The
Ultimate
Modem
Handbook

ISBN 0-13-849415-0

90000

9 780138 494155

The
Ultimate
Modem
Handbook

Your guide to Selection, Installation, Troubleshooting, and Optimization

Cass R. Lewart

To join a Prentice Hall PTR Internet mailing list,
point to http://www.prenhall.com/mail_lists.

Prentice Hall PTR, Upper Saddle River, New Jersey 07458

 © 1998 Prentice Hall PTR
Prentice-Hall, Inc.
A Simon & Shuster Company
Upper Saddle River, New Jersey 07458

Editorial/production supervision: *Joe Czerwinski*
Acquisitions editor: *Bernard Goodwin*
Manufacturing manager: *Julia Meehan*
Cover design director: *Jerry Votta*
Cover design: *Design Source*

Prentice Hall books are widely used by corporations and government agencies
for training, marketing, and resale.

The publisher offers discounts on this book when ordered in bulk quantities.

For more information contact: Corporate Sales Department, Phone: 800-382-3419;
Fax: 201-236-7141; e-mail:corpsales@prenhall.com.
Or write: Prentice Hall PTR, Corp. Sales Dept., One Lake Street, Upper Saddle River, NJ 07458

Printed in the United States of America

10 9 8 7 6 5 4 3 2 1

ISBN 0-13-849415-0

Prentice-Hall International (UK) Limited, *London*
Prentice-Hall of Australia Pty. Limited, *Sydney*
Prentice-Hall Canada Inc., *Toronto*
Prentice-Hall Hispanoamericana, S.A., *Mexico*
Prentice-Hall of India Private Limited, *New Delhi*
Prentice-Hall of Japan, Inc., *Tokyo*
Simon & Schuster Asia Pte. Ltd., *Singapore*
Editora Prentice-Hall do Brasil, Ltda., *Rio de Janeiro*

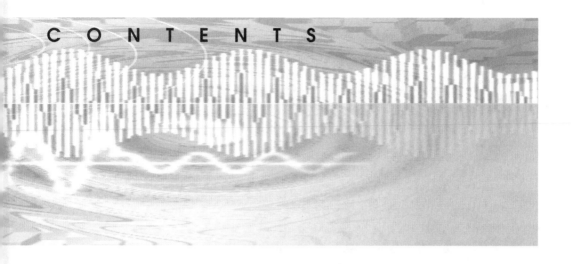

CONTENTS

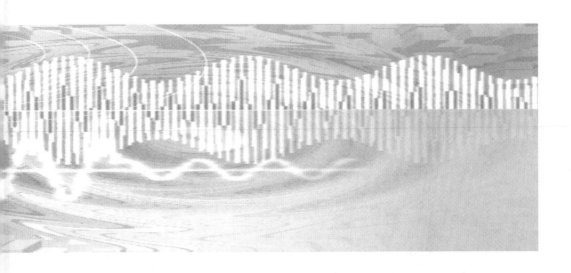

Preface

Ten years have gone by since the first version of this book was published. For an observer of the data communications field, these ten years seem more like a lifetime. The explosive growth of personal computing and the Internet made the whole world easily accessible not just to the selected few who work for the government or large corporations with financial resources and technical expertise, but to anyone willing to buy an inexpensive computer and modem.

Selection of a modem used to involve just a quick check to ensure agreement with applicable standards. In fact, the selection was frequently suggested and performed by a representative of the common carrier—the local telephone company. Today, the decision of which modem to choose and how to install it is mostly done by the PC user with help from obscure manuals. This creates the need for books such as this that can explain the apparent mysteries of data modems and communications software.

While installing assorted components of a personal computer such as a mouse, keyboard, monitor and printer can be relatively straightforward, installing a modem still presents a considerable challenge to most PC users. Modems and sound cards generate the majority of calls to a modem manufacturer's telephone support lines. No wonder—confusing manuals use mysterious terms such as COM ports or IRQs, and specify initialization strings in terms of AT commands! The main purpose of this book is to explain such mysteries by pointing to the basic technical aspects of data communications in general and of modems in particular.

Just like a modern automobile, a modem can be purchased with many options. It is easy to make a wrong decision if you don't know and understand the many modem features that are available, their relative importance, and how they interact with each other. Sometimes a standard solution—connecting two computers equipped with modems via a telephone line—may be the least desirable. Use of less common modem types, such as limited-distance modems or a special device called a null modem may result in both simplification and savings.

This book is written primarily for the personal computer user who would like to learn how to select and install a modem in order to connect to the Internet, to a bulletin board, or to another computer. If something goes wrong, this book should help in analyzing the problem and hopefully solving it.

I also wrote this book for the communications professional, who has been working in the field for many years dealing with common carriers and equipment suppliers. I assume that this person would like to get a better understanding of various elements of the data communications networks in order to make cost-effective specification and planning decisions in modem acquisitions.

This book should also be of interest to those who are not necessarily active in the data communications field, but who would like to expand their knowledge of the latest developments in modem technology.

As an electrical engineer and computer hobbyist active for over 40 years in the data communications field, I have always found modems to be a fascinating subject. There is more to a modem than just a name for a device that lets one computer talk to another computer over a telephone line. Modem technology combines various disciplines of electrical engineering and computer software, such as circuit design, large scale integrative microprocessors, switching, data transmission and programming.

Writing a book like this requires gathering information from many equipment and software sources. This can often be a daunting and harrowing experience. Most companies today appear to be run by their legal departments. The legal department, while not understanding what information can be released without harming the company or helping the competition, considers the only safe action is to not divulge anything. The only information that is easy to obtain are glowing press releases, the main purpose of which seems to be to boost the company's stock price.

Fortunately I found a few memorable exceptions to this rule. In particular, I would like to acknowledge help from a number of individuals who assisted me in understanding certain arcane areas of data communications and of modem design. I would like to thank Dr. Adam Lender, the former president of the IEEE Communications group; Mike Pellegrini of the TAS Corporation; Chuck Hartley of the GRI Corporation; Steve Edwards of Lexis-Nexis; Patrick Chris of Caere Corporation; Sharon Karl of U.S. Robotics; Hannes Kristinsson, formerly of Bell Laboratories; Cynthia Connell of Hewlett Packard; and Naomi Bulock of Mathworks Inc.

My thanks go to my publisher Bernard Goodwin of Prentice Hall for much encouragement, and to copy editor Dick Girard and production editor Joe Czerwinski for the thorough editing of the book.

Final thanks go to my wife Ruth for coming up with many new ideas for the book, and to my son Dan for much general and technical advice.

If, after finishing the last chapter, you find that your understanding of data communications in general and of modem technology in particular has improved, then the purpose of this book will have been fulfilled.

Cass Lewart
Holmdel, NJ
Fall 1997

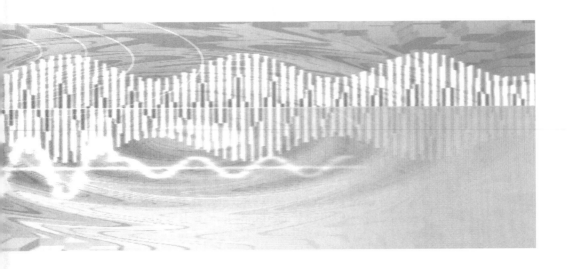

Introduction

The need for the exchange of digital information between computers over analog transmission facilities—in particular over the telephone network—led to the development of modems. Information in a digital computer is stored, processed and transmitted as a sequence of logical 0s and 1s. In terms of voltage, a "0" may correspond to -5 Volts and a "1" to +5 Volts. Unfortunately, the analog telephone network, which was originally designed to carry voice, is not suited to carry such DC voltages.

In order to carry the sequence of digital signals over an analog telephone network, a signal converter is needed. A modem performs a digital-to-analog (D/A) conversion between a digital computer and an analog transmission facility for outgoing signals. It also performs an analog-to-digital (A/D) conversion between an analog transmission facility and a computer for incoming signals.

The D/A conversion is performed by modulating the digital signal on a carrier signal. The A/D conversion is performed by demodulating, or extracting, the digital information from the modulated analog carrier. The combination of the two words—modulator and demodulator—forms the word *modem*.

Modems are always used in pairs—one at each end of a data transmission path. The user has to ensure that the receiving and transmitting modems use the same protocols, which are rules describing precisely which data format, modulation scheme, and transmission speed to use in order to properly transmit and receive information.

The above definition of a modem as a D/A and A/D converter is not always true, however. With the current availability of digital end-to-end transmission, the word *modem* is also used to describe adapters that handles the digital data stream coming from a computer onto digital telephone line such as to an Integrated Services Digital Network (ISDN) line or to a Dataphone Digital Service (DDS) line.

Though this book deals only with data and fax modems, modulators and demodulators are present in every branch of voice and data communications. If you shout, your voice will only carry a few hundred feet at most. However, when a radio transmitter modulates a carrier frequency with speech or music, it will carry the signals over hundreds or even thousands of miles. A radio receiver can then demodulate the carrier signal such that the original speech or music can be heard. In general, modulation and demodulation is used when a signal contains information that is not capable of being propagated by itself over the chosen transmission medium.

This book explains, in considerable depth, exactly how modems perform their task of preparing data for transmission over the telephone and wireless networks. The current modem market is very much alive, thanks to technological advances and competition that started with the deregulation of the United States telecommunications industry. The first part of this book introduces the data communications field, with an emphasis on modems. The second part takes a closer look at modems and communications software for the personal computer user. You will learn the various rules (protocols) that these modems must follow in order to be compatible with each other, which software can be used to operate modems, and what the modems connect to. Global communication—connecting to the Internet and to bulletin boards—is the main reason why most of us bought a modem in the first place.

The third part of this book describes special types of modems used in commercial applications, as well as wireless, cable, ISDN, LAN, cellular, and 56K modems. The fourth part of this book covers how to test modems, transmission facilities and other parts of the data communications channel using diagnostics and troubleshooting aids. If your modem did not seem to operate correctly before you started reading this part of the book, you should have a much better understanding of how to operate it correctly when you finish.

Finally, in the fifth part, you will get a short glimpse into the future of data communications by looking into the data communications crystal ball to see what developments can be expected in the next few years.

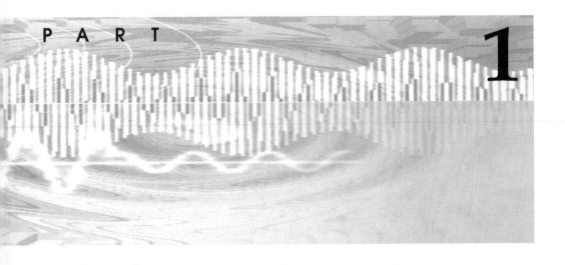

MODEMS AND DATA COMMUNICATIONS

To understand how modems work, you should have a good grasp of data communications. The first part of this book will give you the basics of data communications. The terms and concepts you will come across in the remainder of this book should start to make sense, even for a person who is not familiar with the subject. If you have worked in the data communications field, you might want to just flip through the first part and consider it as a refresher course.

Chapter One starts with an introduction to data communications. It explains why we need modems, and describes basic terms of data communications such as two- and four-wire transmission, half- and full-duplex transmission, transmission impairments, and other aspects of the data channel. It also explains how various parameters affect the transmitted data signal as it is carried over the Public Switched Telephone Network (PSTN). A discussion of the PSTN, which is the backbone of most voice and data communications, is followed by a short description of the Open System Interconnection (OSI) model of the data network. Discussing the impact of the deregulation of the telecommunications industry in the United States, which started a revolution in the spread of data communications tools to the personal computer market, concludes the first chapter.

Chapter Two discusses the theoretical limitations imposed on data transmission by a transmission facility—primarily the telephone line. Terms such as bits, bytes and Baud are also explained.

Chapter Three describes modulation methods used by modems at various transmission rates, and how these methods relate to the transmission error rate.

Chapter Four provides more detail about communications protocols, which are rules of data transmission.

Chapter Five explains the differences between asynchronous and synchronous transmission, and various ways in which a modem can be connected to a computer.

Chapter Six concludes the first part of this book with a detailed description of the serial interface, which is the most common interface used between a modem and a computer.

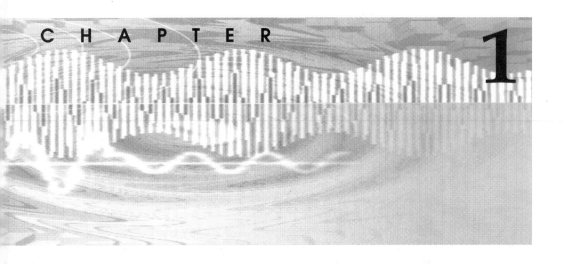

C H A P T E R

1

Introduction to Data Communications

This chapter gives an overview of the relevant aspects of data transmission, and presents terms used throughout the remainder of this book. You should read this chapter to get a better understanding of modem applications, and of modem technology.

You will learn why we need modems to connect a computer to the telephone line. Short definitions of many important data communications terms are presented, with more detail provided in later chapters. Next comes a description of the public telephone network and the variations in the quality of calls made over that network. This chapter concludes with a description of the Open System Interconnection (OSI) model of the data network and the impact caused by the deregulation of the United States telecommunications industry.

Why Do We Need Modems?

If an ideal transmission medium existed, a medium that would accurately pass the digital signals to their destination, there would be no need for modems. The receiver and transmitter would be connected by this ideal medium: the received pulse would be identical to the transmitted pulse, and the life of a communications consultant would be very simple.

There *is* a condition that approaches this ideal situation, and it is called end-to-end digital transmission, as shown in Figure 1-1. Its current implementation in the consumer market is the Integrated Services Digital Network (ISDN), the Dataphone Digital Service (DDS), and the Asymmetric Digital Subscriber Line (ADSL). Although most of the long distance telephone network today is based on digital transmission, the final links between the local telephone office and the customer remain mostly analog, and will probably remain so at least in the near future. Although it is possible to order an end-to-end digital connection now, its cost is considerably higher than that of an analog connection.

Figure 1-1 Digital Transmission, End-to-End

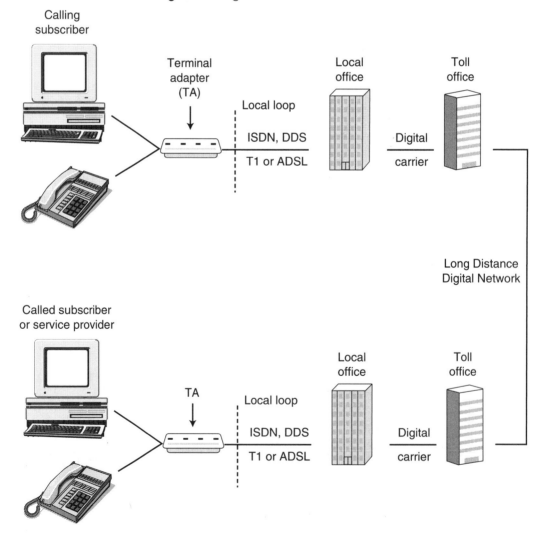

Until end-to-end digital connections become common, the average personal computer and telephone user will encounter a mixed analog/digital network, as shown in Figure 1-2. Because the telephone interface to the subscriber is mostly analog, the digital signal from the computer must be modulated—that is, changed into an analog signal before it is sent over the telephone line. The reverse process takes place at the receiving end, where the analog signal is demodulated, or changed back into a digital signal so it can be understood by the computer. Combining the two words—modulation and demodulation—led to the word *modem*, which is the subject of this book.

Figure 1-2 Analog/digital transmission network

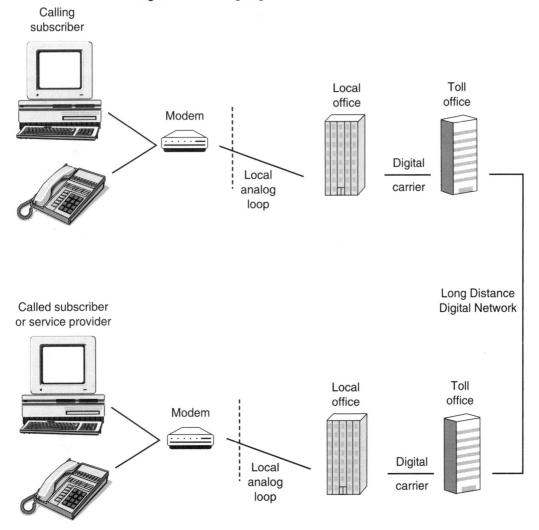

Basic Data Communications Terms

Voice Band

The French physicist and mathematician Jean Baptiste Fourier was the first to recognize that even the most complex time-varying analog signals can be decomposed into separate frequency components, each one being a simple sinusoid of a different frequency and phase. The human ear in young individuals is capable of recognizing sound frequencies between 20 and 20,000 Hz (1 Hz = 1 cycle/sec). This frequency range unfortunately drops, particularly at the high end, as one ages or is exposed to loud noises for extended periods.

Still, most of the information in intelligible speech is concentrated in a more narrow frequency range between approximately 300 and 3000 Hz. Therefore a standard telephone channel was designed to have a bandwidth, or range of frequencies, of approximately 300 - 3300 Hz. Such a channel is usually referred to as a voice-band channel, because it can pass the most important frequencies associated with the human voice.

Transmission Medium

An electrical transmission medium, which can carry analog or digital information, can be anything from a pair of wires to a communications satellite link. In general, we will not concern ourselves with the physical aspects of a transmission medium, but rather with its transmission characteristics, such as bandwidth, circuit loss, and various forms of electrical distortion associated with the transmission of information.

Two- and Four-Wire Transmission

The local telephone loop, which are the wires that connect your telephone to the local telephone office, traditionally consists of two copper wires. This was the most economical solution when the telephone was first introduced. Though telephone companies are now starting to introduce digital facilities such as fiber optics, the copper loop (meaning copper wires) is still the norm. Because the same two-wire physical loop carries both directions of transmission—to and from a subscriber—the signals would normally interfere with each other. The result is that some part of the original signal that is sent out from the subscriber is reflected back to the subscriber. To minimize this from occurring, the telephone company keeps in the local telephone office a so-called *hybrid circuit* as the termination for each two-wire subscriber loop.

A two-wire transmission loop is adequate for voice and certain types of data over short distances. However, a four-wire transmission loop, where each direction of transmission is carried on a separate pair of wires, is frequently required for high-speed data transmission. Four-wire transmission is also used by the telephone company for long-distance transmission between their toll offices.

An analogy would be a divided highway versus a country road. Examples of two-wire and four-wire telephone connections are shown in Figures 1-3 and 1-4.

Figure 1-3 Two-wire telephone connection

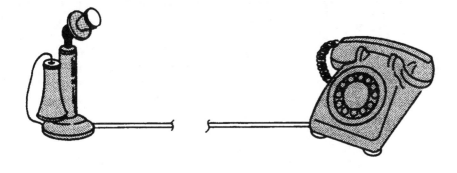

Figure 1-4 Four-wire telephone connection

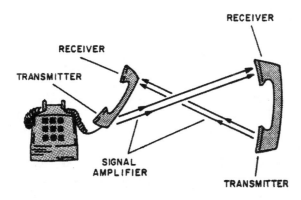

Asynchronous and Synchronous Operation

When transmitting data, there is always a need for synchronization between the transmitting and the receiving computer. Because each transmitted character might consists of seven or eight bits, the receiving modem and its associated computer or data terminal must know exactly where each character begins and where it ends. This information is extracted in a slightly different way when using an asynchronous and synchronous transmission method.

The asynchronous method is used mostly for communications between personal computers and their modems. In this case, each bit in a character is not synchronized with each other; rather, each character starts its own synchronization process. The asynchronous transmission is a series of bits that consist of a start bit, seven or eight data bits that correspond to the American Standard Code for Information Interchange (ASCII) representation of a character, one or two stop bits, and a parity bit that is used for error detection, all depending on the agreed-upon protocol. For example, the ASCII equivalent of the letter A is 65 (decimal) or 1000001 (binary). The seven-bit binary equivalent for the letter A is 1000001, while the eight-bit binary representation is 01000001. The ASCII equivalents of characters and symbols are shown in Appendix C.

The synchronous operation method, in which a continuous stream of data bits is being transmitted, is used mostly for commercial data exchange. Synchronization is performed on larger blocks of data rather than for each character. The advantage is that start and stop bits are not needed for each character sent, thus saving some overhead.

Transmission Impairments

A voice or data signal traveling through a transmission medium, such as a telephone line, is affected by a number of distortion sources associated with the type of medium, the length of the transmission line, and the environment. The major factors affecting data transmission are attenuation and delay distortion, phase jitter, and electrical noise.

Attenuation distortion is the variation in gain or loss of a transmission medium as a function of frequency. It transforms a rectangular data pulse into a rounded and distorted shape. The delay distortion is the difference in velocity at which various frequency components of a signal travel along a transmission line. It causes data pulses to "smear"—or interfere with succeeding pulses. Phase jitter is fast changes in the phase of the received signal compared to the transmitted signal. Its effect is similar to delay distortion.

Finally, electrical noise is a collection of random short pulses and periodic power-line interference, which gets superimposed onto the real signal. The combination of each of these parameters can make the signal unrecognizable at the receiver—for instance, a binary 0 may be interpreted as a 1, or vice versa.

An example of a data signal distorted after passing over a transmission facility is shown in Figure 1-5. Notice that the effect of delay distortion is often more pronounced than that of the amplitude or attenuation distortion. The purpose of a modem is to translate the digital binary signals into analog signals (electrical or light) in a form that will be least affected by the distortion introduced by the transmission medium. As we will see in the following chapters, the modem, when working properly, will satisfy these requirements while being transparent to the user.

Figure 1-5 Effects of transmission distortion on data

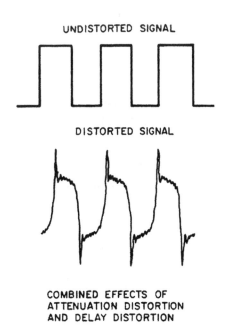

UNDISTORTED SIGNAL

DISTORTED SIGNAL

COMBINED EFFECTS OF
ATTENUATION DISTORTION
AND DELAY DISTORTION

Full-Duplex, Half-Duplex and Simplex Operation

When a modem operates in a two-way connection, where data is passed in both directions simultaneously, it is called *full-duplex* mode. It can also operate in a *half-duplex* mode, where only one direction of transmission is active at one time. Or it can operate in one direction only, called *simplex* mode.

The three modes of operation—full-duplex, half-duplex and simplex—are shown in Figure 1-6. All personal computer modems and many commercial modems use full-duplex mode. However, there is a number of commercial half-duplex modems that trade full-duplex operation for a higher transmission speed. Various schemes are then used to reverse the direction of transmission, when required. There may be a separate low-speed *attention*-calling reverse channel, which is a separate circuit used to activate the reversal of transmission direction, or the modem may reverse the direction of transmission automatically upon detection of a carrier loss. The simplex mode is less common, but it is used for special-purpose applications such as polling of electrical meters by the power utility.

Figure 1-6 Simplex, half- and full-duplex transmission

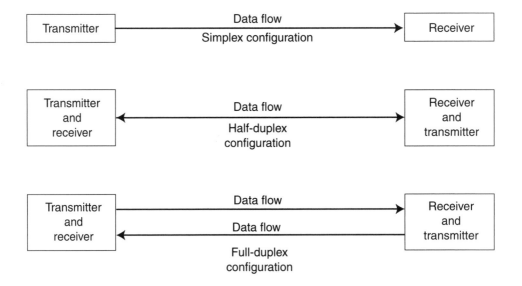

Transmission of Digital Information

To transmit digital information, which is the binary stream of 0s and 1s generated by a computer or a data terminal, one needs a transmitter, a transmission medium, and a receiver. Both the transmitter and the receiver are part of a modem.

The function of a transmitter is to convert the digital signals into an analog voltage, current or light, depending on the transmission medium used. The function of the receiver is to convert the analog signals back into a digital form that can be read by the computer or data terminal. Depending on the properties of the transmission medium—for example, a pair of copper wires, a telephone line or optic fibers—a different type of conversion will be used by the modem.

A generalized form of a combined transmitter and receiver for two-way communication is called a modem, short for modulator (transmitter) and demodulator (receiver). The term modulator has long been associated in telephony with a device that converts the low-frequency voice signals into higher frequencies suitable for long-haul transmission, and for *stacking* of multiple voice-frequency channels on a single pair of wires or a coaxial cable. The telephone company uses either the frequency-division multiplex (FDM) or the time-division multiplex (TDM) scheme for combining multiple voice channels. In the FDM scheme, voice channels are stacked into groups of 12 channels in the so-called A channel banks, super groups of 60 channels, and so on.

Frequency assignments used in the FDM scheme at the group level are shown in Figure 1-7. Notice the many filters, modulators and demodulators involved in the frequency translation, each introducing a certain amount of attenuation and delay distortion to the data signal. Voice channels are stacked in multiples of 24 in TDM schemes, starting with the DS1 channel bank (24 channels at 1.544 Mbps), then continuing with the DS1C channel bank (48 channels at 3.152 Mbps), DS2 channel bank (96 channels at 6.312 Mbps), DS3 channel bank (672 channels at 44.736 Mbps) and the DS4 channel bank (4032 channels at 273.176 Mbps).

As mentioned earlier, a modem converts a digital string of 0s and 1s into an analog form by modulating frequencies suitable for the transmission medium. For example, the early Bell 103A type modem would convert a binary 0 into a frequency of 1070 Hz or 2025 Hz, and a binary 1 into 1270 Hz or 2225 Hz, depending on whether the modem was in the Originating or in the Answer mode. These four frequencies are all well within the frequency bandwidth of a telephone line (300 to 3300 Hz). Higher-speed modems in current use still convert the digital pulses into frequencies within the bandwidth of the telephone line.

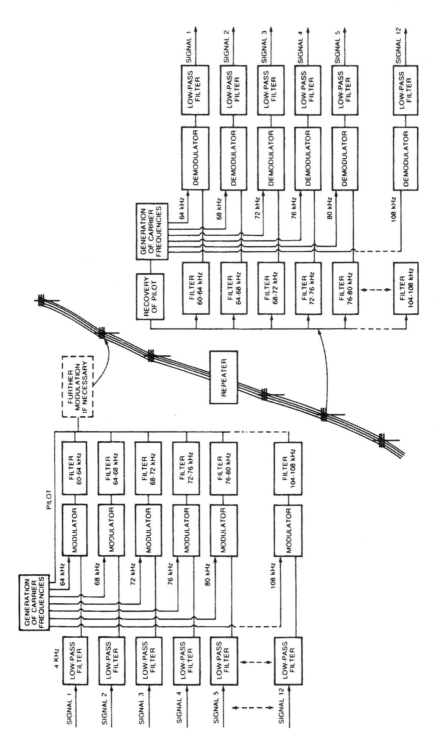

Figure 1-7 Frequency division multiplex

Public Switched Telephone Network (PSTN)

The *public switched telephone network,* or PSTN for short, is the backbone of data communications for most PC users. It was originally developed for voice and not for data transmission. It is therefore primarily designed and optimized for satisfactory voice communication between telephone subscribers. As most of the voice energy is concentrated between 300 Hz and 3000 Hz, the bandwidth of a telephone circuit in the United States and in Canada is between approximately 300 Hz and 3300 Hz in order to satisfy the average telephone voice subscriber. The upper cut-off frequency is a few hundred Hz less in other countries.

Telephone subscribers are, in general, connected to the local telephone office, also called a Class 5 office, with a pair of copper wires, called the local loop. If the calling subscriber and the called subscriber are connected to the same local office, the call will be switched in the local office to establish a connection. If the call is destined for a distant city, the local loop will be switched to a toll-connecting trunk, which used to be another pair of wires but is now mostly a digital multiplexed carrier system. The toll-connecting trunk from the local Class 5 office terminates in the toll office, which is the gate to the world-wide telephone network. From there, the call will be carried on inter-toll trunks, which are four-wire facilities that provide a separate path for each direction of transmission. Depending on traffic conditions, the call may be routed via switched transmission facilities through various hierarchical levels of the telephone network.

The traditional hierarchy of the public switched telephone network is shown in Figure 1-8. On top of the hierarchy are Class 1 Regional offices. Then come the Class 2 Sectional offices, Class 3 Primary offices, Class 4 Toll offices, and finally Class 5 Local or End offices. Long-distance providers, such as AT&T, MCI and Sprint, to name the major ones, connect to the PSTN at local or toll offices and carry the call through their own separate or shared networks.

Calls either follow the hierarchical ladder up and down, or are shunted by means of high-usage trunks with the technique called Dynamic Hierarchical Routing (DHR). DHR relies on Stored Program Control, which provides current information about the telephone traffic and which trunks are busy. While traveling from switch to switch, the call is combined with other voice or data calls traveling to the same intermediate destinations.

At each switching point the signals are combined with other signals (multiplexed) or separated from other signals (demultiplexed). Each of these operations introduces some additional transmission impairments, which, though hardly noticeable on a voice connection, may affect the data signal. While PSTN establishes a permanent connection between the caller and the called subscriber for the duration of the call, the signaling information, which carries dialing instructions, busy and hang-up information, is stripped from the voice call at the local office. It then travels separately from the voice call as data packets.

Figure 1-8 Hierarchical structure of the U.S. telephone network

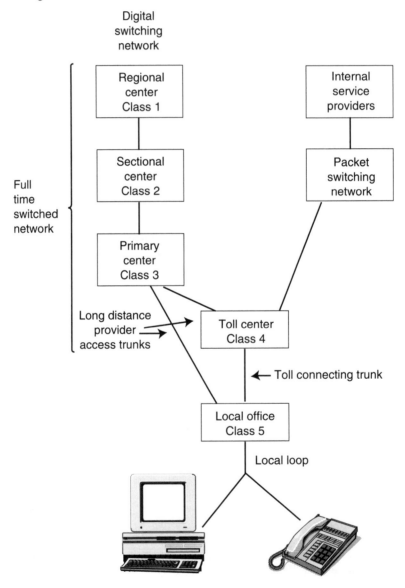

In addition to full-time connections on a switched voice or data call, and packetized signaling connections, the PSTN also supports other packetized data calls and networks such as the one used for the Internet. In this type of network, packets of data, each containing a destination address in its header are routed to its destination by— what else—a device called a *router*. Address information stored in each router propagates the packet to another router until it reaches its final destination.

It is clear that wrong address information stored in a router can easily lead to an endless loop. To prevent such an occurrence, which could overload and clog the packetized network, each packet of data carries information about where it has already traveled. If a router determines that a packet is going in circles, the packet will be deleted by the router.

Local Loop

The local loop is for most telephone subscribers a pair of copper wires that connect a subscriber to the local telephone office. Dozens and even hundreds of local loops from different subscribers are combined into a single telephone cable. The built-in electrical capacitance of the telephone cable increases the transmission loss as the frequency gets higher.

To improve voice transmission, the telephone company inserts a so-called loading coil in a long local loop. The inductance of the coil combines with the capacitance of the cable, and results in a low-pass filter. The combination features a fairly flat attenuation up to about 2,500 Hz, then a steep rise in attenuation at higher frequencies. The frequency response of a *loaded* and *non-loaded* loop are shown in Figure 1-9.

Because a modulated data stream carries energy at frequencies spread over the entire bandwidth of the voice channel up to 3300 Hz, loading coils should be removed from a loop that will be used for data transmission. This will result in expanding the bandwidth and improving the data transmission at the expense of attenuation distortion and of inadequate voice transmission. New telephone installations may also consist of a high-capacity optical fiber or coaxial cable leading from the local telephone office to the customer premises. Similar to combining multiple subscriber loops into one cable, several subscriber access lines can also be combined into a single optical fiber or coaxial cable. A demultiplexer then splits them back into individual circuits.

Figure 1-9 Effect of line "loading" on loop attenuation

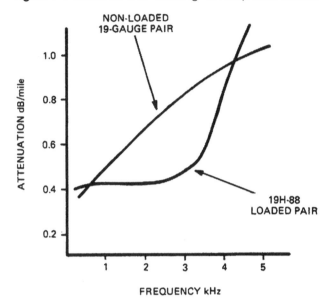

Hybrids and Echo

Most local loops consist of only one pair of wires that carry two-way simultaneous communication. It is referred to as a two-wire full-duplex transmission path. On the other hand, the toll network between Class 4 and higher offices is a full-duplex four-wire path. The transfer from a two-wire to a four-wire transmission facility takes place in a circuit called a *hybrid*. Such a circuit terminates each local loop in the local telephone office.

The effectiveness of a hybrid circuit is measured in terms of the return loss, which is equal to the attenuation of the reflected signal. A higher return loss means a lower reflected energy and a better forward transmission. The hybrid circuit, as shown in Figure 1-10, is somewhat similar to a Wheatstone bridge, except that it uses special transformers instead of discrete resistors, capacitors and coils.

The most important part of the hybrid circuit is the impedance-matching balancing network, Z_b in Figure 1-10. The impedance of the simple matching network roughly approximates the complex impedance of the local loop terminated by the telephone set over the voice frequency range of 300 - 3300 Hz. The better the approximation, the higher the return loss. Because of the cost involved, the actual impedance-matching network is a simple compromise circuit consisting of only a resistor and a capacitor. The mismatch caused by the compromise network returns some of the voice or data energy back to the source instead of the destination.

Figure 1-10 Hybrid circuit telephone loop termination

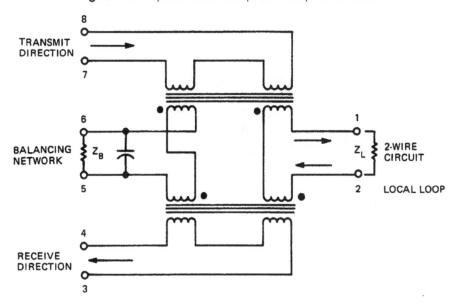

The energy reflected by the hybrid in the local telephone office is called the near-end echo. The energy reflected by the hybrid near the destination is called the far-end echo. The near-end echo is hardly noticeable on a voice call, but it affects data transmission because of its strong signal. The far-end echo affects data to only a limited degree, because of circuit loss along the transmission line, but it is very objectionable on a voice call because of the associated delay. The annoying effect of the far-end echo in a voice conversation increases with the round-trip delay, and decreases with the attenuation of the reflected signal.

The way the telephone network protects its voice subscribers from far-end echo is by introducing extra circuit loss, making it proportional to the length of the connection and hence to the delay of echo. This approach to echo/loss compensation is called the via net loss (VNL) method. The extra loss introduced in a telephone connection affects echo more than it affects speech, because echo makes a round trip on the network and is therefore attenuated twice. The annoying effect of the increasing delay, approximately 20 ms per 1000 km (625 miles), is compensated by this extra circuit loss.

To achieve full-duplex operation over two-wire facilities, modem manufacturers also employ various echo-cancellation techniques such as self-adjustable delay equalizers and echo suppressors, which make a full-duplex modem appear as a half-duplex modem.

On calls exceeding 2400 km (1500 miles), the delay amounts to approximately 45 ms and the extra loss introduced to compensate for the far-end echo would impair the telephone connection and would no longer be acceptable. Therefore, instead of increasing the circuit loss, the telephone company on such long circuits provides *echo suppressors*, which are shown in Figure 1-11.

An echo suppressor puts high loss in the non-active direction of transmission by constantly monitoring energy in both directions of an active circuit. Instead of a full-duplex circuit, we now have a half-duplex circuit, which reverses its direction as each party starts talking. In a voice connection, echo suppressor—being the least evil— are tolerated. On a full-duplex data connection, however, they cannot be accepted because they would not allow for two-way communication.

In addition, the far-end echo, though objectionable on a voice call, will hardly affect a data connection, just slightly decreasing the signal-to-noise ratio of the circuit. The telephone company recognizes this problem and provides the data subscriber with a means to disable echo suppressors on a telephone circuit. This is done by the subscriber sending a tone of 2100 Hz on the United States telephone network (1800 Hz on European telephone networks). Modems frequently send such tones, called the *guard* tones, during the initial handshake sequence, which always precedes the exchange of data.

Figure 1-11 Echo suppressors in a telephone connection

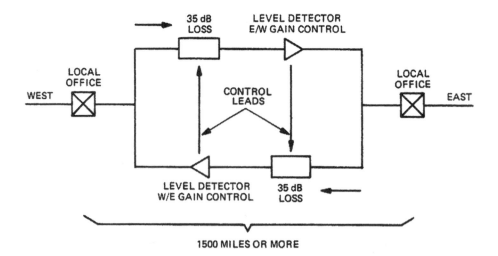

OSI Model of the Data Network

In order to exchange data between different computers located in different cities, countries or continents, some form of standardization is needed. This standardization occurs in a piecemeal way. There are standards for modulation for specific modem types, standards for computer-to-modem interfaces, and error correction and data compression standards. To make development of standards more consistent and uniform, an international standard for data communications was adopted in 1983.

The standard, developed by the International Standards Organization (ISO), defines the by-now famous conceptual seven-layer model of the international data network, the so-called Open System Interconnection (OSI) model. The openness of the standard means that each layer of the model is open to any process or communication conforming to such a standard. A diagram of the OSI model is shown in Figure 1-12. By adopting this model, standards can be developed for the clearly delineated layers, and an assurance exists that, if layer standards agree with each other, data will successfully flow between the layers.

In general, the modem user will only deal with Layer 1 (Physical Layer) and Layer 2 (Data Link Layer). However, understanding the higher layers of the OSI model may sometimes help in diagnosing problems related to data transmission.

Figure 1-12 OSI model of the communications network

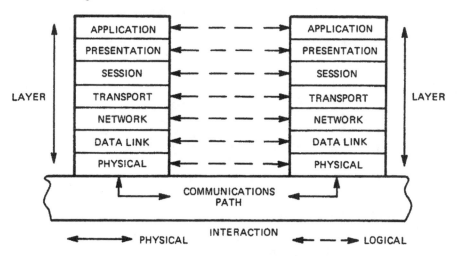

OSI Layer 1— Physical Layer

The physical layer consists of the hardware interface between the computer, modem and the telephone line. The layer standards include functional, electrical and physical specifications, as described in RS-232-C, V.24, X.21, and other interface recommendations. Other standards for this layer include the Local Area network (LAN) standards that are being developed by the IEEE 802 committee. These LAN recommendations will likely follow the OSI model.

OSI Layer 2—Data Link Layer

The data link layer is primarily responsible for error detection and correction. The MNP error correction protocols work at this level. Data transmission protocols used at this level include SDLC, HDLC, and others (discussed in later chapters). The protocols take care of data-block framing and of data synchronization, error control and recovery, message sequence control, link initialization, connection, disconnection, and addressing.

OSI Layer 3—Network Layer

The network layer provides a means to transmit data through the network from the originator to the receiver. It provides the mechanism to establish, maintain and terminate data connections between separate computer networks. The CCITT X.25 Packet Switching protocol is one example of the network layer standard. The standard describes how a packet of data is to be assembled, framed and addressed.

OSI Layer 4—Transport Layer

The transport layer allows end users to communicate independently of network constraints imposed by layers 1 to 3. The appropriate standards specify priority levels, security, response-time expectations, error rates, and recovery strategies. The layer takes data from the higher layer (layer 5), splits it into smaller units if necessary, passes the units to the network layer, and assures that they reach their destinations. The transport layer also provides guidelines for data-flow management.

OSI Layer 5—Session Layer

The session layer is the true user's interface to the data network. It handles the log-on and log-off procedures and user authentication, including password verification. This layer may also handle communications parameters to other local devices, such as the choice of local disk storage units. It also handles recovery from unreliable transport layer connections.

OSI Layer 6—Presentation Layer

The presentation layer is approaching the user's programs. It handles the display of data, file formatting, code encryption and decryption, and data compression and expansion. The layer also assures that the syntax of the originator is understood by the receiver. If the originator uses ASCII and the receiver uses EBCDIC codes, a translation procedure would occur at this layer.

OSI Layer 7—Application Layer

The application layer contains recommendations for the programs used at each end of the path. Some of the recommendations for this layer are suggested standards for electronic funds transfer, point-of-sale terminals, automated teller machines, airline reservation systems, and other large-scale applications.

Application of the OSI Model to a Data Call

Layers 2 through 6 of the OSI model affect the data by adding successive headers to the blocks of data originated by the user in layer 7. Each header adds some additional control bytes so the network can successfully transmit the data from source to destination. For most users, layers 2 through 7 are purely conceptual and are part of the user's computer and associated software. However, in the following example from the Hewlett Packard Data Communications Tutorial, we will assume that the OSI model is fully implemented, and that a data call is made between two locations using the OSI model network.

Let us examine in detail how an end-to-end data transfer might take place. Our Los Angeles computer, which is running an application program, finds a reference to a database that is located in New York. Assume that the communications link has been established, and that a file transfer is about to take place. The data to be transferred may look like this:

```
<Region Sales = 50% of Quota>
```

The New York application process has pulled this data from a file and passes it into layer 6 of the OSI model. The presentation layer adds some overhead to the message it received from the file. Does our system operate in ASCII-8 (8 bits per character)? Do we use text compression? Do we utilize encryption? This information is attached to the header of the message, perhaps in the form of "ASC8" (8-bit ASCII) and "NTC" (No Text Compression), such that the resultant information appears as follows:

```
<ASC8 NTC><Region Sales = 50% of Quota>
```

Each angular bracket delimits a sub-block of data, which can be recognized by the equipment. The composite message is passed on to layer 5, which treats the data as a complete message. Here is one of the most important concepts of the OSI—the information passed from layer to layer is treated by any one layer as *user* information, which is not to be changed in any way.

Next, the session layer takes our composite message from the presentation layer and adds the session layer header. What is important here is that the LA/NY link is now established, perhaps with some session keyword, that validates the session in progress. The resultant package of information passes to the next lower layer, and looks like this:

```
<LA NY KEY><ASC8 NTC><Region Sales = 50% of Quota>
```

The transport layer adds whatever information is important to its peer. In this case, it may be only the network number through which communication is established. It may add the identifier to the growing information frame, resulting in:

```
<NET1><LA NY KEY><ASC8 NTC><Region Sales = 50% of Quota>
```

The network layer recognizes what it receives from the higher layer as data, and prefaces to that data the information important to its peer on the receiving end. In this case, it merely passes back to the network the virtual circuit number (VCN) it was assigned when communication was established. The data frame that gets passed to layer 2 then appears as:

```
<VCN><NET1><LA NY KEY><ASC8 NTC><Region Sales = 50% of Quota>
```

The data link layer recognizes this as user data and frames it as such, conveying it to the local network node to which it is physically connected. It passes the address and control information, as well as a frame-check sequence. Our frame keeps growing and now looks like this:

```
<FAC><VCN><NET1><LA NY KEY><ASC8 NTC><Region Sales = 50% of Quota>
```

This outgoing frame is passed to the physical link layer, which converts all the data to a bit stream going across the RS232-C interface to a modem, or whatever device is connected to the transmission medium.

An important point to recognize here is that a protocol analyzer, like the one discussed in Chapter 17, can actually "see" the data only at the physical link layer. To analyze the data, the instrument must then convert the received data string using the appropriate protocol.

Current Status of the OSI Model

OSI layers 1, 2, and 3 are currently implemented in the X.25 standard, and work continues in various international committees on the development of standards for the remaining layers. The International Standards Organization and the ITU have several study groups defining these standards. Still, the OSI model is more of a template for future development than a set of currently existing standards. It is an important step, which had to be taken to achieve more consistent standards for data transmission

In the future, application layer standards for specific areas of industry, such as banking, retailing, manufacturing, medicine, education, and government, will be available in addition to generic application layer standards that are used across all industries. Still, progress in this area is rather slow, as most large applications use proprietary software and protocols.

An organization called Global Information Center in Englewood, Colorado keeps track of new developments in the area of international standards and recommendations and provides many publications for interested users.

Deregulation of the U.S. Telecommunications Industry

Prior to 1975, common carriers—AT&T, and independent telephone companies in the United States and government-operated Post, Telegraph and Telephone organizations (PTTs) in most other countries—owned, maintained and operated all communication equipment attached to the public telephone network. The rationale for this approach was protection of the fragile telephone network and its users from unauthorized equipment that might interfere with the network. To show the need for such protection, one can consider a frequency-multiplexed carrier system that in its first stage, stacks twelve voice channels into a so-called channel bank. The assumption used in the design and operation of a carrier system is that adjacent channels will not interfere with each other, as long as their bandwidth and power are limited. Due to the nonlinearity of the system, a signal that is more powerful than the design limit could generate modulation products falling into other channels, resulting in crosstalk and distortion.

Before the deregulation of the United States telecommunications industry, all equipment operated by common carriers and PTTs was thoroughly tested before being put into service. This process assured proper operation of the telephone network. Because the equipment was not owned by the user—it was leased from common carriers—the cost of maintenance was included in the monthly lease. The equipment was conservatively designed and reliable, and considering the overhead cost of large organizations, it was also expensive. A typical charge for a low-speed modem leased from a telephone company used to be around $50 per month.

What further added to the cost of communications equipment was cost allocation by common carriers. In United States, the common carriers provide Universal Telephone Service, assuring that a farmer in a remote location could be provided with telephone service at a reasonable cost. In other countries, postal and telephone expenses were pooled together. In many instances the telephone service was subsidizing the money-losing postal service.

The path to proliferation of communication equipment was opened in the United States with the famous 1968 Carter phone decision. The Federal Communications Commission decided that the telephone user has the right to own and maintain equipment, as long as it does not interfere with the public telephone network. The initial approach of the common carriers was to lease to the end users, for a fee of around $7 per month, the so-called Data Access Arrangement devices (DAA). The DAA was an amplitude and frequency limiter to protect the telephone network from interference from privately-owned equipment. The economical problem DAA created was that the monthly fee for users with private modems and other attached devices, such as answering machines and memory dialers, was not required on equipment leased directly from common carriers. The common carriers claimed that their equipment had been thoroughly tested and thus did not require protection required by the FCC. This approach created a considerable competitive advantage to common carriers and was strongly opposed by equipment manufacturers selling to the private market.

The next step in the deregulation process was the decision by the FCC to let the equipment vendors build the protective circuitry into their equipment and not require a separate DAA box to be supplied by the common carriers. This decision was made effective in June 1977, after several appeals by AT&T and by the communication equipment manufacturers. From then on, only *type*-certification of the equipment was required, in addition to individual registration of the equipment by the user with the local telephone company.

These two measures were then supposed to protect the public telephone network. Even though the last provision—registering the equipment with the local telephone company—has been mostly ignored, the FCC policy is apparently working well. I cannot recall over the past 20 years any major outage of the public telephone network caused by privately-owned communication equipment. The major outages since deregulation have been caused by natural disasters such as fires and hurricanes, or by bugs in the increasingly-complex software that runs the telephone network.

The subsequent breakup of the Bell System in the early 1980s, and the competition of United States, European and Far East equipment makers, decreased modem prices and introduced a large variety of equipment available to the data communications user. Advances in large scale integration (LSI) led to the development of complex chip sets, which perform most functions of a modem. The manufacture of high-volume items, such as modems, became more of an assembly than an engineering and development task.

Today's modems consist of between one and three LSI chips and a few extra components mounted on a printed circuit board. A handful of chip makers, such as Lucent Technologies, Motorola and Rockwell, make modem chips used by hundreds of manufacturers all over the world. Changing a mask during the chip fabrication—changing software instructions stored in a PROM (Programmable Read Only Memory) chip—can give a different personality to a particular modem, differentiating it from similar products.

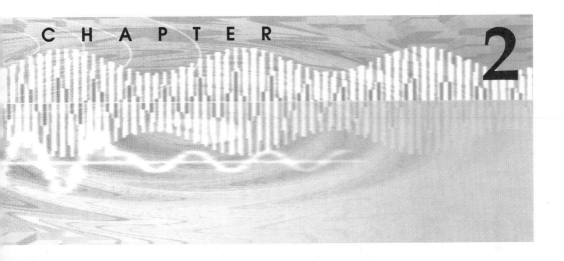

CHAPTER 2

How Fast Will That Modem Go?

There are physical limits to each activity. A man cannot run faster than about 40 km per hour even in the Olympic 100-meter dash. Similarly, there are limits to how fast data can be transmitted over telephone lines or other medium. This chapter discusses such inherent limitations and clarifies certain terms used to describe transmission rate—often referred to as transmission speed.

Basic Limitations—Nyquist and Shannon

A voice-band channel on the Public Switched Telephone Network in the United States has a bandwidth of approximately 3000 Hz; an overseas telephone channel has a bandwidth of approximately 2500 Hz. Typical measured characteristics of voice-band telephone channels with and without equalization are shown in Figures 2-1 and 2-2.

In 1928, a Bell Laboratories mathematician, Harry Nyquist, derived a relation between channel bandwidth and the number of signal changes that a channel can carry. His theorem states that the maximum number of signal changes in a channel of bandwidth B is $2 \times B$. In practical applications, it was found that the maximum number of signal changes is actually slightly less than B. Therefore the application of the Nyquist theorem leads to an apparent limitation of the maximum transmission rate

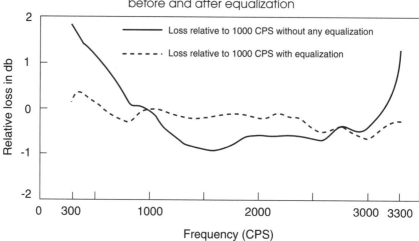

Figure 2-1 Voice band channel attenuation vs. frequency before and after equalization

Loss of a typical L multiplex circuit consisting of two pairs of type A channel banks. (51R / 54R WP2-BUFF ch.2/8)

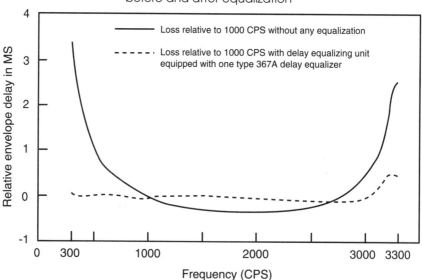

Figure 2-2 Voice band channel group delay vs. frequency before and after equalization

Relative envelope delay of a typical L multiplex circuit consisting of two pairs of type A channel banks. (51R / 54R WP2-BUFF ch 2/8)

for a voice-band channel to less than its bandwidth in Hertz. The bandwidth would be typically less than 3000 Hz for a half-duplex channel, and less than 1500 Hz for a full-duplex channel, with signals flowing in opposite directions. Therefore it appears that the highest full-duplex transmission rate is less than 1500 bps (bits per second). A 1200 bps modem was considered state-of-the-art in 1985 and it cost over $500. So how does this explain the 33,600 bps and even faster full-duplex modems announced by various manufacturers in 1997?

Fortunately, there is an easy way to get around the Nyquist theorem. The theorem refers explicitly to signal changes and not to the transmission rate (measured in bits per second). By assigning more than one bit of information to each signal change, one can reach a much higher transmission rate than the number of signal changes per second.

The old measure of telegraph transmission—Baud—is the unit that specifies signal changes per second. It is also referred to as the rate of modulation at which signals are being transmitted. If the signals can only assume two states, e.g., 5 Volts for a 1 and 0 Volts for a 0, then the modulation rate in Baud is equal to the transmission rate in bits per second. The 1970's Bell 103A type modem, which used frequency-shift keying (FSK) as the type of modulation, is an example of such a system. It was a simple two-state modem operating at a transmission rate of 300 bps. The modem was therefore limited to the maximum theoretical transmission rate of 1200 bps in full-duplex mode and could comfortably operate at its native transmission rate of 300 bps.

The formula to find the maximum transmission rate R of a modem in bits per second (bps), assuming that you know the modulation rate B in Baud, and assuming that a signal can assume D distinct states, is:

$$R = B \times (D / 2)$$

A 1200 bps full-duplex modem such as the Bell 212A assigns two bits (called a dibit) to each transmitted signal, for a total of four distinct states, effectively doubling the transmission rate. The full-duplex transmission rate of 1200 bps is equivalent to a modulation rate of only 600 Baud (1200 bps divided by 2 bits per signal element). Similarly, for higher-rate modems, up to six bits, which result in 64 distinct states, can be assigned to each signal change, increasing the transmission rate while still retaining a low modulation rate. Increasing the number of bits per signal element therefore increases the effective transmission rate.

It is somewhat similar to the transmission of characters and words rather than bits, when speaking to another person. Saying the word "Hello" takes considerably less time than giving a long sequence of zeros and ones, corresponding to the binary representation of the same word.

What, then, is the theoretical limit of transmission rate—the channel capacity—for a channel of bandwidth W in Hertz? It is clear that by increasing the number of bits per signal element one can not increase the transmission rate indefinitely. With less differentiation between distinct signal states, the detection and discrimination process becomes more and more difficult as the redundancy of the transmitted signal decreases. The maximum attainable transmission rate was postulated in 1949 by another Bell Laboratories mathematician, Claude Shannon. Shannon's theorem relates the channel bandwidth, the signal-to-noise ratio of the transmission channel, and the capacity of the transmission channel into the following equation:

$$C = W \times \log_2 (1 + S/N)$$

Where

1. C = Maximum channel capacity in bps
2. W = Channel bandwidth in Hz
3. S = Signal power in Watts
4. N = Noise power in Watts over the channel bandwidth
5. $\log_2$ = Logarithm to base 2

This formula finds the maximum theoretically possible transmission rate for a given transmission channel without consideration for circuitry, modulation type, error correction or coding required to approach this limit. It is somewhat similar to Albert Einstein saying that the speed of light cannot be exceeded without saying how it can be reached by a solid body traveling at high speed.

Figure 2-3 evaluates the previous formula for a voice-band channel with a bandwidth of 3000 Hz. A typical end-to-end dial-up voice connection in the United States will have a bandwidth of approximately 3000 Hz and a signal-to-noise ratio of 30 to 40 dB (decibels). Looking up the signal-to-noise (S/N) ratio of 35 dB yields a theoretical channel capacity of approximately 35,000 bps. Each additional 5 dB improvement in S/N increases the channel capacity by an additional 5,000 bps. Commercially available modems used on dial-up lines now reach 33,600 bps, which is close to that theoretical limit. As many telephone connections experience S/N ratios of less than 30 dB, it is not surprising that a 33.6K modem often connects at a lower rate than 33,600 bps. The x2[©] scheme pioneered by U.S. Robotics and discussed in more detail in Chapter 4 pushes the transmission speed above 50,000 bps by taking advantage of low-noise local loops and low-noise digital transmission through the backbone of the telephone network.

Figure 2-3 Maximum channel capacity for an ideal 3,000 Hz facility based on Shannon theorem

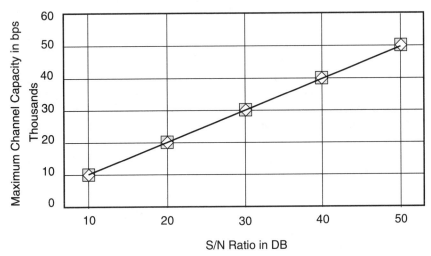

It should be noted that the signal-to-noise ratio of the Public Switched Telephone Network (PSTN) increased by about 10 dB over the last ten years. This is due mainly to a shift from analog transmission and switching equipment to digital transmission and switching equipment. With analog transmission, noise increases linearly with distance. An analog transmission facility 2000 miles in length will have twice the noise of a 1000 mile facility. If the noise is of a random nature, doubling the distance will result in deterioration of the signal-to-noise ratio by 3 dB. If the sources of noise are coherent, then the deterioration will be 6 dB.

On properly designed digital transmission facilities, noise can be made nearly independent of distance. Properly located digital repeaters restore the signal to its original shape. The main source of noise is then the quantizing noise resulting from the analog-to-digital (A/D) conversion of the signal originating with the sender, and digital-to-analog (D/A) conversion at the receiving end. This noise is independent of distance and depends only on the number of A/D and D/A conversions and the number of bits in the converted digital signal.

A larger number of bits results in a finer approximation of the original analog signal and a lower signal-to-noise ratio. It also makes the conversion process more complicated and requires a larger bandwidth to transmit the digital signal. The actual values are an economic compromise. Still, as mentioned earlier, the change to digital facilities over the last 10 years resulted in a considerable improvement in the quality of the long-distance telephone network.

Figure 2-4 shows the effect of quantization on an analog signal. An analog signal being transmitted over a digital circuit can only assume one of several specific discrete values, which is determined by the analog-to-digital conversion process. You can see the analog sine wave—the input signal to the A/D converter, the superimposed staircase output at the receiving end after passing through a D/A converter, and the difference between the output and input signals, which is the quantizing noise. The quantizing noise can be reduced by assigning more bits, resulting in a larger bandwidth in the A/D and D/A conversion. The quantizing noise can also be reduced with non-linear conversion by assigning proportionally more bits to weaker signals.

Figure 2-4 Quantizing noise due to A/D and D/A conversion

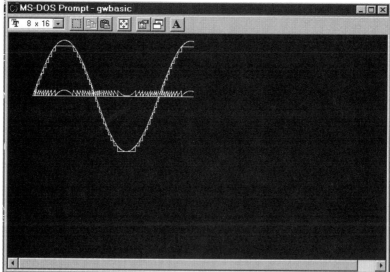

A "back-of-envelope" type computation shown in the next paragraph will find an approximate S/N ratio resulting from A/D and D/A conversions in the popular T1 digital carrier system.

1. The T1 carrier combines 24 voice channels, each 4 kHz wide, into a 1.544 Mbits per second data stream. The number of bits per voice channel is then:

1,544,000 bps/24 channels/(2 x 4000) samples per second = 8 bits/sample
2. The resulting quantizing/noise is $1/2^8 = 0.00391$
3. The effective noise is half of 0.00391. The signal-to-noise ratio, when compared with a signal of value 1, is then:

-10 log (.00391/2) = 27.1 dB

The actual S/N ratio of the T1 carrier will be a few dB better than 27.1 dB due to various circuit enhancements such as pre-compression of the analog signal before quantization.

As the theoretical transmission limit is approached, the transmitting and detecting circuitry is becoming exceedingly complex. It is much more difficult to detect amplitude and phase relations between successive signal samples using the quadrature amplitude modulation (QAM) technique than it is to simply distinguish between two frequencies using an FSK type of modulation. When transmission facilities that have better signal-to-noise ratios are developed (probably based mainly on fiber optics transmission), then both the theoretical and the practical transmission rate limits will be pushed up even further.

Measures of Transmission Rate—Baud, Bits and Bytes

The transmission rate, or speed of a modem, is alternately given in bits per second (bps), in characters per second (cps), in words per minute (wpm), and in Baud. This section explains the relations between these four units of measure.

The basic transmission rate of a modem is given in bits per second. It is typically 14,400, 28,800, or 33,600 bps. A modem rated at 28,800 bps will transmit a maximum of 28,800 binary digits of information—or 0s and 1s—each second. This is equal to transmitting one pulse, or absence of a pulse, every 0.035 ms. The effective data transmission rate of a synchronous modem will be close to this rate. The effective data transmission rate for an asynchronous modem will be about 30% less due to start, stop and parity bits. If data compression is in effect and some of the information is redundant, such as when a picture is being sent, the transmission rate between the computer and the modem can be up to four times higher.

When a computer is not sending any binary data, a synchronous modem will send a fixed sequence of bits, such as 010101..., to synchronize the receiving modem. An asynchronous modem will transmit a continuous sequence of 1s, which are interpreted as stop bits—such as 111111...

The data fed into the modem consists typically of ASCII characters, translated into a sequence of 1s and 0s. At least seven bits are required to transmit one ASCII character. Seven bits will produce 2^7, or 128 different combinations of 0s and 1s, which corresponds to the first half of the ASCII alphabet, and includes upper- and lower-case letters, digits and control codes. In most cases, when ASCII characters are transmitted as 7 or 8 bits, one byte (8 bits) is assigned to each character. The ASCII Table is presented in Appendix C. The eighth bit is required to transmit graphic data, so one of 2^8 (or 256) different combinations are possible.

In addition to the seven or eight data bits, asynchronous transmission also requires a start bit to indicate the start of a character, one or two stop bits to indicate when the transmission of a character is completed, and optionally one parity bit to check for errors in transmission. To transmit one character in asynchronous mode therefore requires 10 to 12 bits, including data, start, stop, and parity bits. Table 2-1 shows the number of bits required for each character, depending on the protocol used.

Table 2-1: Asynchronous Transmission Parameters

Data Length	Parity	Stop Bits	Total Length
7	None	2	10
7	Odd	1	10
7	Even	1	10
7	Mark	1	10
7	Space	1	10
8	None	1	10
8	Odd	1	11
8	Even	1	11
8	None	2	11
8	Odd	2	12
8	Even	2	12

To transmit an ASCII character in synchronous mode requires 7 to 8 bits per character and an occasional frame synchronizing byte.

The ratio between bits-per-second (bps) and characters-per-second (cps) is approximately ten to one. To convert characters per second into words per minute (wpm), multiply cps by 60 and divide by 6. This accounts for 60 seconds in a minute and assumes an average word length of five characters per word followed by a space. A 28,800 bps modem can therefore transmit approximately 2800 characters per second, or 28,000 words per minute. Transferring a 1 Megabyte file with a 28,800 modem working at full speed will therefore take about 1,000,000/2800 = 357 seconds, or approximately 6 minutes.

The other measure of transmission rate is the modulation rate in Baud, named after Emil Baudot, who invented the first constant-length teleprinter code in 1874. One Baud is equal to one code or signal element transmitted each second. If one code element represents one bit of information, like in the FSK 300 bps modem, then bps and Baud are the same. The now-obsolete Bell System 103A type modem using FSK operated at 300 bps and its modulation rate was 300 Baud. Higher-speed modems use the amplitude and the relative phase position of each transmitted signal to represent successive bits. The equivalent modulation rate in Baud is therefore only a fraction of the transmission rate in bps. Modem literature and specifications frequently confuse bits per second and Baud. They incorrectly use the same numbers for transmission rate and for the modulation rate in either units without consideration for the type of modulation.

To summarize, the primary measure of modem transmission rate is bits per second (bps), which is directly related to characters per second (cps) and to words per minute (wpm). The modulation rate in Baud is related to bps by a factor determined by the type of modulation. The modulation rate is mostly irrelevant to a modem user.

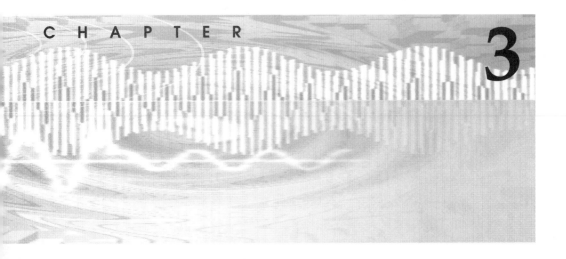

CHAPTER 3

Modulation Techniques

Modulation is a process where two input signals—a constant carrier frequency signal and a variable signal that carries the frequency information—combine to produce a unique output signal. The *black box* shown in Figure 3-1 can be a simple multiplier or the most complicated signal processor.

Figure 3-1 Signal Modulation

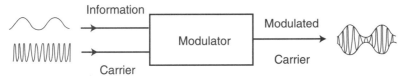

A digital signal transmitted over a voice-band telephone circuit contains significant energy components between DC and about three times the basic modulation rate, which is equal to the number of signal changes per second. For a modulation rate of 1000 Baud, the signal energy would therefore span the frequency range between 0 and 3000 Hz. Most of the signal energy would be concentrated at low frequencies below the modulation rate, or below 1000 Hz. The telephone network over which the digital signal is to be transmitted unfortunately will not pass frequencies below 300 Hz and above 3300 Hz.

The cutoff limits on circuits, which include underwater cable and satellite links, are even more stringent. The energy distribution of a digital signal and the bandwidth of a telephone network are superimposed in Figure 3-2. To fit the digital signal into the 300 to 3300 Hz band pass of the telephone network, the signal has to be modulated on carrier frequencies located in the flat portion of the bandwidth of the transmission facility.

Figure 3-2 Energy spectrum of a digital signal vs. bandwidth of a telephone channel

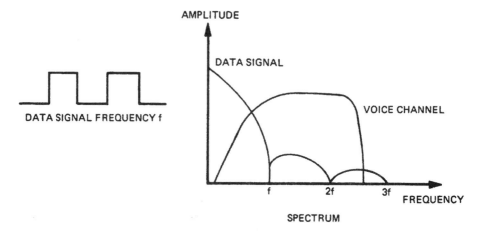

The instantaneous value of a carrier frequency Y can be represented as shown in Formula 3-1 at a given time t by a sinusoid having three parameters—the maximum amplitude A, the frequency F, and the phase P:

Formula 3-1 Instant value of carrier frequency signal

$$Y = A \times \mathrm{Sin}\,(2\pi F \times t + P)$$

Consequently, if the carrier signal is to carry digital information, all three parameters—amplitude, frequency and phase of the carrier wave—must be separately or jointly modulated by the digital signals. By proper processing of the modulated output signal, it should be possible to derive the information included in the input signal.

In modem technology, frequency modulation is only used at transmission rates of 300 bps or less; phase modulation is used at 1200 bps. The method used exclusively by the current modem technology at transmission rates above 1200 bps is a combination of amplitude and phase modulation. The type of modulation used by a particular modem depends on what protocol the modem adheres to—for instance, the Bell 212A or 202 standards, the CCITT/ITU V.34 standard, and so on. Most modems on the market today will support a wide range of protocols that become active by sending an appropriate software command to the modem.

A description of the four basic modulation methods—amplitude, frequency, phase, and amplitude/phase modulation—follows next in more detail. For a complete description of appropriate modulation methods, refer to the CCITT/ITU V. series modem recommendations listed in Chapter 4.

Amplitude Modulation (AM)

The simplest method of modulation is amplitude modulation (AM). It has a limited use for data transmission because of its susceptibility to noise and the need for a wide frequency bandwidth—twice the highest modulating frequency—for proper transmission. The main (but quickly declining) application for AM is in carrier telephony, specifically the frequency division multiplex (FDM) system, where one of two sidebands generated during the modulation process is suppressed in order to conserve the frequency spectrum.

Amplitude modulation is of course still widely used in the form of AM radio and television signals. The main advantage of amplitude modulation is the simplicity of the receiver demodulation circuit—a diode and a low-pass filter. However, this advantage has lost its importance with advances in large scale integration, the technology that packs complex circuitry into a very small package at minimal cost. An example of an amplitude-modulated signal is shown in Figure 3-3.

Figure 3-3 Simulated picture of an AM modulated signal

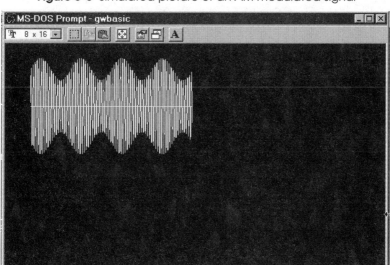

The fast-changing signal is the carrier wave, and the slow peaks and valleys represent the information content—voice or data.

By the year 2000, the amplitude modulation method will probably disappear from the telephony industry, as nearly all new facilities will be digital. In a few years, amplitude modulation will also disappear from TV transmission with the introduction of digital TV.

Frequency Modulation (FM)

Frequency modulation is also a relatively simple modulation technique that translates the digital 0s and 1s into preassigned discrete frequencies. The term used for this type of modulation is frequency shift keying (FSK). Figure 3-4 shows a simulated oscilloscope picture of an FSK-modulated signal. Unlike an AM signal, the amplitude of an FSK signal remains constant, but its frequency changes whenever data changes from a mark to a space, and vice versa. Frequency assignments used for the now-obsolete 300 bps Bell 103A FSK modem are shown in Table 3-1, while Figure 3-5 displays how these frequencies fit in the voice-band spectrum.

Figure 3-4 Simulated picture of an FSK modulated signal

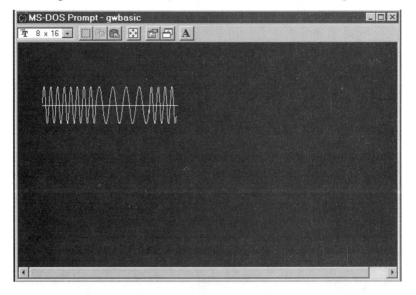

Figure 3-5 Bell 103 frequencies vs. bandwidth of a telephone channel

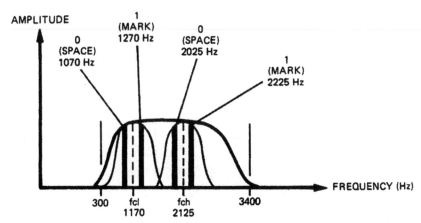

Table 3-1: Bell 103A Frequency Assignments

Direction of Transmission	Data Signal	Frequency
Originating	Mark (1)	1270 Hz
Originating	Space (0)	1070 Hz
Answering	Mark (1)	2225 Hz
Answering	Space (0)	2025 Hz

The circuitry required for FSK modulation and demodulation is relatively simple by today's standards, which makes this type of modem very inexpensive. The modulator requires only an electronic switch to select one of two frequencies for the originating or answering mode. The demodulator requires a phase-locked loop or four band pass filters, followed by a detector, to determine which frequency is being received.

No synchronization between the receiving and the transmitting modem is required for proper demodulation. An example of an early implementation of a Bell 103A-equivalent modem with an acoustic coupler is shown in Figure 3-6. The reason for acoustic coupling rather than a direct connection to a phone line was that in the pre-deregulation days, a customer-owned device was not allowed to be electrically connected to the telephone network. The design was based on a single-chip modem made by National Semiconductor Corporation.

In the 1990s, the only application for FSK technology was in some point-of-sale (POS) terminals and in process control applications.

Figure 3-6 Bell 103 implementation

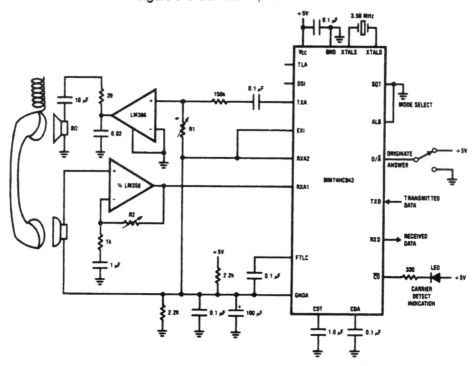

Phase Modulation (PM)

The only modem that used this modulation technique was the venerable 1200 bps Bell 212A. Although it is no longer manufactured, it can still be found in some older commercial installations. The modulation process used by the Bell 212A modem was the precursor of the modern modulation techniques. It was a variation of the simple phase modulation, called differential phase-shift keying (DPSK).

This technique uses phase changes relative to the previously-sent signal as indicators of the signal value. For example, in two consecutive sequences of data—00 and 11—the dibit 00 will give a phase change of 90 degrees, while the dibit 11 will change the current phase by 270 degrees.

This technique relies on incremental phase changes rather than on the absolute phase value of the carrier signal, as would be the case in pure phase modulation (PM). This method to detect the transmitted signal obviates the need for keeping an absolute time standard at the transmitting and at the receiving terminals. But a synchronization procedure is still required for all phase modulation techniques to ensure that the initial conditions are correct and that the received signals are properly interpreted.

The synchronization between the transmitting and receiving terminal is performed by sending a fixed pattern of 1s and 0s, which can be recognized by the receiving terminal during the modem-handshake sequence. This sequence occurs at the beginning of the transmission and at periodic intervals throughout the transmission of data.

An example of a simulated DPSK signal is shown in Figure 3-7. Notice that neither the maximum amplitude of the signal nor its frequency changes with modulation. The only parameter that changes is the phase of the signal.

Figure 3-7 Simulated picture of a DPSK modulated signal

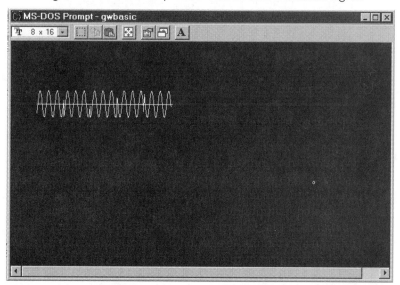

The two methods to detect incremental phase changes are the *comparison* and the *coherent detection* methods. The comparison method requires a circuit introducing a delay equal to the time period between two successive signal elements. The circuit compares the two signals—one of the signals is delayed, and the other is not delayed. The comparison of two signals that differ by a certain phase angle (delay) is accomplished by measuring the timing of the zero crossings of the respective sine waves.

The coherent detection of phase-modulated signals consists of comparing the received signal with an internal reference carrier that is kept in phase-lock with an internal clock. A decision is then made for each received signal element based upon which phase position it represents, depending on its relation to the local clock. The local clock receives its synchronization from the received data during the handshake phase and during the periodic synchronization sequences, and continuously adjusts itself based upon some error criteria.

Constellation Diagrams

The differential phase-shift keying (DPSK) modulation and the combined phase/amplitude modulation (QAM) discussed in the following section can best be described by means of the so-called *constellation diagrams*. A constellation diagram is a vector diagram that shows the phase and amplitude relations associated with each combination of bits. The constellation diagram shown in Figure 3-8 for the Bell 212A modem shows phase assignments for each dibit combination. The frequency and relative phase assignments for 212A modems are shown in Table 3-2. Figure 3-9 displays how these frequencies fit into the voice-band spectrum.

Figure 3-8 Constellation diagram
for the Bell 212A modem

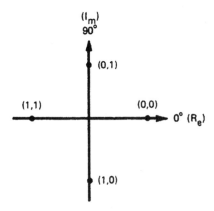

Another important interpretation of the constellation diagram is to consider two sinusoids, the second shifted by 90 degrees relative to the first. The amplitude of the first sinusoid is the projection of the

Figure 3-9 DPSK Bell 212A frequencies vs. bandwidth of a telephone channel

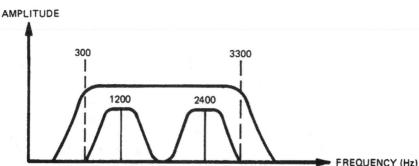

Table 3-2: Bell 212A Frequency and Phase Assignments

Direction of Transmission	Frequency (Hz)
Originating Modem Send	1200
Originating Modem Receive	2400
Answering Modem Send	2400
Answering Modem Receive	1200
Dibit Phase	**Shift (degrees)**
01	0
00	90
10	180
11	270

data point in the constellation diagram on the X axis; the amplitude of the second sinusoid is the projection of the data point on the Y axis. The modulated signal is then the sum of the two sinusoids.

This interpretation of the constellation diagram can be used to visualize the transmitted signal and also to design the modulating and demodulating circuitry. The modulator sets the proper amplitude of the two sinusoids before combining them and sending the resulting wave shape to the transmission line. The demodulator shifts the received modulated sine wave by 90 degrees, then recovers the original information from the shifted and unshifted components of the received signal.

Quadrature Amplitude Modulation (QAM)

The technique of simultaneous amplitude and phase modulation is called Quadrature Amplitude Modulation (QAM). It is being used in most modern modem designs at transmission speeds above 1200 bps. To be able to send data at higher transmission rates over the telephone network, several bits must be combined into a signal element. The number of possible states that the signal can assume is equal to 2^n (n = number of bits in a group). For instance, when five bits are combined into a group, 32 states are possible ($2^5 = 32$).

As explained in Chapter 2, the transmission rate in bits equals the modulation rate in Baud multiplied by the number of possible states in each group, divided by 2. The modulation rate should not exceed about 2000 for a full-duplex connection, or 2400 for a half-duplex or simplex (one way) connection. Advances in modem design made the full-duplex modem appear to the transmission line as a half-duplex modem, thus increasing the acceptable modulation rate above 2000 Baud. Combining five bits in a group would be acceptable for a full-duplex modem with a maximum transmission rate of 33,600 bps.

Figure 3-10 shows constellation diagrams for various 2400 bps, 4800 bps, and 9600 bps modulation standards. The constellation diagram of a 33,600 bps modem would look similar but would have over 100 points. Since the phase-changes of the signal are relative to the preceding time period, the same bit combinations appear in all four quadrants. For example, a 90-degree phase shift will move a phase vector from quadrant 1 to quadrant 2, from quadrant 2 to quadrant 3, and so on.

Figure 3-11 shows a modulated carrier wave according to the 2400 bps V.22 bis protocol. Notice the changes in both amplitude and phase due to the QAM modulation. Demodulation of a QAM signal is a complex operation, consisting of detection of phase differences, followed by a decision about the value of the received amplitude. Only the recent advances in very large scale integration (VLSI) made a QAM modem economically feasible, in particular for the personal computer user.

Figure 3-10 Constellation diagrams for 2400 bps,
4800 bps and 9600 bps modulation

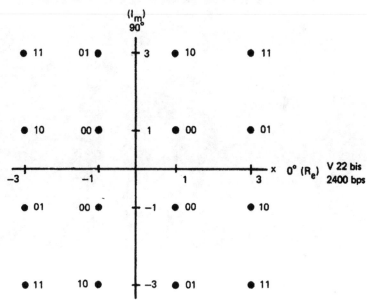

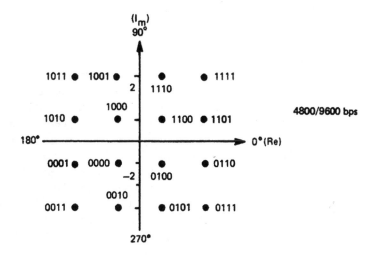

Figure 3-11 Simulated picture of a QAM modulated signal

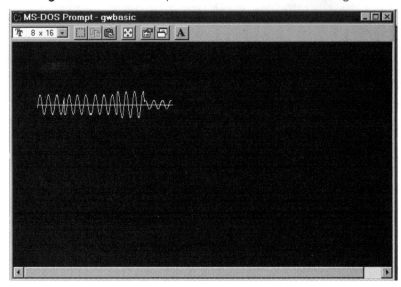

Trellis Modulation

Another trick up the sleeve of modem designers is the trellis modulation technique. Though the technique was first described in a 1967 paper by A.J. Viterbi, it was not implemented in modem design until the mid-1980s. All current modems transmitting at rates of 9600 bps and higher use this technique. In the trellis modulation scheme, the data stream is grouped into several bits at a time, and an additional redundant bit is generated for error correction from each group.

The trellis constellation diagram shown in Figure 3-12 shows the phase and amplitude values for each combination of five bits in each group. In this example, the fifth bit is used for error detection and correction by the receiving modem. The receiving modem uses the Viterbi decoder, which performs the so-called *soft-bit* detection.

In the more common *hard-bit* detection, the circuit decides—after reception of each bit—whether that bit represents a binary 0 or a binary 1. This decision depends on the amplitude and phase of the received bit when compared to a threshold value of the adjacent bit. After detection in the hard-bit detector, the bit value—as it was received—is discarded. In a soft-bit detection scheme, as implemented in the Viterbi decoder, the value of the received bit is compared with other bits in a group, and the actual decision whether a received bit is a 0 or a 1 is made only after a number of bits (up to 32) are received.

The name *trellis* comes from the picture of the decision tree used by the Viterbi decoder, which looks like a wooden trellis. The trellis modulation was first implemented in the V.32 modulation protocol. The actual decision tree is, however, not described in the V.32 protocol. The decision trees of various V.32 modems using trellis coding is proprietary and was independently arrived at by modem manufacturers.

Fortunately, in the last few years, de facto standards arose, such that modems from various manufacturers using trellis modulation are compatible with each other. The trellis modulation scheme is also referred to as *forward error correction* and provides 3 to 5 dB of additional noise immunity. As current high-speed modems use modulation rates of up to 2,400 Baud, the full-duplex signal occupies the entire voice-band spectrum.

Figure 3-12 Constellation diagram of the 9600 bps trellis coded signal

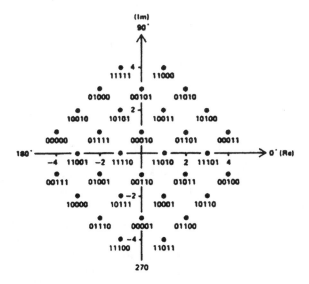

Susceptibility to Error

As the number of amplitude and phase levels increases with each successive—and more complex—modulation method, both the transmitting and the detecting circuitry become more complex. In the previous edition of this book, I used to say that they also became more costly. This is no longer true, as the price of high-speed good-quality modems always seems to hover between $100 and $200.

Simple FSK modulation requires only frequency discrimination; DPSK modulation requires detection of successive relative phase positions QAM requires detection of both phase and of amplitude changes—a formidable undertaking, in particular as the transmission rate increases in each successive modem generation. Finer distinctions between detected data signals in terms of phase and amplitude make it more difficult for the receiving modem to decide whether the signal is a 0 or a 1.

As discussed in Chapter 2, Shannon's theorem gives the upper bound of the transmission speed as a function of the signal-to-noise ratio for a given transmission facility. In the last 10 years the average S/N ratio of the telephone network—in particular, on long haul connections—improved by about 10 dB (from less than 30 dB to over 40 dB). This was due to the replacement of most analog facilities with digital facilities, and by the replacement of microwave transmission with fiber optics cable.

The obvious advantage of digital over analog systems is that random noise adds to the signal as transmission progresses in analog systems, and cannot be removed. In digital systems, it is possible, through proper facility design and error correction, to transmit a signal with an infinitesimally-small probability of error.

Modem design slowly approaches the Shannon limits, as the PSTN network improves, by introducing successfully more and more complex modulation schemes.

Computing Bit Error Rate

The bit error rate (BER) of a specific modulation method can be computed. Because the thermal noise is assumed to have a Gaussian distribution of its energy as function of time, BER values—as shown in Formula 3-2—use the Gaussian error function, or the complementary error function, as their arguments. These functions can easily be computed even on a simple programmable calculator. The theoretical bit error rates, as a function of signal-to-noise ratio, for various modulation methods are given by the following formulas:

Formula 3-2 Probability of error Pe vs. signal-to-noise ratio

1. On-Off Keying (OOK), Coherent

$$Pe = \frac{1}{2} Erfc\left(\frac{1}{2}\sqrt{\frac{S}{N}}\right)$$

2. Frequency Shift Keying (FSK), Coherent
Amplitude Shift Keying (ASK), Coherent
Pulse code Modulation (PCM), Unipolar

$$Pe = \frac{1}{2} Erfc \sqrt{\frac{S}{2N}}$$

3. PCM, Polar Phase Shift Keying (PSK)

$$Pe = \frac{1}{2} Erfc \sqrt{\frac{S}{N}}$$

4. FSK, Noncoherent

$$Pe = \frac{1}{2} e^{-\frac{S}{2N}}$$

5. Differential Phase Shift Keying (DPSK)

$$Pe = \frac{1}{2} e^{-\frac{S}{N}}$$

6. ASK, Noncoherent

$$Pe = \frac{1}{2} Erfc \left(\frac{1}{2} \sqrt{\frac{S}{N}} \right)$$

The function Erfc (complementary error function) related to the Gaussian noise distribution is defined as follows:

$$Erf\,(x) = \frac{2}{\sqrt{\pi}} \int_0^x e^{-t^2} dt \qquad Erfc(x) = 1 - Erf(x)$$

Actual BERs measured on a 3002 unconditioned line are shown in Figure 3-13. The instrument setup used for this measurement is shown in Chapter 17. The actual BER is not only a function of the modulation method and the type of telephone line used, but is also determined by the modem circuit design of a particular manufacturer. Modems from different manufacturers—and even different models from the same manufacturer—will have different noise sensitivities.

Modem design and manufacture involves many decisions where cost of production, reliability and skill of the designer all affect the final product. Tests performed on different modems (described in Chapter 16) let the consumer compare various modem implementations.

It should be noted that, for equivalent designs, the sensitivity to noise increases with the transmission rate.

Figure 3-13 Typical bit error rate curves
as a function of S/N ratio

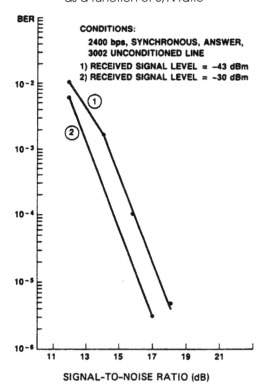

CONDITIONS:

2400 bps, SYNCHRONOUS, ANSWER, 3002 UNCONDITIONED LINE

1) RECEIVED SIGNAL LEVEL = −43 dBm
2) RECEIVED SIGNAL LEVEL = −30 dBm

SIGNAL-TO-NOISE RATIO (dB)

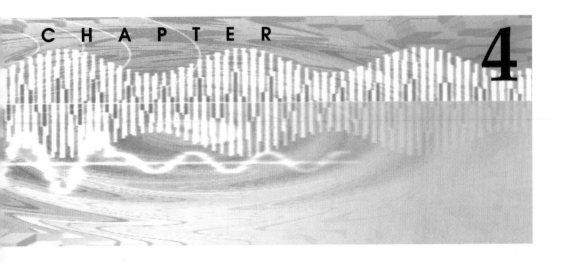

Data Communications Protocols

When you buy a new modem, the User Manual will usually list the modem specifications. In a modem purchased in 1997, you will typically see a set of specifications similar to that shown in Table 4-1.

Table 4-1: Typical Data/Fax Modem Specifications

Data Specifications	
V.34 bis, Plus	33,600 bps
V.34, V.FC	28,800 bps
V.32 bis	14,400 bps
V.32	9,600 bps
V.23	1,200/75 bps and 75/1200 bps
V.22 bis	2,400 bps
V.22	1,200 bps
V.21	300 bps
Bell 212A	1,200 bps
Bell 103	300 bps

Table 4-1: Typical Data/Fax Modem Specifications (continued)

Fax Specifications	
V.17	14,400 bps
V.29	9,600 bps
V.27 ter	4,800 bps
V.21 Ch. 2	300 bps
Error Correction: V.42/MNP 2-4	
Data Compression: V.42 bis/MNP 5/MNP 10	

What do all these cryptic words mean? That is the subject of this chapter. Modems, like people, need a common language to understand each other. Most modems, except for a few that use proprietary standards, follow one of the established standards called *protocols*. In the 1970s, the Bell system was the largest producer and designer of modems in the U.S., and its modem designs became the de-facto standards that were used by other modem makers.

These standards were later adopted as recommendations by the world standards organization, the Comité Consultatif International de Telegraphie et Telephonie, or CCITT for short. This standard-making organization was subsequently renamed the International Telecommunications Union - Telecommunication Standard-ization Sector, or ITU-T for short. Member organizations send their delegates to periodic meetings, mostly in Geneva, Switzerland. The meetings in this beautiful city are not just for fun, vacation and sightseeing. They require a lot of hard work and persuasion, arguing differences among members to arrive at mutually-acceptable telecommunications standards.

ITU-T V. Series Standards

The standards for data communication over the telephone network, and specially for modems, were developed by the ITU-T . These standards are named with the letter V as the first character, and are therefore known as the V series standards and recommendations. They describe data protocols, data translation, data compression, and so on. All standards in force as of December 1996 are listed next. The standards that are currently most important to the modem user are discussed in more detail later in this chapter.

V.1 - V.8 bis	General description of coding, symbol rate and power levels for modems
V.1 (12/72)	Equivalence between binary notation symbols and the significant conditions of a two-condition code
V.2 (11/80)	Power levels for data transmission over telephone lines
V.4 (11/88)	General structure of signals of international alphabet No. 5 code for character-oriented data transmission over public telephone networks
V.7 (11/88)	Definitions of terms concerning data communication over the telephone network
V.8 (9/94)	Procedures for starting sessions of data transmission over the general switched telephone network
V.8 bis (8/96)	Procedures for the identification and selection of common modes of operation between data circuit-terminating equipment (DCE) and data terminal equipment (DTE) over the general switched telephone network, and on leased point-to-point telephone-type circuits
V.10 - V.34	Interface and modulation standards for modems using voice-frequency telephone lines
V.10 (3/93)	Electrical characteristics for unbalanced double-current interchange circuits operating at data signaling rates nominally up to 100 Kbit/s
V.11 (10/96)	Electrical characteristics for balanced double-current interchange circuits operating at data signaling rates up to 10 Mbit/s
V.12 (8/95)	Electrical characteristics for balanced double-current interchange circuits for interfaces with data signaling rates up to 52 Mbit/s
V.13 (3/93)	Simulated carrier control
V.14 (3/93)	Transmission of start-stop characters over synchronous bearer channels
V.15 (10/84)	Use of acoustic coupling for data transmission
V.16 (10/76)	Medical analog data transmission modems

V.17 (2/91)	A 2-wire modem for facsimile applications with rates up to 14,400 bit/s
V.18 (10/96)	Operational and interworking requirements for DCE operating in the text telephone mode
V.19 (10/84)	Modems for parallel data transmission using telephone signaling frequencies
V.21 (10/84)	300 bits per second duplex modem standardized for use in the general switched telephone network
V.22 (11/88)	1200 bits per second duplex modem standardized for use in the general switched telephone network and on point-to-point 2-wire leased telephone-type circuits
V.22 bis (11/88)	2400 bits per second duplex modem using the frequency division technique standardized for use on the general switched telephone network and on point-to-point 2-wire leased telephone-type circuits
V.23 (11/88)	600/1200-Baud modem standardized for use in the general switched telephone network
V.24 (10/96)	List of definitions for interchange circuits between data terminal equipment (DTE) and data circuit-terminating equipment (DCE)
V.25 (10/96)	Automatic answering equipment and general procedures for automatic calling equipment on the general switched telephone network, including procedures for disabling of echo-control devices for both manually- and automatically-established calls
V.25 bis (10/96)	Synchronous and asynchronous automatic dialing procedures on switched networks
V.25 ter (8/95)	Serial asynchronous automatic dialing and control
V.25 ter An A (8/96)	Procedure for DTE-controlled call negotiation
V.26 (10/84)	2400 bits per second modem standardized for use on 4-wire leased telephone-type circuits

V.26 bis (10/84)	2400/1200 bits per second modem standardized for use in the general switched telephone network
V.26 ter (11/88)	2400 bits per second duplex modem using the echo-cancellation technique standardized for use on the general switched telephone network and on point-to-point 2-wire leased telephone-type circuits
V.27 (10/84)	4800 bits per second modem with manual equalizer standardized for use on leased telephone-type circuits
V.27 bis (10/84)	4800/2400 bits per second modem with automatic equalizer standardized for use on leased telephone-type circuits
V.27 ter (10/84)	4800/2400 bits per second modem standardized for use in the general switched telephone network
V.28 (3/93)	Electrical characteristics for unbalanced double-current interchange circuits
V.29 (11/88)	9600 bits per second modem standardized for use on point-to-point 4-wire leased telephone-type circuits
V.31 (12/72)	Electrical characteristics for single-current interchange circuits controlled by contact closure
V.31 bis (10/84)	Electrical characteristics for single-current interchange circuits using opto-couplers
V.32 (3/93)	A family of 2-wire duplex modems operating at data signaling rates of up to 9600 bit/s for use on the general switched telephone network and on leased telephone-type circuits
V.32 bis (2/91)	A duplex modem operating at data signaling rates of up to 14,400 bit/s for use on the general switched telephone network and on leased point-to-point 2-wire telephone-type circuits
V.33 (11/88)	14,400 bits per second modem standardized for use on point-to-point 4-wire leased telephone-type circuits

V.34 (10/96)	A modem operating at data signaling rates of up to 33,600 bit/s for use on the general switched telephone network and on leased point-to-point 2-wire telephone-type circuits
V.36 - V.38	Signal convention standards for wideband modems
V.36 (11/88)	Modems for synchronous data transmission using 60-108 kHz group band circuits
V.37 (11/88)	Synchronous data transmission at a data signaling rate higher than 72 kbit/s using 60-108 kHz group band circuits
V.38 (10/96)	A 48/56/64 kbit/s data circuit terminating equipment standardized for use on digital point-to-point leased circuits
V.41 - V.42	Error control conventions
V.41 (12/72)	Code-independent error-control system
V.42 (10/96)	Error-correcting procedures for DCE using asynchronous-to-synchronous conversion
V.42 bis (1/90)	Data compression procedures for data circuit-terminating equipment (DCE) using error-correction procedures
V.50 - V.80	Testing methods and noise measurement specifications
V.50 (10/68)	Standard limits for transmission quality-of-data transmission
V.51 (11/88)	Organization of the maintenance of international public switched telephone circuits used for data transmission
V.53 (10/68)	Limits for the maintenance of telephone-type circuits used for data transmission
V.54 (11/88)	Loop test devices for modems
V.55 (11/88)	Impulsive noise-measuring equipment for telephone-type circuits
V.56 (11/88)	Comparative tests of modems for use over telephone-type circuits

V.56 bis (8/95)	Network transmission model for evaluating modem performance over 2-wire voice-grade connections
V.56 ter (8/96)	Test procedure for evaluation of 2-wire 4-kHz voice-band duplex modems
V.58 (9/94)	Management information model for V-Series DCEs
V.61 (8/96)	A simultaneous voice-plus-data modem, operating at a voice-plus-data signaling rate of 4800 bit/s, with optional automatic switching to data-only signaling rates of up to 14,400 bit/s, for use on the general switched telephone network (GSTN) and on leased point-to-point 2-wire telephone type circuits
V.70 (8/96)	Procedures for the simultaneous transmission of data and digitally-encoded voice signals over the GSTN, or over 2-wire leased point-to-point telephone type circuits
V.75 (8/96)	DSVD terminal control procedures
V.76 (8/96)	Generic multiplexer using V.42 LAPM-based procedures
V.80 (8/96)	In-band DCE control and synchronous data modes for asynchronous DTE
V.100 - V.230	Inter-networking specifications, including ISDN
V.100 (10/84)	Interconnection between public data networks (PDNs) and the public switched telephone networks (PSTN)
V.110 (10/96)	Support of data terminal equipment with V-series type interfaces by an integrated services digital network (ISDN)
V.120 (10/96)	Support by an ISDN of data terminal equipment with V-series type interfaces with provision for statistical multiplexing
V.130 (8/95)	ISDN terminal adapter framework
V.230 (11/88)	General data communications interface layer 1 specification

Copies of these standards can be ordered from:

ITU - Sales
Place des Nations
CH-1211
Geneva
Switzerland

ITU-T publications can also be ordered at prices typically between $25 and $35 and downloaded on-line through the ITU web page at:

http://www.itu.int/publications

The de-facto standards for modems operating at 9600 bps or less were developed before the breakup of the Bell system. As seen from the list of specifications in Table 4-1, these older standards are still supported in the current modem design. Therefore if a modern 33,600 bps modem encounters an extremely poor telephone connection, it may fall back to the Bell System 103A protocol operating at only 300 bps. In addition to these de-facto transmission standards, all modems follow interface standards of either the U.S. Electronics Industries Association (EIA) or the ITU-T. These standards apply to electrical connections and types of plugs used between computers, terminals, printers, modems, and other assorted devices. The prime example of an interface developed by the EIA is the ubiquitous serial RS-232-C interface. The ITU-T equivalent corresponding to the RS-232-C interface is the V.24 standard. The ITU-T standards apply to current modem designs in the United States and in the rest of the world.

Adherence to standards is a necessary—but not sufficient—requirement to ensure that a modem made by one manufacturer will communicate adequately with a modem made by another manufacturer. Adherence to a standard only means that a standard is not violated and that the manufacturer follows a certain subset of requirements expressed in the standard. To differentiate between modems, manufacturers often enhance their products to make them unique and hopefully still be able to communicate with other modems that follow the same standard.

Another problem that affects modem compatibility is the length of time it takes before a standard—originally a recommendation—is adopted by the international body. It could take years. In the fast-paced unregulated commercial market, every modem manufacturer tries to be the first with a new product. Though a specific standard may only exist as a preliminary recommendation, someone will start manufacturing modems to fit that recommendation. Another manufacturer may follow a later recommendation of the same standard, which may be slightly different. Both will market their modems as *adhering* to the same standard, although that standard has not yet been formally adopted.

The result is that modems from different manufacturers, supposedly following the same standard, may not be able to talk to each other. Once the standard is adopted and implemented by modem chip makers, the situation clears up: the more reputable modem manufacturers provide customers with software or hardware patches, and modems from different manufacturers become compatible. Less reputable modem manufacturers simply disappear, turning their products into expensive door stops.

In the last few years, this was the case with the V.32 bis 14,400 bps modems, and later with the V.34 and V.FC 28,800 bps modems. Typically for about a year after a new modem introduction, a U.S. Robotics modem would not communicate with a Hayes modem at the top transmission rate, and vice versa. The same is currently happening with the V.34 bis and V.34+ 33,600 bps modems and with the proprietary 56,000 bps designs—the x2$^©$ design from U.S. Robotics, and the K56Flex$^©$ from Lucent Technologies and Rockwell International Corp. The compatibility between these designs will not be resolved until final standards are adopted. To confuse the situation even more, there is an obscure FCC regulation on the books that limits the maximum transmission rate for data calls made over analog loops to 53,000 bps. It was probably written many years ago to avoid competition with the Dataphone Digital Service (DDS) and to protect the telephone network from interference.

Many modems provide several standards from which a user can choose prior to establishing a connection. This is done by sending the appropriate commands to the modem. The selection of the best standard for a particular connection can also be done automatically by the modem based on responses of the remote modem during the handshaking sequence.

The few types of modems that do not follow established standards can be found in proprietary commercial applications. Examples of these types of modems are used in secure bank transactions, airline reservations, automatic teller machines (ATMs), power company control applications, and short-haul connections with Limited Distance modems. Such special applications may require private or leased transmission facilities. Furthermore, proprietary modems fill a certain small niche of high-speed, high-performance (and high-priced) devices using techniques such as multiplexing, data compression, and specialized error correction. As these devices are mostly used in point-to-point or multi point connections over leased transmission facilities, compatibility with other modems is of limited importance.

To illustrate how an international standards organization such as the ITU-T operates, this chapter starts with a description of how new modem standards are born. This is followed by a description of the importance of modem transmission speed. Then an introduction to data compression, error detection and error correction methods used in modem protocols is provided.

The next two sections, gives an overview of the ITU-T and the older Bell system standards currently used for data and fax transmission. These standards, though of only limited importance to the PC user, help in understanding how a specific modem works. The last section of this chapter describes a set of modem protocols developed by Microcom Corporation. These protocols have been widely adopted as de-facto standards by modem designers, and are called MNP protocols.

Birth of a Standard

Development of a new standard or recommendation is a slow, painful and circuitous procedure. Though this report was written in the early 1980s, it still describes well the current process of technical arguments and negotiations. This report is followed with a description of more recent negotiations about standards for high-speed modems. The development of the V.22 bis 2400 bps standard was extracted from a report by one of the CCITT delegates, Dale Walsh of U.S. Robotics.

> The 1200 bps V.22 recommendation was completed during the April 1980 meeting in Geneva. The session was particularly tough for U.S. manufacturers because of several hot issues—would the standard be compatible with Bell 212A or with Vadic 3400 modems? The 212A scheme won, but Vadic extracted an 'Alternative C' recommendation, which shares some of the features of Vadic 3400 but is neither compatible with Vadic 3400 nor with Bell 212A. An 'Alternative B,' compatible with Bell 212A, was then also adopted.
>
> At the same time groundwork began on 2400 bps full-duplex two-wire modems and was given a tentative name FDX 2400. The echo cancellation scheme was proposed to lower the cost of the modem. Two groups then started working on specifications—one on the low priced echo cancellation scheme and the other one

on a more expensive frequency division solution. The two groups with totally different approaches became firmly entrenched. The frequency division group also split into two subgroups suggesting different modulation/coding schemes, one using a modulation rate of 600 Baud, the other 800 Baud. During 1981 the U.S. Modem Working Party met periodically and agreed on the 600 Baud approach, which became the V.22 bis Recommendation, the current world-wide standard. To satisfy the first group, working on the echo cancellation scheme, that approach became the V.22 ter Recommendation. Both recommendations were formally adopted at the Plenary Assembly meeting of CCITT in October 1984.

Things did not change much in the following 15 years. The convoluted negotiation process of establishing standards continues. In the late 1980s, ITU-T established the TR-30 Group to work on V.Fast for modems faster than 14,400 bps. The V.Fast was to be renamed V.34 after adoption. In the meantime, modem chip manufacturers were divided into two camps—those supporting a public domain high-speed standard (V.32 terbo)—and those supporting a proprietary standard (V.FC) developed by Rockwell International. In 1994, modems appeared on the market claiming to be compatible with the not-yet-ratified V.34, and with V.FC, V.Fast and V.32 terbo. It took another year until the V.34 standard was approved. The confusion cleared, modem manufacturers offered upgrades and replaced some chips, and most 28,800 bps modems started talking to each other again.

High Speed Saves Money

In the previous edition of this book, a *high-speed modem* was defined as a modem with transmission rate higher than 4800 bps. Because the technology has advanced in the past 10 years, this edition defines a high-speed modem as one with transmission rate of over 28,800 bps. No doubt that in future editions, this "high speed" number will keep increasing.

Transmission speeds higher than 28,800 bps can be achieved in several ways. As discussed in previous chapters, the transmission line noise and other transmission impairments, such as attenuation and delay distortion and bandwidth, can all affect the transmission speed. In the beginning of 1997, the maximum modem speed that could be achieved using analog transmission over a switched telephone network and an analog copper loop between subscriber and the telephone office was 33,600 bps. The increase from 28,800 to

33,600 was achieved by improvements in line noise and by "tweaking" the V.34 standard.

Several new technologies are in the process of breaking this barrier. One approach is to rely on asymmetry of data traffic—using a heavy load from the network to the subscriber and a light load from the subscriber to the network. While the upload traffic from subscriber to the network consists mostly of short alphanumeric slowly-typed queries and requests, the download traffic from the network to the subscriber consists often of large audio-visual files. More and larger files are downloaded than are uploaded.

To speed up the download traffic, a service provider can, at a moderate cost per subscriber, arrange for digital transmission to the nearest central office. This will improve the noise level, as the main contributor will now be the short local loop from the central office to the subscriber. This new technology, announced first by U.S. Robotics at the end of 1996, allows for downstream transmission speeds of up to 56 kbps. An older but quickly-expanding technology, based on a digital connection between the subscriber and the central office, is the integrated services digital network (ISDN), which allows for even faster speeds of up to 118.2 kbps.

The continuing expansion of cable TV is leading to another quantum leap of transmission speed due to the higher bandwidth of coaxial cable compared to the traditional copper-based telephone network. A number of trials have been conducted with thousands of cable subscribers to test the feasibility of providing Internet service with cable modems. The first commercial service was introduced in December 1996 by a cable company on Long Island.

The cost of a phone call does not depend on how much information is being transmitted. Therefore there is a strong incentive to speed up the information flow, in particular for file transfers. The size, in bytes, of a file containing text is no longer the number of characters of plain text. Formatting and embedded images can easily convert a memo of a few pages into a 1-Mbyte file. Speeding up transmission of information will not only save time but it will also reduce telephone charges. Table 4-2 shows how long it will take to transmit a 1-Mbyte file and how much it will cost assuming a 10-cent-per-minute rate. As mentioned in Chapter 2, it takes approximately 10 bits to transmit one character (or one byte). Therefore the formula to find time t in minutes to transmit a file of n bytes with a modem operating at b bits per second is:

$$t = n \times 10/b/60$$

Table 4-2: Time required and cost of transmitting a 1 Mb file

Trans. Speed (bps)	Trans. Time (minutes)	Cost at $0.10/min. (dollars)
300	555	55.50
1200	139	13.90
2400	69	6.90
4800	35	3.50
9600	17	1.70
1440	12	1.20
28800	6	0.60
33600	5	0.50
56000	3	0.30

Note:

Table 4-2 assumes no time-loss due to error correction, block re-transmission, and so on. The actual data throughput, in particular at higher transmission rates, will typically be 10% to 20% slower. On the other hand, a substantial throughput improvement can be obtained by using data compression.

Data Compression and Error Detection

Various techniques have been developed to speed up the transmission of information. The transmission medium—the switched telephone network—puts inherent limitations on the transmission rate. As discussed in previous chapters, noise and distortion of the analog voice-band telephone lines limit the transmission rate to approximately 35,000 bps. Use of fully-digital transmission schemes such as ISDN, or use of cable modems and coaxial cable, can increase this rate. Use of partially-digital transmission and of asymmetric transmission (high rate from the network and low rate to the network), as in the x2 (c) technology, can raise the transmission rate on the switched telephone network to 56 kbps. While the transmission medium and technology limit the transmission rate, a faster transfer of information can be achieved by using data compression.

The concept of data compression is quite simple. A file being transmitted, be it text, video or sound, contains much redundant information. For example, 100 spaces between two paragraphs can be transmitted as 100 separate bytes of value 32—the ASCII-equivalent of a space—or can be transmitted as the number 100, followed by a repeat symbol, followed by 32. The only requirement is that both the sending and the receiving modem, or the communications software, understand this shorthand language. Even associating individual characters with single bytes involves redundancy and waste. A byte can assume one of 256 values, while ASCII characters, including control characters, can only assume one of 128 values. Therefore sending ASCII text with proper encoding could double the rate of information transfer.

Often, the variable-length Huffman code instead of the fixed-length ASCII code is used to represent individual characters. This approach is similar to the Morse code, which assigns a single bit— "."—to the most common letter in the English alphabet—E. Frequently-occurring words, such as "the," "and," "yours," "invoice," or single digits, can also be compressed by smart algorithms, which analyze the data before transmitting it.

A further enhancement of compression techniques is the Lempel Ziv Welch encoding scheme, implemented in the V.42 bis data compression protocol, as one of the alternatives (the other alternative is the MNP4 protocol). The heart of the Lempel Ziv Welch algorithm is a dynamically-generated dictionary of words encountered most frequently during data transmission. The same dictionary is generated at the receiving and the sending modem. When a dictionary word is transmitted for the second time, only its dictionary entry code, rather than the word itself, is sent.

There are two possible approaches to data compression. The first is to perform data compression offline on files residing in the local computer before data is sent; the second is to compress data by the local modem in real time while it is being transmitted.

Offline compression can be done by means of archiving and de-archiving utility programs such as the popular PKZIP/PKUNZIP from the PKWARE Inc. in Brown Deer, Wisconsin. A file is first "zipped" on the local computer, generating a new file that is in general substantially smaller than the original file. Then the zipped file is transmitted to the remote computer. At the receiving computer, the zipped file is decompressed by the same archiving/dearchiving utility and restored to its original form.

In real-time compression, assignment of specific redundant data strings to shorter compressed strings is a dynamic process and is repeated during a data call. To perform data compression in real time requires buffering and analyzing transmitted data by the sending and receiving modems. This method of data compression became practical with the development of inexpensive and fast microprocessors that handle the digital signal processing in the modem chip sets.

Data is transmitted one block at a time. To further maximize data throughput, the length of a block of data can also be changed dynamically and made to change with the quality of the transmission facility. A quiet line, which requires few re-transmissions, would have long blocks; a noisy line would have short blocks. In some compression schemes, a modem may also operate in synchronous mode by stripping the start and stop bits.

Compression of data always requires decompression at the receiving end. If your modem is capable of data compression and this feature is enabled by sending a proper initialization string to the local modem, your modem will query the remote modem about its capabilities during the handshaking sequence. Only if the handshaking sequence shows that the remote modem supports the same data compression protocol as your modem, data compression in your modem will be activated.

Typical gains in throughput transmission obtained through data compression are 50% to 100%. Even larger gains can be obtained when transmitting graphic image files, which contain much redundancy. Though some modem manufacturers claim gains of 300% to 400% due to compression, such claims should be taken with a grain of salt.

The major modem developer that uses built-in data compression protocols is the Microcom Corporation, with their proprietary MNP protocols. In general, these protocols perform both data compression and error correction. The protocols have been licensed to most modem and modem-chip manufacturers, and have been implemented in nearly all personal computer modems in use today. In addition to MNP protocols, many modems also support the ITU-T-promoted V.42 and V.42 bis data compression and error correction protocols. As MNP protocols became de-facto standards in the industry, the V.42 protocols are compatible with MNP protocols and include them as acceptable options.

To take advantage of automatic real-time data compression, not only the local and the remote modem must have the same protocols embedded in their firmware, but the serial port of the local computer must also be capable of transmitting uncompressed data to the modem at a rate higher than the transmission rate. This higher rate is often called the Local Transmission Rate. For this reason, terminal and browser programs allow you to select a transmission rate higher than that of the modem itself. For example, for a 28,800 bps modem, the suggested Local Transmission Rate is 115,200 bps. Certain older UARTs (Universal Asynchronous Receiver Transmitters)—the chips that convert data for transmission—will not support this higher rate. Therefore you should ensure that your computer is equipped with one of the higher-speed UARTs such as the 16550. You will find a more detailed discussion of UARTs in Chapter 6. Modem diagnostic programs, such as Microsoft's MSD or the WhatCom program on the enclosed disk, will display the UART type in your system. MSD and WhatCom are described in detail in Chapter 16.

A direct file transmission without an error detection protocol occurs at the full modem transmission rate. Therefore a 28,800 bps modem will transmit approximately 2,800 characters per second. However, once an error-controlling protocol is in effect, the transmission will slow down, depending on the line conditions and the need to re-transmit blocks of data to correct errors. Even without block re-transmission, the throughput will slow down. The reason is that, in addition to actual data, a checksum is transmitted with each block of data. At the receiving end, the checksum is compared with the checksum computed from the block of received data. If the two checksums do not agree, then a special code called Automatic Resend Request (ARQ) is issued by the receiving modem and the block of data is re-transmitted.

A typical loss in data throughput due to error detection and correction is 10% to 20%, but can get considerably worse. A recent study by the Software Digest Ratings Newsletter showed the ratio of effective transmission speeds between the fastest and the slowest communications software package was 3 to 1. This information is usually not found in the software specifications and can only be determined by an actual test that consists of sending a file of several pages over a telephone line to a computer equipped with a similar modem and software. Such a test may also show one software package to be better on a high-quality line, since it might use large blocks of data, and another package might be better on a poor, noisy connection if it uses short blocks. The best overall transmission speed should be obtained from a communications protocol and software using variable block sizes. Similarly, different modems—even from the same manufacturer—can exhibit different sensitivity to noise and other transmission impairments.

Current Standards for Data Transmission

A description of the data transmission protocols developed by the former Bell System, the CCITT and the ITU-T, which will be found in the current crop of modems used in personal computers, follows. Though some of these standards are now obsolete, they are still implemented to maintain backward compatibility.

Table 4-3 summarizes the most common Bell and ITU-T data transmission standards implemented by modem manufacturers who sell their products to the personal computer industry. A description of transmission standards and modems used in some commercial applications is found in Chapter 13.

Table 4-3: Voice-Band Modem Data Transmission Standards

Standard	Speed (bps)	Remarks
Bell 103	300	US standard
Bell 212A	1,200	US standard
ITU-T V.21	300	European low speed standard
ITU-T V.22	1,200	European low speed standard
ITU-T V.22 bis	2,400	World low-speed standard
ITU-T V.23	1,200	Old European standard
ITU-T V.32	9,600	World Medium speed standard
ITU-T V.32 bis	14,400	World Medium speed standard
V.FC	28,800	Temporary de-facto standard
ITU-T V.34	28,800	World high-speed standard
ITU-T V.34 bis	33,600	Temporary de-facto standard
x2©, 56Flex©	56,000	Proprietary standards

Table 4-4 shows the approximate equivalents between Bell and ITU-T modem standards. It should be remembered that although Bell and ITU-T modems may use the same modulation protocols and bit rates, they may differ in other parameters, such as timing of handshake sequences, fall-back transmission speeds, and so on.

Table 4-4: Equivalent Bell and ITU-T Voice-Band Modems

Speed	Bell Standard	ITU-T	Mode
300	103	V.21	Full-duplex
300/1200	212A	V.22	Full-duplex

Following is a short description of the Bell and ITU-T modem standards implemented in personal computer modems.

Bell 103/103A/108/113

The grand-daddy of personal computer modems, the 300 bps asynchronous, full-duplex Bell 103/103A/108/113, uses frequency shift keying (FSK) modulation. Though obsolete, it can still be found in some older installations and in specialized applications. The FSK modulation scheme used by this modem was described in detail in Chapter 3. The assorted versions of this modem, such as 103J, 108F and 113B, differ only in specific implementations for the public switched telephone network, TELEX, TWX, or for two- or four-wire leased facilities. Most of these modems are equipped with voice/data switches, automatic-answer capability, and can work with the standard 500/2500-type telephone sets. The modem is basically transparent to the computer/terminal and it can operate at any asynchronous speed not exceeding 300 bps. It is implemented in most modem chip sets as the lowest fall-back alternative when the quality of the transmission facility is extremely poor.

Bell 212A

The Bell 212A-type modem developed in the 1970s was one of the most popular standards in the United States for both commercial and personal computer modems. The modem operates in full-duplex mode over the public telephone network facilities. Though the Bell modem standards are not used in Europe, a Bell 212A or a compatible modem should still be able to communicate with a European V.22 type modem operating at 1200 bps.

The Bell 212A-type modem features 300/1200 bps transmission speed and is, by now, obsolete. When operating at 300 bps, the 212A modem is equivalent to the Bell 103 standard. At 1,200 bps, the modem uses differential phase shift keying (DPSK) modulation, which provides excellent results on the public switched telephone network. Two bits are encoded in each signal element, resulting in a modulation rate of 600 Baud in full-duplex mode. Separate carrier frequencies are used for both directions: 1,200 Hz for the originating modem, and 2,400 Hz for the answering modem.

At 1,200 bps, the 212A-type modem can operate in either synchronous or asynchronous mode. At 300 bps and slower transmission speeds, the modem operates only in asynchronous mode. In asynchronous mode, the modem can use any character format with an arbitrary number of data, start and stop bits per character. In synchronous mode, it uses between 8 and 11 bits per character. The number of bits per character is selected with software, jumper or strapping options.

The 212A-type modems used in personal computing field were all equipped with pulse and touch-tone dialers, which could be operated by sending appropriate ASCII commands from the computer or terminal to the modem. The modems were also used with an automatic dialer to save the computer operator the effort of manually dialing and re-dialing and saving the cost of a telephone set.

It may be difficult to believe, but in the 1970s and even in the 1980s, modems without dialers were quite common. They were usually provided with voice/data switches and RJ11 jacks for a telephone set connection. The 212A-type modems operated in the Originate mode until a ring signal was detected. Because a ring signal is usually not provided on leased two-wire lines, an option switch was required for such an operation to force the receiving modem into Answer mode.

ITU-T V.21

The currently-obsolete European version of the Bell 103 300 bps full-duplex modem was also designed for use on the public telephone network. The modem was using the same modulation scheme (FSK) and carrier frequencies as the Bell 103 modem, but was not fully compatible with it.

ITU-T V.22

This currently-obsolete 1200 bps modem standard covered a whole family of full-duplex devices for use on the public telephone network and on two-wire private-line circuits. The standard is very similar to the Bell System 212A. In general, a V.22 and a Bell 212A modem can exchange data with each other. However, the Bell 212A modem could not be officially sold in Europe. The principal characteristics of the V.22 modem are channel separation by frequency division, with carrier frequencies at 1200 Hz and 2400 Hz, inclusion of a scrambler, and DPSK modulation at 600 Baud (2 bits/transmitted signal). The V.22 (but not the 212A) modem was also capable of transmitting a guard tone of 1800 Hz used for echo suppression on many European telephone networks.

The two major versions of this modem are the V.22A for synchronous operation and the V.22B for synchronous or asynchronous operation.

ITU-T V.22 bis

This now-obsolete design incorporated many modern modem features. The standard is similar to V.22 in that it is designed for use in full-duplex mode on the public switched telephone network and on two-wire private-line circuits. The principal characteristics of the V.22 bis modem are channel separation by frequency division, with carrier frequencies at 1200 Hz and 2400 Hz, inclusion of a scrambler, QAM at 600 Baud (4 bits/transmitted signal), inclusion of self-test facilities, inclusion of an adaptive and a compromise equalizer, and 1200/2400 bps operation in synchronous or asynchronous mode. The V.22 bis standard and its various subsets were, for many years in the 1980s, the leading standard in the personal modem field.

ITU-T V.23

The now-obsolete 1200 bps low-speed modem was used mostly in Europe on the public telephone network in half-duplex mode. The modem employs FSK modulation at 1200 bps for its main channel and the ON/OFF Keying (OOK) modulation for the slow-speed 75-bps reverse channel. The modem has two modes of operation. Mode 1 uses modulation rates of up to 600 Baud with mark/space carrier frequencies of 1300 Hz and 1700 Hz. In Mode 2, which corresponds to higher modulation rates of up to 1200 Baud, the corresponding carrier frequencies are 1300 Hz and 2100 Hz. The ability to operate at lower frequencies in Mode 1 was helpful on local loops that were equipped with loading coils, which would otherwise interfere with higher carrier frequencies.

ITU-T V.32

The 9600-bps standard provides for synchronous operation at 4800 bps and at 9600 bps. For several years in the 1980s, it was the highest transmission speed standard designed for operation over two-wire public switched telephone network. A V.32-type modem can operate in half-duplex or full-duplex mode with a single carrier frequency located in the middle of the voiceband. The full-duplex operation is similar to voice conversation on a two-wire circuit. A hybrid coil separates the two directions of transmission, although they both use the same frequency spectrum. An important adjunct circuit for full-duplex operation is an echo canceler, which partially cancels both the high amplitude near-end echo, and the delayed low-amplitude far-end echo, which would otherwise be returned to the transmitting modem and be treated as a received data stream.

The method of echo cancellation was not described in the V.32 standard and was left to the individual modem manufacturer to implement. The methods used were proprietary and made modems from different manufacturers exhibit different bit error rates on the same call. The carrier frequency for V.32 is 1800 Hz for both directions of transmission and each signal element consists of 4 bits. The modulation rate is 2400 Baud.

At 9600 bps, the V.32 standard mentions two alternative modulation methods. As both modes are usually implemented in commercial modems, care should be taken that the receiving and the transmitting modem follow the same modulation method. The first method assigns 16 states in the constellation diagram; the second method assigns 32 states. In the second method of operation, the V.32 modem uses trellis coding and modulation, and an error correcting scheme employing a special algorithm that generates a fifth parity bit based on the four data bits. The trellis coding and modulation scheme gives the extra performance edge to the V.32 standard such that modems that support the trellis scheme could satisfactorily operate in the 1980s at 9600 bps over the public telephone network. The cost of the V.32 modem at its inception was around $2500.

ITU-T V.32 bis

The V.32 bis was the first in the series of high-speed modems that are still in use in 1997. When it was first announced in the early 1980s, its speed of 14,400 bps was considered a major technical breakthrough. Its constellation diagram features 128 points and trellis coding is one of the options.

V.FC

This proprietary standard was developed by Rockwell International while waiting for adoption of the V.34 standard. The standard supports modems operating at speeds of up to 28,800 bps over the PSTN. At the time of introduction in late 1993, the cost of a V.FC modem was around $500. The V.FC standard is not compatible with the current V.34 28,800 bps standard. If a strictly V.FC modem encounters a V.34 modem, they will connect after handshaking at the fall-back intermediate V.32 rate of 14,400 bps.

ITU-T V.34

This 28,800 bps standard is currently supported by most modem manufacturers. It was adopted by ITU-T in late 1995. The standard provides many advanced features such as non-linear encoding and trellis coding to improve immunity to noise. The standard provides over 100 options, which may or may not be implemented in a particular modem chip set.

V.34 bis/V.34-1996/V.34+

In the mid 1990s, the pace of modem development increased considerably and prices of modems started falling at nearly the same rate as the prices of personal computers. Only a year after approval of the V.34 28,800 modem standard, ITU-T was considering for approval an improved version of that standard, named V.34 bis. This standard extends the maximum speed at which modems can operate on the PSTN to 33,600 bps. At the writing of this book, there were several competing standards for 33,600 bps operation with names such as V.34+ and V.34 Plus. Although many modems currently on the market claim a 33,600 bps speed, the actual rate established on a connection is often less. It is expected that by the end of 1997, modem chip manufacturers will agree on the common subset of features such that a 33,600 bps connection with a V.34 bis modem becomes a standard rather than an exception.

The following overview of the 33.6 kbps modem standards is based on a white-paper from Circom Corporation.

> Before initial ratification of the V.34 standard during the summer of 1995, Study Group 14 of The International Telecommunications Union-Telecom (ITU-T) began developing extensions and improvements to this important standard. Today, a new version known as V.34-1996 provides for two additional, optional data transmission speeds of 31.2 and 33.6 kbps. Further enhancements to supporting protocols allow devices implementing V.34-1996 to deliver more robust and more frequent 26.4 and 28.8 kbps connections.
>
> **The Need For Speed: Faster Internet and Intranet Access**
> With additional, optional speeds of 31.2 and 33.6 kbps, modems implementing the V.34-1996 standard can communicate at speeds of up to 16.6% faster than existing V.34 modems. To the end user, this means improvements when transferring files, downloading information from online services, accessing the Internet, or corporate intranets (these sites must also use compatible modems that support the optional higher speeds).
>
> **The Need For Reliability: Better High-Speed Connectivity**
> All modems are susceptible to noisy phone lines, often resulting in the negotiation of a connection speed less than the speed that is printed on the box. Users purchasing V.34 modems who expect consistent 28.8 kbps performance are often disappointed, rarely experiencing connections at that speed. Rather, they observe that the two modems will frequently "downshift" and negotiate a lower speed in the 21.6 or 24.0 kbps range.
>
> Improvements in software algorithms specified in V.34-1996 provide more frequent, stable connections at speeds of 26.4

and 28.8 kbps than with previous-generation V.34 modems. During initial trials using real-world phone lines, tests indicated that about 60 % of V.34 connections would increase by 2.4 to 4.8 kbps.

Haves vs. Have Nots: Analog vs. Digital Phone Lines

Many so-called Internet experts proclaim that if you really want to surf the Web, then you must subscribe to expensive ISDN service and purchase or lease expensive terminal adapters (TAs) to efficiently access graphics-laden Web pages. However, in many geographic locations, it is still difficult, tedious, and expensive to obtain reliable ISDN service to a home or remote office. If you are able to get good ISDN service, most ISDN TAs still communicate with your system through a serial port limited to speeds of 115.2 kbps. Additionally, in some cases bottlenecks accessing the Web are not between the user's modem or TA and that of their Internet service provider (ISP). Instead, the bottleneck is elsewhere on the Internet backbone, and therefore out of the direct control of the user and their equipment.

In these situations, both Internet neophytes and power-surfers can still get a lot of bang for their buck by upgrading to V.34-1996 modems. When compared to earlier V.34 implementations, users should reliably see a 2.4 to 4.8 kbps improvement in modem connections.

What's In A Name? V.34, The Sequel

Several different names were used to describe this new revision of the V.34 standard. Rockwell originally suggested V.34+ or V.34 Plus, but withdrew this proposal in May 1996. Other companies, such as Lucent Technologies (formerly AT&T Microelectronics) have referred to it in terms of "extended rate V.34." Some modem manufacturers also refer to the term "V.34 bis."

In October 1996, Study Group 14 of the ITU-T standards committee finalized the naming of the new standard as V.34-1996.

V.34-1996: Superior Performance Over Original V.34

There are four areas of improvement that distinguish devices implementing V.34-1996 from those using the initial version of the standard.

Testing by various manufacturers indicates that on about 60% of networks currently supporting 26.4 kbps data transmission, the enhancements in V.34-1996 offer 2.4 to 4.8 kbps improvement in connection speeds.

V.8 bis

The original V.34 standard includes a component protocol known as V.8 bis. This protocol specifies the negotiation startup, or handshaking, procedures used between modems before a data exchange. The V.34-1996 proposal includes an updated startup protocol, V.8 bis, which provides a faster connection initialization.

Additionally, while certain types of echo-canceling equipment previously caused V.8 to fall back to V.32 bis auto-mode negotiation (limiting speed to a 14.4 kbps maximum), V.8 bis delivers a true V.34-protocol connection. V.8 bis also improves faxing, reduces connection delays, and provides more reliable support when switching between fax and telephone operation.

Signaling System 5 Problem Resolved

Most modern telephone networks in the United States use Signaling System 7 (SS7) protocols to manage data transmission between central office (CO) switches. However, some older COs still use an earlier version known as Signaling System 5 (SS5). Two first-generation V.34 modems communicating between COs using SS5 occasionally experience connection failures. In V.34-1996, the startup algorithms are modified, allowing successful operation on older networks using SS5.

While not a panacea, the improvements detailed in the V.34-1996 standard will prove valuable in many modem applications. Users disenchanted with the performance of first-generation V.FC or V.34 modems over moderately-impaired phone lines can reasonably expect improved connectivity and data throughput using V.34-1996 devices. Extensive testing by Xircom, Lucent Technologies, and other companies ensures that this latest version of V.34 provides recognizable benefits in real-world usage.

The x2© Design

The x2 is a proprietary design, possibly a future standard, announced by U.S. Robotics in mid 1996 and brought to market in the first quarter of 1997. It is an asymmetric design, in that the data transfer rate from the computer user to the service provider—typically an Internet site or a bulletin board—is only 28,800 bps or 33,600 bps, while the data transfer rate from the service provider to the computer user is supposed to reach 56,000 bps. The asymmetry should be acceptable to users, as in general data transmission from the personal computer user to the service provider consists of short queries, while the much heavier traffic from the service provider to the computer user usually consists of large text, graphics and multimedia files.

Based on limited information released by U.S. Robotics, the x2© approach to faster modem operation, which overcomes the theoretical limitations imposed on standard, analog modems, is as follows:

1. It is made possible by the fact that Internet service providers (ISPs) usually connect digitally with a T1 or an ISDN line to the telephone company trunks. Therefore in the direction from the ISP to the subscriber, the only analog portion of the phone network is the analog loop that connects the subscriber

to the telephone company's local Class 5 office. Over the past two decades the telephone companies have been replacing portions of their original analog network with digital facilities. But the slowest portion of the network to change has been the local connection to the local Class 5 telephone office. That connection will likely remain analog for some years to come.

2. The x2 design also takes advantage of the fact that the service providers encode data with the 256 PCM codes used in the digital portion of the telephone network. Therefore there is no additional quantization noise associated with converting discrete PCM codes to analog-type signals. The PCM codes are converted by the service provider to discrete analog voltages and sent to the subscriber modem via an analog low-noise loop circuit. The analog loop between the telephone office and the subscriber is therefore converted into a quasi-digital line. Finally, the subscriber's modem decodes the discrete PCM codes from the analog signals it receives, reconstructing what the transmitting modem has sent.

The x2 scheme will only work on non-multiplexed low-noise analog local loops. Unfortunately, on local loops where the telephone company multiplexes several loops onto one pair of copper wires by means of the subscriber line concentrator (SLC) scheme, x2 will not work. U.S. Robotics provides a test discussed in Chapter 17 to determine if SLC is present. If it is detected, the only hope for a faster transmission speed is to ask your telephone company for a separate loop to the local telephone office. As the telephone company is not required to do it under the current tariffs, you may not have much luck.

The x2 will also not work on extensions to a private digital branch exchange (PBX). A modern PBX has its own multiplexing scheme, which includes A/D and D/A conversions that would interfere with the x2 transmission.

The x2 design will generally require changes to the wiring and equipment at service provider sites, all the subscriber needs is an x2 modem.

A recent study of x2 performance was reported in various Usenet news groups. The study summarized in Table 4-5 shows the distribution of connection rates on calls made to Internet Service Providers equipped with x2 modems. Though no connection reached 56 kbps, the increase in speed over the previous generation of modems is quite impressive.

Table 4-5: Connection Speed Distribution of x2 Calls

Connection Rate	# Calls	Percentage
52000	2	0.1%
50666	10	0.5%
49333	71	3.7%
48000	159	8.3%
46666	567	29.6%
45333	317	16.6%
44000	132	6.9%
42666	189	9.9%
41333	82	4.3%
40000	39	2.0%
38666	24	1.3%
37333	72	3.8%
36000	45	2.3%
33333	126	6.6%
32000	20	1.0%
29333	60	3.1%

It should be noted that Lucent Technologies and Rockwell International Corporation, as mentioned earlier, announced a competing high-speed 56,000 bps modem design called K56Flex. It is not compatible with x2. For further discussion of 56K operation, see "What Makes 56K Modems Tick," in Chapter 14.

Current Standards for Fax Transmission

Fax transmission, which is similar to data transmission, requires a set of common rules adopted by equipment manufacturers, to allow the receiving and transmitting modems to set up a connection, exchange data corresponding to the image of a transmitted page, and correctly interpret the transmitted data. Data modems codify these rules in sets of protocols starting with the letter V, which were first developed by CCITT and later by the ITU-T organization. To make fax transmission more challenging and confusing, the fax transmission is governed not only by V-series protocols, but also by Class and by Group protocols. Though the distinction between the three types of protocols is somewhat muddy, the Class protocols mainly describe commands and handshaking between the computer and fax modem; the Group protocols describe the details of page translation and handshaking between receiving and transmitting modems; and the V-series protocols describe the type of modulation and transmission speed.

Dedicated fax machines have been in use since the 1960s. In the last few years, modems with fax capability have begun to take their place. Fax modems are considerably smaller and cheaper than fax machines but will not read a hard copy of a document without a scanner. The early fax machines were classified as Group 1. Group 1 fax machines using analog signals were very slow, taking up to six minutes to send a single page. In the late 1970s, Group 1 fax machines were replaced by Group 2 fax machines. These machines converted the image into digital signals, which were then transmitted by built-in modems. These digital signals were much more reliable, since they were less affected by line noise than the analog signals sent by Group 1 machines. A mathematical compression formula was added as well, bringing the transmission time down to about three minutes per page.

In the early 1990s, the current standard—Group 3 faxes—were given the option of two resolutions: 200 x 200 dots per inch (dpi) and 200 x 100 dpi. With the V.17 protocol, Group 3 faxes are capable of transmission at speeds up to 14,400 bps. The T.30 protocol is the method by which Group 3 equipment manages fax sessions and negotiates the capabilities supported by each fax in the connection. The T.4 protocol controls page size, resolution, transmission time, and coding schemes for Group 3 faxes. The less common Group 4 faxes are designed for ISDN (digital) lines.

Class Protocols

These protocols describe the set of AT commands (ATtention commands are part of the Hayes command set) that can be sent from a computer, and which will be understood by the fax modem. Chapters 9 and 10 explain how to send a command from a computer to a modem and how to monitor the response. To find out which Classes are supported by your modem, send it the following AT command:

```
AT+FCLASS=?
```

A typical response from a full-featured fax modem should be:

```
0,1,2.0
```

This indicates that the modem supports Class 0 (data), as well as fax Classes 1 and 2.0. Notice that Class 2.0 is a newer version of Class 2, which was not officially approved by the ITU-T. Most modems today will support both classes of operation, 2 and 2.0. Which Class is used depends on the communications software package used for fax transmission.

Class 1

This protocol, approved in 1990, supports only six AT commands and makes the computer's CPU perform most of the fax processing. Both Class 1 and 2.0 follow the Hayes convention of prefixing each command with AT. Class 1 relies on the computer to handle both data and communications. Class 1 fax modems have AT language extensions to their command sets that allow them to act as Group 3 fax machines. The official standard for Class 1 fax modems is EIA/TIA-578. All fax commands start with "AT+F". For example:

```
AT+FRM=?
```

Class 2.0

This more up-to-date protocol supports over 50 AT+F-type commands. These commands instruct the modem to perform functions that would be otherwise handled by the computer's software for a Class 1 modem. Some of these functions include checking and conversion of resolution mismatches, reporting of caller and sender ID, checking of line quality, and many other housekeeping chores. Because the modem rather than the computer handles most of the communication tasks, a Class 2.0 modem is preferable if using a slower microprocessor with low resources, such as a 386SX with 4 MB of RAM. On a fast Pentium computer with lots of memory, there will not be much difference in performance between Class 1 and Class 2.0 modems.

As it took a very long time for Class 2 to be approved, many fax modems were produced before the standard was official. Therefore Class 2.0 is used for fax modems that apply strictly to the standards, while Class 2 applies to those fax modems that were produced before the standard was official.

The proposed Class 3 is expected to handle the conversion of data streams into images. Class 4 fax modems, which are in limited use, have buffers that allow the CPU to go for a short period of time without processing fax data.

Group Fax Protocols

There are four Group protocols used in fax transmission. They are appropriately named Group 1, 2, 3 and 4. These four Groups define modem capabilities in terms of coding, page scanning and type of transmission. Group 1 and 2 modems use analog transmission and are currently obsolete. They were popular in early analog fax machines and are not implemented in current PC modems.

For personal fax communication, the most popular protocol, which is used nearly exclusively today, is the Group 3 protocol, adopted by CCITT in 1980. The more-recent Group 4 modems, adopted in 1984, are used only for special applications. In general, Group 4 modems will not operate on the switched public telephone network. Both Group 3 and 4 are digital designs and can produce high-resolution output. Both use data compression algorithms, which reduce the amount of transmitted data by a factor of between 5 and 10. A typical transmission time for Group 3 is 30 to 60 seconds per page. The Group 3 resolution is selectable, and is either 100 vertical by 200 horizontal dots per inch (dpi) in Standard mode, or 200 x 200 dpi in Fine mode. The fax software lets you select which mode to use. Group 4 fax modems are capable of 400 x 400 dpi resolution.

V-Series Fax Protocols

Similar to data modem protocols, ITU-T approved a number of V-series protocols for fax transmission. There are currently four V-series protocols dealing with the modulation and transmission speed of fax messages.

Protocol	Maximum Speed (bps)
ITU-T V.17	14,400
ITU-T V.29	9600
ITU-T V.27ter	4800
ITU-T V.21 Ch.2	300

The V.17 protocol is implemented in most modern data chips. To achieve a high fax transmission rate, it uses the trellis coded modulation (TCM) scheme, the same method of forward error correction as used for high-speed data transmission.

You can easily find out which V-series protocols and transmission speeds are supported by your modem. To do so, issue the following AT command to the modem:

```
AT+FRM=?
```

A typical response might be:

`3,24,48,72,73,74,96,97,98,121,122,145,146`

Table 4-6 interprets these results.

Table 4-6: Fax Protocol Implementations

Modulation Code	V Series Protocol	bps
3	V.21 Ch.2	300
24	V.27 ter	2400
48	V.27 ter	4800
72, 73, 74	V.29	7200
96	V.29	9600
97, 98	V.17	9600
121,122	V.17	12000
145,146	V.17	14400

Higher transmission speeds require better transmission facilities, meaning quieter telephone lines with less distortion. A fax modem will automatically switch to a lower transmission speed if the telephone line will not support the specified higher speed.

The majority of facsimile modems are based on the Rockwell chip set called RC288. The chipset satisfies the CCITT/ITU-T Group 2 and Group 3 requirements and follows the V.17 Recommendation at 14,400 bps, V.29 at 9600, 7200, or 4800 bps, V.27 ter at 4800 or 2400 bps, and V.21 at 300 bps. Table 4-7 shows the line signal-to-noise requirements to achieve an error rate of 1 in 100,000 for these transmission speeds and standards.

The values in Table 4-7 are typical, but may vary from modem to modem. As most telephone lines have a signal-to-noise ratio in excess of 30 dB, reliable fax transmission can easily be achieved at speeds of up to 14,400 bps.

Table 4-7: S/N Ratios (dB) for a Fax Modem

Bit Error Rate: 1 Error in 100,000 Bits		
Standard	**Transmission Speed (bps)**	**S/N (dB)**
V.17	14400	26
V.29	9600	23
V.29	7200	20
V.27	4800	18
V.29	4800	15
V.27	2400	11
V.21	300	5

Microcom Network Protocols (MNP)

The Microcom Corporation developed a series of firmware and software data compression and error correction schemes that are known by the collective name *MNP*. MNP protocols Levels 1 through 4 are in the public domain, while higher levels are licensed to most modem manufacturers. The lowest level is MNP Level 1, which is no longer in use; the highest is currently Level 10.

Each higher level MNP protocol is more complex than the preceding one. The first three levels are based on cyclic redundancy checks (CRC) and block re-transmission.

The higher levels of the MNP protocols are Levels 4 through 10, except Level 8, which was never implemented. The error correction and compression algorithms are implemented in software embedded in modem firmware. Following is a short description of each MNP protocol.

MNP Level 1

This lowest level of MNP protocol is no longer being implemented. It is a set of software algorithms, developed for early processors, that provides CRC checks for error correction. It poses minimum demand on processor speed and memory storage. It is based on an asynchronous byte-by-byte half-duplex transmission of data. The error correction slows data transmission by approximately 25%.

MNP Level 2

Level 2 differs from Level 1 in that it uses full-duplex transmission—meaning data can be exchanged in both directions at the same time. An improved CRC algorithm slows data transmission by approximately 15%

MNP Level 3

This protocol increases effective transmission speed by approximately 15% by converting asynchronous data, which is received from the computer's UART, into synchronous data. The local modem strips the start and stop bits from each byte and generates clock signals, which are added to the synchronous data stream. The remote modem converts the synchronous data stream to asynchronous data stream by adding start and stop bits to each byte.

MNP Level 4

Level 4 introduces variable packet size to data being transmitted. In the event of an error, the suspected packet is re-transmitted. The process of sending and receiving each packet introduces overhead consisting of control and error-checking bits. Therefore in order to optimize the effective transmission speed, these packets should be kept small on noisy transmission lines, and should be kept large on good-quality lines. Furthermore, in MNP Level 4, repetitive control information is removed from the data stream to further improve efficiency. The protocol increases the effective transmission speed by approximately 20%.

MNP Level 5

Level 5 protocol is being replaced by MNP Level 7 and is only implemented for backward compatibility. MNP5 uses real-time adaptive compression algorithms based on data being transmitted, and optimally doubles data throughput. In particular, MNP5 breaks data into small chunks and checks for repetition of these chunks in a long sequence. If more than three repetitions occur, then the chunk count rather than the actual data is re-transmitted.

MNP Level 6

Level 6 protocol includes a link negotiation protocol to determine the highest transmission speed acceptable to the remote modem. In addition, Level 6 provides the so-called *statistical duplexing* on older modems that implement the V.29 half-duplex protocol. The V.29 modem with Level 6 therefore appears to the user as a full-duplex modem.

MNP Level 7

Level 7 protocol replaces Level 5 in new designs with its improved compression algorithms. In addition to frequency tables of individual characters, the algorithms in Level 7 also try to predict succeeding characters. For example, a "q" is most probably followed by a "u." The frequency tables are dynamically adjusted for the data being transmitted. Therefore data compression can be optimized for any language, not just for English text.

MNP Level 9

Level 9 protocol is optimized for its house-keeping chores, such as acknowledging that a block of data has been received, and for re-transmitting data in the event of a detected error.

MNP Level 10

Level 10 protocol is the latest in the MNP series. It is optimized for poor transmission facilities that may be encountered on cellular networks and on international calls. Some of the features implemented in this protocol include negotiation of optimum transmission speed at the start of a call and during transmission of data, dynamic adjustment of size of data blocks, and dynamic adjustment of transmission power on cellular calls.

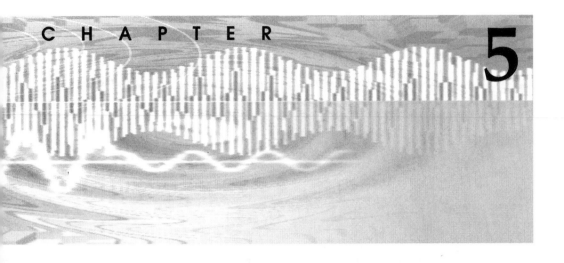

Asynchronous and Synchronous Transmission

This chapter explains the two methods of synchronization of data during the transmission process—asynchronous and synchronous. It explains how a character is transmitted and what the purpose of the start, stop and parity bits are in asynchronous transmission. It also explains what a frame looks like in a synchronous transmission, and what the advantages and disadvantages are of each method.

When two computers talk to each other, their respective timing must not differ. If, to take an extreme example, the sending computer transmits a 0 or a 1 every second, and the clock in the receiving computer runs twice as fast as that of the sending computer, then the sequence 00101 might be interpreted in a stuttering way as 0000110011, which is an obvious error. Even if clocks in the two computers run at the same speed, there is still a need for synchronizing the transmitted events such that the beginning of a sentence or of a word is interpreted correctly.

In the asynchronous method, extra data pulses are sent before and after each character to alert the receiving computer that a character is on its way. In the synchronous method, the transmitted data is divided into frames containing many characters. Each frame is then preceded and followed by extra data pulses, which alert the receiving computer that a frame is on its way or that it is finished.

Start, Stop and Data Bits in Asynchronous Transmission

Each ASCII character, when expressed as a binary number, is composed of a sequence of seven or eight bits—0s and 1s. The receiving computer or terminal therefore needs some indication to determine where the bits that belong to one character end, and where the bits that belong to the next character begin. In the asynchronous transmission method, when no data is sent, a steady stream of 1s (called a *mark* in the language of telegraphers) is transmitted. To alert the receiving terminal that a character will immediately be sent, a start bit (a 0, which corresponds to a space) is transmitted. It is then followed by seven or eight data bits, a parity bit used for error detection, and one or two stop bits (a 1, or mark). Figure 5-1 shows an example of the character A being transmitted. The sequence begins with a start bit, followed by seven data bits, which correspond to the ASCII code for A (1000001 = decimal 65), a parity bit, and a stop bit.

A simple and crude way of checking for transmission errors is to include a parity bit (P) at the end of each character. If the parity is odd, then the parity bit will be 1 if the total number of 1s in the previous seven bits is odd. It will be 0 otherwise. Similarly, if the parity is even, then the parity bit will be 1 if the total number of 1s in the previous seven bits is even. It will be 0 otherwise. Most data transmission protocols provide their own error checking: they do not rely on the parity bit following each character, but instead compute a checksum of larger blocks of data.

After the last bit associated with the A is sent, either more 1s will follow, or another start bit will announce the next packet of data bits associated with the next character.

Figure 5-1 Diagram of the serial transmission of the letter "A"

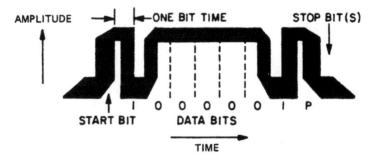

Figure 5-2 shows the decision process at the receiving modem during the reception of individual bits that comprise a single character. As each character in asynchronous mode of transmission is surrounded by start, stop and parity bits, both the transmitting and the receiving modems are always synchronized by those bits on a character-by-character basis. No synchronization is required between characters. The asynchronous mode is particularly suitable for manual exchange of data, where the two participants type at their respective terminals. The hardware implementation for the asynchronous mode of transmission is relatively simple and the method is deeply ingrained in communications software and modem design.

Figure 5-2 Decision tree for detection of a serially transmitted character

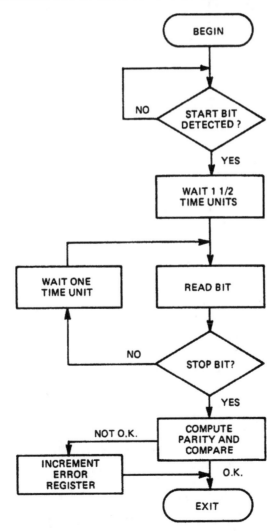

Synchronous Transmission

In the synchronous mode of transmission, no start or stop bits precede or follow each character. This alone can speed up the transmission by up to 30% (7 to 8 instead of 10 to 11 bits per character). The data can then be sent in a steady, non-stop stream. The local and remote modems synchronize themselves by means of special framing signals that surround blocks of data, and by extracting timing information from the data itself. In some commercial applications, a clock signal is sent separately from the data on a special circuit. Figure 5-3 shows a block of data (TEXT) with synchronizing, error checking, and framing characters surrounding it.

Figure 5-3 Block of synchronous data with synchronizing and framing bits

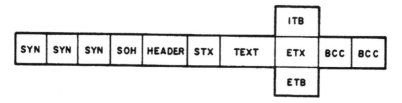

Many medium- and high-speed modems—in particular, those used in commercial environments—are equipped with the synchronous transmission option. Although modem chip manufacturers provide both asynchronous and synchronous options in most of their modem chip sets, the synchronous option—due to limited public interest—is usually not implemented in lower-priced modems. Therefore only a few modems used in personal computers implement the synchronous option. In the commercial environment, synchronous modems are often used in polled and multi-point private networks. Synchronous operation permits higher transmission speeds and sophisticated error recovery systems.

As in other areas of data transmission, there is also a need here for common protocols, so that synchronous terminals can communicate with other synchronous terminals and understand each other's signals. The number of framing characters and their composition is determined by the protocols used by both the sending and receiving modem. The most commonly-used synchronous transmission protocols are the binary synchronous communication protocol (BISYNC) and several high level data link protocols (HDLC). The BISYNC protocol, developed by IBM, is a half-duplex character-oriented procedure most often used on polled multi-point networks. The HDLC protocols are attempts to improve on BISYNC by providing full-duplex operation, and being bit-oriented rather than character oriented. IBM's own synchronous data link control protocol (SDLC) belongs in this category. The SDLC protocol independently manages each link between terminals. SDLC can be used to control a link between a terminal and a concentrator, or used between a concentrator and the host computer.

The overall network control is performed by additional software in the concentrators and the host computer. Other HDLC protocols were developed by Digital Equipment Corporation and other manufacturers. Another entry is the Hayes synchronous interface (HSI) protocol introduced by Hayes Corporation. Hayes placed the protocol in the public domain. However, the company charges developers a license fee for use of drivers required by the protocol in order to interface with various Hayes-compatible modems. The HSI protocol, unlike other HDLC protocols, is specifically designed to operate on personal computer modems. The HSI protocol should allow synchronous transmission through a standard RS232-C serial port.

Network protocols are a big subject in themselves, and their detailed discussion is beyond the scope of this book.

Table 5-1 provides a short summary of control symbols used in synchronous transmission protocols and their equivalents in hexadecimal notation (when available).

Table 5-1: Control Characters for Synchronous Transmission

SYN	Used to establish and maintain synchronization
16H	Fills time in absence of data. Between two and five SYN characters are required in each frame.
SOH	Start of Header; user defined
01H	Includes source/destination codes, parity, date, and so on. Terminated by STX
STX	Start of Text segment block
02H	Text may be divided into several blocks
ITB	Used to separate messages into multiple blocks. No reply required until ETB or ETX sent. Followed by BCC
ETB	End-of-Text Block for multi-block messages
17H	Requires response from receiving station. Followed by BCC
ETX	End-of-Text; terminates a block of text
03H	Requires reply. Followed by BCC

Table 5-1: Control Characters for Synchronous Transmission (continued)

BCC	Block Check Character used for error detection. Computed by Cyclic Redundancy Check from block data
EOT	End-of-Transmission; resets all stations on the line
04H	Response to a poll if nothing is sent
ACK	Acknowledge, positive response, message received correctly
06H	(Is there something missing here?)
NAK	Negative response, message not received or incorrect
15H	Determined by agreement of received data with BCC
WACK	Waiting for acknowledgment

Hardware Considerations

For proper operation in synchronous mode, both the receiving and transmitting terminals must use the same kind of data communications hardware and software. In addition, the serial interface (on the computer) and the serial cable should have pin 15 (Transmit Signal Element Timing), pin 17 (Receiver Signal Element Timing), and pin 24 (Transmit Signal Element Timing) all connected through and implemented. Both terminals must also agree on a common timing option, namely whether the modem provides its own clock, whether the clock will be generated by the computer or derived from the receiving modem through the so-called *slave* operation. They also must agree on the number of bits used for each character. These options are usually selected by switch or strap settings.

One of the problems inherent in synchronous transmission is that dialing and other modem commands (the AT commands) must be sent from the computer to the modem in asynchronous mode. The communication between the computer and the modem, because of hardware considerations, is always asynchronous, even when the modem-to-modem transmission is synchronous. Without it, one could not use the escape code (+++) in the synchronous mode to change from the online to offline command mode. The escape code would be embedded in a frame and would not be recognized by the modem.

The final option to be selected is the dialing mode. In dialing Mode 1, the call is placed in the standard asynchronous fashion. When the connection is established, the modem switches automatically to the synchronous mode. In dialing Mode 2, the modem automatically dials a stored number when the DTR (Data Terminal Ready) signal lead is high. This number must be previously stored in the modem by means of asynchronous data transmission. In dialing Mode 3, the operator manually dials using a standard telephone set, or dials a stored number using one of the modem panel buttons.

There are only three conditions under which the modem will change from synchronous to asynchronous operation, each of them terminating the previously-established connection:

1. One can terminate a synchronous connection by pressing a front- panel disconnect button normally provided on a modem.
2. When the computer drops the DTR line, the connection is terminated.
3. When the modem detects a loss of carrier, the connection will also be terminated.

Considering all the above options and settings, it is no wonder that synchronous transmission is used mostly for specialized applications in commercial environments. However, if implemented, it can speed up data transfer by approximately 20% by avoiding start and stop bits required in asynchronous transmission before and after each character.

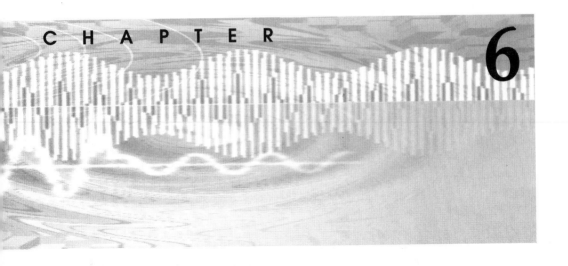

The Serial Interface

A computer needs a set of formalized low- and high-level procedures in both hardware and software to be able to communicate with the outside world through a modem.

The low-level procedures generally describe the hardware interfaces such as plugs and jacks, the electrical interfaces such as voltage or current levels, timing interfaces, and signal interfaces such as whether synchronous or asynchronous transmission should be used.

The high-level procedures, in the form of transmission protocols described in Chapter 4, provide rules for establishing a connection between the sender and the receiver by specifying the handshake protocol, modulation and error-correction method, granting the sender access to the physical channel, defining the message format and blocking rules, and providing message acknowledgment and error-recovery methods.

A set of low-level procedures is called an *interface protocol*. The most common interface protocol for communication between a computer and a modem is the serial interface. The word *serial* means that bits are transmitted in sequence, or serially, one bit at a time, over a single wire with a ground return, or over a pair of wires balanced to ground. This differs from the parallel interface, commonly used to transfer data from a computer to a printer, but also used in some modem implementations, where eight bits of data are transmitted simultaneously over eight separate wires on a data bus with ground returns.

The most common serial interface protocol is the RS-232-C standard. Because of the technical obsolescence of the RS-232-C standard, other standards have also been proposed. Still, due to the large installed base of equipment using the RS-232-C standard, these newer standards have not taken over the market. Quite the contrary, except for some special applications, including the military or high-speed data transmission, the RS-232-C still reigns supreme. A short review of other serial interface standards besides RS-232-C is provided later in this chapter.

RS-232-C Serial Interface

The serial interface, as implemented in the RS-232-C standard, is the most common way of connecting a computer to a modem. The RS-232-C standard was developed by the Electronic Industries Association (EIA) in the stone age of computing—in August, 1969. The full name of the standard is *Recommended Standard 232 Version C, Interface Between Data Terminal Equipment and Data Communications Equipment Using Serial Interface*. The EIA standard describes functions of 25 leads connecting the Data Terminal Equipment (DTE)—typically a terminal or a computer, to the Data Communications Equipment (DCE)—typically a modem. The serial port on most computers and on most communications devices uses only between nine and 11 leads out of the 25 leads shown in Figure 6-1. The nine to 11 leads are sufficient for most control applications and are also sufficient for operating a modem.

Figure 6-1 RS-232-C DB-25 connector and lead/pin description

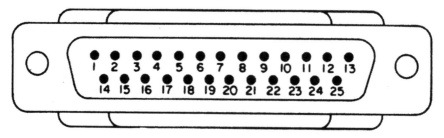

Abbrev.	Pin No.	Circuit	Description	Signal Type	Function	Direction To:
PG	1	AA	Protective Ground	Ground	Ground connected to the equipment frame or power cord ground.	
TD	2	BA	Transmitted Data	Data	Information sent from the local terminal.	DCE
RD	3	BB	Received Data	Data	Information received at the local terminal.	DTE
RTS	4	CA	Request to Send	Control	A control signal from the local terminal to determine if a transmission can be sent. This control is most useful for half duplex operation where data are transmitted in both directions, one at a time, over a single communication line. It is not necessary on full duplex (communication in both directions simultaneously) but is often implemented.	DCE

Abbrev.	Pin No.	Circuit	Description	Signal Type	Function	Direction To:
CTS	5	CB	Clear to Send	Control	A control signal to the local terminal which indicates that data transmission to the remote terminal may commence. This is in response to a Request-to-Send-command.	DTE
DSR	6	CC	Data Set Ready	Control	A handshake signal to the local terminal indicating that the data communication equipment is connected and ready to operate.	DTE
SG	7	AB	Signal Ground (Common Return)	Ground	The ground connection to which other RS-232 signals are referenced.	
RLSD	8	CF	Received Line Signal Detector	Control	Carrier Detect is a handshake to the local terminal indicating that an acceptable signal is being received on the communications line between DCEs.	DTE
	9	—		—	Reserved for data set testing.	DTE
	10	—		—	Reserved for data set testing.	DTE
	11	—	Unassigned*	—		
SRLSD	12	SCF	Secondary Received Line Signal Detector	Control	Auxiliary carrier detect used when signals on pins 14 and 16 are implemented.	DTE

Abbrev.	Pin No.	Circuit	Description	Signal Type	Function	Direction To:
SCS	13	SCB	Secondary Clear to Send	Control	An auxiliary handshake used when signals on pins 14 and 16 are implemented.	DTE
STD	14	SBA	Secondary Transmitted Data	Data	An auxiliary channel by which low speed data or special control functions are sometimes transmitted.	DCE
TSET	15	DB	Transmission Signal Element Timing (DCE Source)	Timing	Used for synchronous transmission.	DTE
SRD	16	SBB	Secondary Received Data	Data	A low speed data or control function received from the remote DCE.	DTE
RSET	17	DD	Receiver Signal Element Timing (DCE Source)	Timing	Used for synchronous transmission.	DTE
	18		Unassigned*	—		
SRS	19	SCA	Secondary Request Send	Control	An auxiliary handshake used when signals on pins 14 and 16 are implemented.	DCE
DTR	20	CD	Data Terminal Ready	Control	A handshake from the terminal to prepare the DCE for communication.	DCE

Abbrev.	Pin No.	Circuit	Description	Signal Type	Function	Direction To:
SQD	21	CG	Signal Quality Detector	Control	Indicates to the local terminal that there is a low-grade signal on the communications line which could result in a high error rate. This signal is normally used on long distance telephone communications.	DTE
RI	22	CE	Ring Indicator	Control	A handshake sent to the local terminal indicating that a ringing signal is being received on the communication line. This signal would be used on an auto answer telephone communication system.	DTE
DSRS	23	CH/ CL	Data Signal Rate Selector (DTE/DCE Source)	Control	A handshake used on a telephone communication system to indicate and/or select one of two bit rates. This signal gives the user the option to operate with different speed communications equipment at the remote end.	DCE
TSET	24	DA	Transmit Signal Element Timing (DTE Source)	Timing	Used for synchronous transmission.	DCE
	25		Unassigned*			

*While unassigned in RS-232-C, some of these pins are used for specialized functions by the manufacturers of DTE and DCE equipment

Most modems, printers, controllers and other devices in the data communications market follow selected subsets of the RS-232-C recommendation. There is also a European equivalent of the RS-232-C, namely the CCITT/ITU-T recommendation (standard) V.24. The V.24 recommendation does not describe the electrical interface like the RS-232-C does, but it specifies more than the 25 leads listed in the EIA standard.

Though many communications devices claim to adhere to the RS-232-C standard, they may still be incompatible with each other. The primary reason for incompatibilities is that the RS-232-C standard, as written in the EIA specifications, covers only some aspects of the communication protocol. For example, though most RS-232-C devices use a 9-pin (DB-9) or a 25-pin (DB-25) male or female D-shell connector, the type of connector is not mentioned in the EIA standard. The standard also does not say which connectors should be male and which should be female. This ambiguity alone spawned a whole family of gender-changing cables and adapters.

The RS-232-C standard assigns specific functions to 22 leads and lets the equipment manufacturers assign the remaining three leads. The pin assignments of the leads to the 25-pin D-shell connector, as used by most manufacturers who comply with the RS-232-C standard designations, are shown in Figure 6-1. However, some companies, such as Apple for their MacIntosh line of computers and IBM with their AT line of computers, use 9-pin D-shell connectors (DB-9) with only a subset of the 25 leads of the RS-232-C standard. The current crop of Pentium computers normally provides one RS-232-C serial port that uses a 9-pin connector and one RS-232-C serial port that uses a 25-pin connector.

As a de-facto standard, all manufacturers provide a male jack for the serial port at the computer side and a female plug on the serial cable. To avoid confusion, IBM wisely decided to use a male DB-25 jack for the serial port and a female DB-25 jack for the parallel port on the computer side, starting with their first PC. This convention prevents accidental use of the parallel port plug instead of the serial port plug.

This wise move compensated for a less wise move to use the same kind of DB-25 connector for both parallel and serial printer ports. It just added to the confusion when IBM introduced the same 25-pin D-shell connector as the parallel port with specifications completely different from the RS-232-C specifications. The parallel port is used mostly for connecting printers to computers. Other connectors for the serial port have also been used. For example, IBM went even further out on a limb by introducing a rectangular Berg connector as the serial interface on the unsuccessful PCjr. As board space became more valuable and computer manufacturers realized that, for most purposes, only a subset of the 25 RS-232-C signals is required, many designs switched from a 25-pin D-shell to a 9-pin D-shell connector. Pin assignments used in the MacIntosh serial interface are shown in Figure 6-2, while pin assignments used in the IBM-compatible AT-type DB-9 connectors are shown in Figure 6-3.

Figure 6-2 Macintosh DB-9 serial interface connector

Pin Number	Macintosh Signal Name	RS-232 Signal Name
1	Chassis Ground	Chassis Ground (1)
2	Plus 5 Volts (+5)	
3	Signal Ground	Signal Ground (7)
4	Transmit Data + (TXD+)	
5	Transmit Data – (TXD–)	Transmitted Data (3)
6	Plus 12 Volts (+12)	Data Terminal Ready (20)
7	HSC Input	Data Carrier Detect (8)
8	Receive Data + (RXD+)	Chassis Ground (1)
9	Receive Data – (RXD–)	Receive Data (2)

Macintosh to Modem Connecting Cable

Macintosh Connector	Modem Connector (DCE)	Function
DB-9P	**DB-25P**	
(Pin Numbers)		
8 --------- 1		Chassis Ground
3 --------- 7		Signal Ground
5 --------- 3		Data Mac to Modem
7 --------- 8		Carrier Detect (HSC)
9 --------- 2		Data Modem to Mac
2 --------- 6		Data Set Ready
6 --------20		Data Terminal Ready

Figure 6-3 AT type serial interface DB-9 connector

DB-9 Pin#	DB-25 Pin#	Signal
1 - - - - - - - - - - - - 8		Carrier detect
2 - - - - - - - - - - - 3		Transmitted data
3 - - - - - - - - - - - 2		Received data
4 - - - - - - - - - - 20		Data terminal ready
5 - - - - - - - - - - - 7		Signal ground
6 - - - - - - - - - - - 6		Data set ready
7 - - - - - - - - - - - 4		Request to send
8 - - - - - - - - - - - 5		Clear to send

By definition, a data communications equipment (DCE) device complies with a standard as long as it does not violate it. If one device adheres to one subset of the RS-232-C standard and another device adheres to a different subset, then although both devices adhere to the RS-232-C standard, they may be incompatible with each other. To confuse things even further, the RS-232-C standard always refers to data terminal equipment (DTE) and to data communications equipment (DCE). This designation is fine if one always attaches a terminal (DTE) to a modem (DCE). But how should one designate a terminal that has a printer attached to it and is connected directly to a computer? It could be considered a DTE or a DCE. Some serial printers are also configured as a DTE device, while some are configured as a DCE device.

The RS-232-C standard does not cover the communication protocols such as type of transmission (asynchronous or synchronous), number of bits per character, and so on. The features, which the RS-232-C standard covers best and that were of importance in 1969 when it was developed, are the electrical characteristics of the data signals. The received data signals, according to the RS-232-C standard, are interpreted as follows:

- -3 to -25 Volts is a mark (binary 1)
- +3 to +25 Volts is a space (binary 0)
- Voltage levels between -3 and +3 Volts are not defined.

Similarly, negative voltages on control leads indicate the OFF (or LOW) condition, while positive voltages indicate the ON (or HIGH) condition. These relations are shown in Figure 6-4. In real-life situations, what is referred to as the RS-232-C interface is really a subset of the above electrical requirements, combined with the asynchronous serial transmission of data.

Figure 6-4 RS-232-C voltage levels corresponding to mark/space

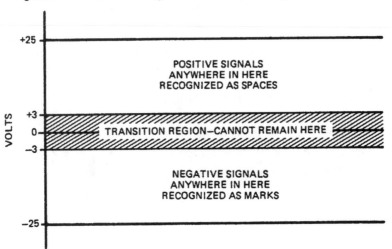

One important safety feature of the RS-232-C standard is that any lead can be connected to any other lead without causing damage to the equipment it is connected to. Cross-connecting various leads may not help in establishing a data connection, but at least it will not fry anything. The null modem described later in this chapter, as well as various cheater cords used to connect serial printers and other devices, all rely on the great non-damage promise of the RS-232-C standard.

At a recent computer convention, I heard about a computer salesman telling a customer how easy it was to interconnect various devices, as long as they follow the RS-232-C standard. The customer then asked whether he needs a 25-pin male or a 9-pin female cable, and whether the data leads should be reversed or lead-through. The salesman's answer was that his company carries every "standard" cable, and if they do not have one, they will make it.

After reading all this, you may wonder how two RS-232-C devices ever communicate with each other. Well, sometimes they do, and sometimes one has to put some extra effort to achieve a successful exchange of information.

In the next two sections you will see how to connect an external modem to a serial port and how to transfer data between serial ports of two computers without a modem—or, to be precise, by means of a *null modem.*

Connecting an External Modem to a Serial Port

Only the basic connections and handshake sequence between the computer and the modem are described in this chapter. External and internal modem installation will be discussed in more detail in Chapter 7.

If the transmitting and receiving computers are separated by a substantial distance, then a data path, which will include a modem on each end and a telephone connection, must be established. The basic connections and typical pin assignments between an external modem and the computer serial port are shown in Figure 6-5.

Figure 6-5 Basic external modem hook-up

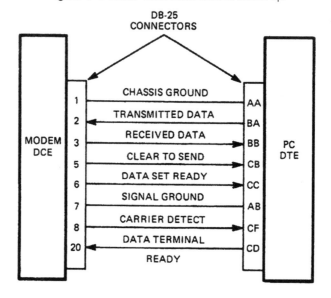

Certain RS-232-C pins such as pin 15 (TC), pin 17 (RC), pin 12 (SCD), and pin 18 (SRS) may not be implemented in the computer or in the modem. The same considerations about choosing COM1 or COM2, as will be described in Chapter 8, still apply. Remember that the DB-25 jack on the computer will be male and the one on the modem will be female. The start-up handshake sequence between the serial port and the modem is shown in Table 6-1.

Table 6-1: Start-Up Sequence (Control Signals)

RS-232-C Signal	Description
DTR	Normally high when DTE (computer) is ON
DSR	Normally high when DCE (modem) is ON
RTS	Goes high to request service
CTS	Goes high after a typical 100- ms delay
Connection is now established (dialing)	
CD	Goes high to indicate received carrier
TD	Starts sending data from DTE to DCE
RD	Starts receiving data from DCE to DTE

To start the handshake sequence, the computer sets the DTR lead to ON. The modem responds by setting the DSR lead to ON. Next, the computer indicates that it is ready to send data by setting the RTS lead to ON. The modem responds, after a specified delay time, that it is ready to accept data by setting the CTS lead to ON. The computer should then send instructions to the modem to dial the remote modem. After a connection is established, the local and remote modems start negotiating the common protocol. The final step before transmission of data from the local computer to the local modem is confirmation by the local computer that the carrier was received from the remote modem by setting the CD lead to ON. Once these hurdles are cleared, the data exchange can proceed by sending data on the TD lead and receiving data on the RD lead.

Connecting Two Computers with a Null Modem

Use of the serial port connected to a modem allows sending files from one computer to another across the globe. If the two computers are located physically near each other, one only needs a so-called null modem. For example, I have used a null modem to transfer text, data, and program files between my new Pentium, an old Unix AT&T PC7300, and an assortment of old and new computers including the Radio Shack TRS-80 Model I and the TANDY Model 100. This technique is particularly useful if the disk formats and operating systems of the two computers are not compatible. For example, one computer may have a CD drive, the other may only use 5.25" or 3.5" diskettes, and one may have a need to transfer data or program files between the two computers.

The reason why a null modem is necessary is that most computers are configured as a DTE device, with data being transmitted on pin 2 and data being received on pin 3, rather than as a DCE device, where data is transmitted on pin 3 and received on pin 2. The null modem is a cable or a small male/female connector that is wired to make one of the computers appear to the other computer as a modem (DCE).

If one connects, for example, serial ports of two IBM-compatible PCs to each other, one will need either two modems back to back, or as a simpler solution, a null modem that interchanges the Receive Data and the Transmit Data signals on pins 2 and 3, as well as all of the appropriate control signals, when used. The control signals indicate if the transmitting terminal is ready to send data, and if the receiving terminal is ready to receive data.

The physical implementation of a null modem might be a cable or might be a large plug with DB-25 connectors at both ends. Figure 6-6 shows a typical implementation of a null modem. Notice in particular the crossed leads, #2 and #3. If the null modem is not available, one can easily put one together by using Figure 6-6. The RS-232-C 25-pin male and female D-shell connectors can be easily found in most electronic and computer supply stores.

When buying a null modem, or when building one from individual components, be sure to check for gold-plated pins on all connectors. Gold makes the best electrical corrosion-proof contact. Figure 6-7 shows connections required to transfer files between a desktop and a laptop computer by means of a null modem. Figure 6-7 would also apply to most other computers equipped with a serial interface.

After making the proper connections, you should be able to exchange files between two computers. The following DOS MODE command demonstrates how easy it is to exchange simple ASCII files. To exchange non-ASCII files, you will have to install a file-sharing program or terminal software on both computers. Otherwise, the receiving computer would get confused by special characters present in non-ASCII files and interpret them as control characters, carriage returns, form feeds, disconnects, and so on. Terminal software is discussed in Chapter 10.

To send an ASCII character or a source-program listing from the desktop to the laptop and vice versa, follow these steps. If you plan to send a BASIC program, for example, first convert it to an ASCII file with the command:

```
SAVE "filename",A
```

Figure 6-6 Diagram of a "null" modem

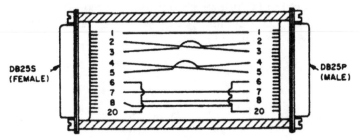

Figure 6-7 Hook-up of a desktop PC and a laptop via "null" modem

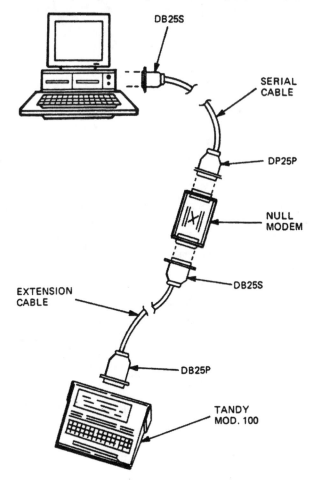

To initiate the file transfer between the two computers, first redirect the PC printer output to the serial port by entering the following DOS commands, or by selecting the printer port as COM1 in the Windows control panel.

```
MODE COM1:1200,n,7,1,p
MODE LPT1:=COM1
```

The display should now look like this:

```
C:\> MODE COM1:1200,n,7,1,p
Resident portion of MODE loaded
COM1: 1200,n,7,1,p
C:\>MODE LPT1:=COM1
Resident portion of MODE loaded
LPT1: redirected to COM1:
C:\>
```

The first command sets the transmission parameters for the first serial port (COM1) to 1200 bps, no parity, 7 data bits, 1 stop bit, and "p" to continuously check for time-out errors. The second command redirects the first parallel printer port (LPT1) to the serial communication port. Change COM1 to COM2 in both commands if you have a mouse, a printer, an internal modem, or some other device already installed in your PC as COM1. Based on experience, the previous MODE statements are foolproof. In particular, 1200 bps seems to be the fastest common denominator for reliable transmission between various computers, serial printers, operating systems and communications programs.

When transmitting to a laptop, call its resident terminal program—Terminal in Windows 3.x, or Hyper Terminal in Windows 95—and set its serial parameters (bit rate, parity, start and stop bits) to the same values as the desktop. Now simply "print" the file that you want to transmit by using the DOS statement PRINT <filename>. If you want to transfer a C, Basic, or other source program, simply type LLIST from Basic or PRINT from DOS. The program listing or other ASCII file will now go to the laptop via the serial interface instead of to the printer. The file will appear on the screen of the receiving computer where it can be captured from the screen and saved.

You can also transfer a non-ASCII file by means of programs that translate a string of characters, including special characters, into a series of ASCII codes. Such programs, with names like UUDECODE and UUENCODE, have been developed for nearly every computer, and are either in the public domain or can be purchased at a reasonable cost. Chapter 10, on data communications software, describes such programs in more detail.

Connecting Modems With a Cheater Cable

Considering the large number of different types of modems and personal computers on the market, it may occasionally happen that no combination of jumper and DIP switch settings results in satisfactory modem operation. In such cases, where everything else fails, a *cheater* cable might be the only solution. A cheater cable internally straps certain serial-interface leads and forces them to be HIGH or LOW to satisfy the handshake protocol.

Fortunately, the RS-232-C standard allows you to connect any interface lead to any other interface lead without causing permanent damage to the equipment it is connected to. Of course an incorrect strap, although it may not result in subsequent damage, will not set up a data connection either. For example, in a handshake sequence, the DTE side of the circuit—the Personal Computer—expects the Carrier Detect (CD) lead to be HIGH when it sets the Request To Send (RTS) lead HIGH. The cheater cord simply straps together the RTS and CD leads on the PC side and therefore forces the CD lead to be HIGH, satisfying the protocol requirement.

Figure 6-8 shows a cheater cable that does away with the handshake protocol between the computer and modem. The only leads carried between the modem and computer are the TD, RD and signal ground. The danger of this arrangement is that transmission will continue, even if the respective terminals are not ready to accept data for some reason. In such a case, the transmitted data will simply get lost.

However, even use of a cheater cable still allows certain error-correcting block protocols to operate properly. Such protocols do not have to rely on the condition of the RTS and CTS leads for re-transmission of blocks of data, but could instead use control characters exchanged between the two terminals to control data flow. Such data-flow protocols are discussed in more detail in Chapter 9 and are referred to as XON/XOFF software flow control. The XON character is ASCII 17 (DC1/Ctrl-Q), and the XOFF character is ASCII 19 (DC3/Ctrl-S).

Figure 6-8 Example of a "cheater" cable

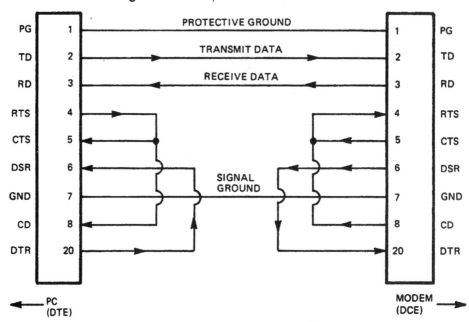

Serial Interfaces Other Than RS-232-C

The RS-232-C standard developed in 1969 has severe limitations in terms of the nominal maximum transmission speed of 19,200 bps and maximum distance of 50 feet. Though the limitations of the RS-232-C are often ignored, a number of other standards have been developed since 1969 to enable reliable operation at higher transmission speeds and at longer distances.

Some of these newer interfaces were developed by the same EIA committee as the RS-232-C and start with letters RS. Some were developed by the European standard body CCITT and later by the ITU-T, and start with the letter V. Others were developed by the U.S. military and start with the letters MIL-STD. These standards are frequently as confusing as the RS-232-C standard, occasionally leaving out the hardware specifications.

Table 6-2 lists lead assignments of the most common non-RS-232-C serial interface standards in current use—RS-449 and the CCITT/ITU-T V.35—and compares them with the RS-232-C standard.

Table 6-2: Non RS-232-C serial interface lead/pin assignments

RS232C/CCITT V.24	CCITT V.35	RS449	
25 Pin	34 Pin	37 Pin	9 Pin
1—Protective Ground	A—Protective Ground	1—Shield 37—Send Common	1—Shield 9—Send Common
2—Transmitted Data	P—Transmit Data (A) S—Transmit Data (B)	4—Send Data (A) 22—Send Data (B)	
3—Received Data	R—Received Data (A) T—Received Data (B)	6—Received Data (A) 24—Received Data (B)	
4—Request to Send	C—Request to Send	7—Request to Send (A) 25—Request to Send (B)	
5—Clear to Send	D—Clear to Send	9—Clear to Send (A) 27—Clear to Send (B)	
6—Data Set Ready	E—Data Set Ready	11—Data Mode (A) 29—Data Mode (B)	
7—Signal Ground	B—Signal Ground	19—Signal Ground	5—Signal Ground (C)
8—Carrier Detect	F—Receive Line Signal Detect	13—Receiver Ready (A) 31—Receiver Ready (B)	
9—Reserved for Testing	m—Reserved for DSU Testing		
		20—Receive Common	6—Receive Common
10—Reserved for Testing		10—Local Loop (A) 14—Remote Loop (A)	
11—Unassigned		3—SPARE 21—SPARE	
12—Sec. Carrier Detect		32—Select Standby	2—Sec. Receiver Ready
13—Sec. Clear to Send			8—Sec. Clear to Send
14—Sec. Transmitted Data			3—Sec. Send Data
15—Transmit Clock (DCE Source)	Y—TX Signal Element Timing o—TX Signal Element Timing	5—Send Timing (A) DCE Source 23—Send Timing (B) DCE Source	
16—Sec. Received Data			4—Sec. Received Data
17—Receive Clock	V—RX Signal Element X—RX Signal Element	8—Receive Timing (A) 26—Receive Timing (B)	
18—Unassigned		18—Test Mode (A) 28—Term in Service (A) 34—New Signal	
19—Sec. Request to Send			7—Sec. Request to Send
20—Data Terminal Ready		12—Terminal Ready (A) 30—Terminal Ready (B)	
21—Signal Quality Detector		33—Signal Quality (A)	
22—Ring Indicator		15—Incoming Call (A)	
23—Data Signal Rate Selector		2—Signaling Rate Indicator (A)	
		16—Signaling Rate Selector (A)	
24—Transmit Clock (DTE Source)		17—Terminal Timing (A) 35—Terminal Timing (B)	
25—Busy		36—Stand by Indicator	

The most important of these standards, the RS-449, defines only the mechanical interface, the 37-pin and the 9-pin connectors. It defines 10 new leads beyond those already specified by the RS-232-C standard, and uses two additional electrical standards—the RS-422 and the RS-423. It is inter-operable with RS-232-C and is equivalent to MIL-188-114. The big disadvantage of the RS-449 standard is the need for two connectors and the associated heavy cables.

The RS-422 standard provides signals that are balanced to ground. Both data and control signals require a pair of wires for each signal. The standard can operate at transmission speeds of up to 100 kbps at distances of 1 km, and at 10 Mbps at distances of up to 10 meters. The inherent advantage of a balanced circuit is that it is less affected by ambient electrical noise. The RS-422 generator has two outputs that connect to two transmission lines. The receiver responds to the voltage difference between these two lines rather than to the voltage difference between a line and ground, as in the unbalanced RS-232-C interface. When the transmission lines are subject to external electrical noise, the noise affects both lines to about the same degree and cancels each other out when the difference in voltages is used. Therefore the noise difference between the two lines, as measured by the receiver, is much lower than the noise on each line. The RS-422 installations are able to use inexpensive twisted-pair telephone cable, which is mentioned in the appendix to the RS-422 standard.

Another interface standard, the RS-423, provides signals that are unbalanced to ground, similar to RS-232-C. Because of this limitation, the interface can only operate at up to 3 kbps at distances of up to 1 km, and at 300 kbps at up to 10 meters.

The V.35 interface, based on the ITU-T recommendations, is popular with high-speed modems and multiplexers. It is specified as the interface for various high-speed digital services. The data and control lines are balanced to ground using multi-pair twisted cable. The V.35 interface uses a rectangular connector.

Universal Serial Bus

Because of the inherent limitations of the RS-232-C specification, another serial interface is being proposed. The so-called Universal Serial Bus, or USB, is a possible future industry standard for the connection of external devices to a PC. Should that standard be adopted by the industry, it would provide a universal connector, high-speed access, and a more robust Plug-and-Play capability.

Currently, there are several ports on a PC to connect external devices:

1. Keyboard port
2. Display port
3. Mouse port
4. Serial port
5. Parallel port
6. Audio port
7. Other proprietary ports

The speed at which data travels through these ports varies from port to port. Still, none of them are capable of the speeds required by new high-speed modems, multimedia, or high-speed storage devices. The USB would provide high-speed data transfer rates of up to 12,000 kbps, compared to today's standard PC serial port rate of up to 115 kbps.

The variety of different ports on a system can make the addition of peripherals confusing for the user. The key aspect of USB is the standardization of a single connector for a wide variety of devices. This would simplify expansion for the user and could save valuable space on a system, because all of the ports listed above could be eliminated by using a single USB port with a daisy-chain cable that goes from device to device.

Acceptance of the USB would provide a method by which external devices could better identify themselves and their system requirements. The operating system and the computer BIOS would then interpret this information and provide a Plug-and-Play capability.

Interface Converters

When two pieces of data communications equipment use different and incompatible interfaces, it might be possible to find an interface converter. The following is a partial list of interface converters listed by a modem and interface manufacturer, the Black Box Corporation of Pittsburgh.

1. RS-232-C to RS-422 Converter
2. RS-232-C to V.35 Converter
3. RS-422 to V.35 Converter
4. RS-232-C to MIL-188 Converter
5. RS-232-C to Current Loop Converter
6. RS-232-C to RS-449 Converter
7. RS-232-C to Burroughs TDI Converter

Each of these converters consists of a small box with jacks to accommodate the two incompatible interfaces. For example, the first converter interconnects the electrically-balanced RS-422 with the unbalanced RS-232-C equipment. All supported leads are converted to their electrical equivalents. Both DCE and DTE equipment are supported, and the interface conversion occurs in both directions. The mechanical connection is to a 25-pin connector on the RS-232-C side and to a 37-pin connector on the RS-422 side. Both synchronous and asynchronous operations are supported.

Nearly every interface comes with its own set of connectors. Some interfaces, such as the RS-232-C, even come with a variety of different plugs and jacks, although the DB-25 connectors are most prevalent. Table 6-3 compares specifications of various serial interfaces, while Table 6-4 shows connector types associated with serial and other interfaces.

Table 6-3: Serial interface specifications

	RS-232-C	RS-422-A	RS423-A	RS-485	RS-449
Recommended maximum distance (feet)	50	4000	4000	—	200
Maximum signaling rate (bits/sec)	20k	10M	100k	10M	20k/2M
Generator levels (V) Open circuit	<25	≤6	≤6	<6	RS-422/423
On, Space, 0	+5 to +15	+2 to +6	+3.6 to +6	+1.5 to +5	RS-422/423
Off, Mark, 1	−5 to −15	−2 to −6	−3.6 to −6	−1.5 to −5	RS-422/423
Maximum receiver levels (V)	±3 to ±25	±12	±12	−7 to +12	RS-422/423
Balanced/ Unbalanced	U	B	U	B	U/B
Connector	25-pin D	—	—	—	37- and 9-pin D

Comparison of UART Chips

The Universal Asynchronous Receiver Transmitter, or UART, is the main component of a serial port. The UART chip translates data coming from the parallel computer bus on eight separate wires and ground returns (one byte, equal to eight bits—one bit per wire) into serial signals (one bit at a time). Simply speaking, the eight bits taken off the parallel bus are put into a buffer and then released one bit at a time. The serial signals are sent to, or are received from, the remote terminal via a single pair of wires, namely the telephone line. The UART is initialized by the communications software, which supplies the UART with information about the transmission rate, number of bits per character, parity information, and other parameters.

The CPU in the computer communicates with the serial port by writing to the UART's registers. UARTs have First-In First-Out (FIFO) buffers through which the transfer of data occurs. First-In First-Out means that the first data that enters the buffer is also the first data to leave the buffer.

Table 6-4: Serial and parallel interface connectors

INTERFACE	CONN. TYPE	NO. OF PINS	APPLICATION
	DB 25	25	RS-232
	DB37	37	RS-449
	DB50	50	DataProducts, Datapoint, UNIVAC, & others
	DB15	15	DSU, CSU, TI, NCR POS, & others
	DB9	9	449 Secondary, ATARI, DAA, & others
	V.35	34	V.35
	M/50	48	Data Products, UNIVAC, DEC, & others
	Champ	36	Centronics, Champ, Printronics, Epson & others
	IEE-488	24	IEEE-488
	MATE-N-LOK	8	Current Loop
	RJ-11	4	Telephone
	RJ-45	8	Telephone

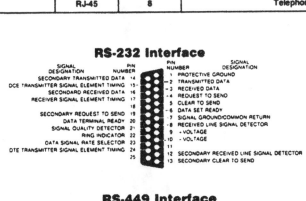

RS-232 Interface

SIGNAL DESIGNATION	PIN NUMBER	PIN NUMBER	SIGNAL DESIGNATION
SECONDARY TRANSMITTED DATA	14	1	PROTECTIVE GROUND
DCE TRANSMITTER SIGNAL ELEMENT TIMING	15	2	TRANSMITTED DATA
SECONDARD RECEIVED DATA	16	3	RECEIVED DATA
RECEIVER SIGNAL ELEMENT TIMING	17	4	REQUEST TO SEND
	18	5	CLEAR TO SEND
		6	DATA SET READY
SECONDARY REQUEST TO SEND	19	7	SIGNAL GROUND/COMMON RETURN
DATA TERMINAL READY	20	8	RECEIVED LINE SIGNAL DETECTOR
SIGNAL QUALITY DETECTOR	21	9	+ VOLTAGE
RING INDICATOR	22	10	- VOLTAGE
DATA SIGNAL RATE SELECTOR	23	11	
DTE TRANSMITTER SIGNAL ELEMENT TIMING	24	12	SECONDARY RECEIVED LINE SIGNAL DETECTOR
	25	13	SECONDARY CLEAR TO SEND

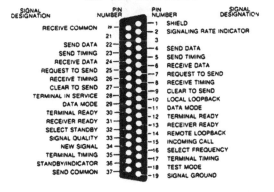

RS-449 Interface

SIGNAL DESIGNATION	PIN NUMBER	PIN NUMBER	SIGNAL DESIGNATION
RECEIVE COMMON	20	1	SHIELD
	21	2	SIGNALING RATE INDICATOR
		3	
SEND DATA	22	4	SEND DATA
SEND TIMING	23	5	SEND TIMING
RECEIVE DATA	24	6	RECEIVE DATA
REQUEST TO SEND	25	7	REQUEST TO SEND
RECEIVE TIMING	26	8	RECEIVE TIMING
CLEAR TO SEND	27	9	CLEAR TO SEND
TERMINAL IN SERVICE	28	10	LOCAL LOOPBACK
DATA MODE	29	11	DATA MODE
TERMINAL READY	30	12	TERMINAL READY
RECEIVER READY	31	13	RECEIVER READY
SELECT STANDBY	32	14	REMOTE LOOPBACK
SIGNAL QUALITY	33	15	INCOMING CALL
NEW SIGNAL	34	16	SELECT FREQUENCY
TERMINAL TIMING	35	17	TERMINAL TIMING
STANDBY/INDICATOR	36	18	TEST MODE
SEND COMMON	37	19	SIGNAL GROUND

Table 6-4: Serial and parallel interface connectors (continued)

Centronics Interface

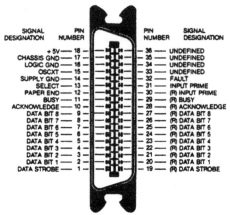

SIGNAL DESIGNATION	PIN NUMBER		PIN NUMBER	SIGNAL DESIGNATION
+ 5V	18		36	UNDEFINED
CHASSIS GND	17		35	UNDEFINED
LOGIC GND	16		34	UNDEFINED
OSCXT	15		33	UNDEFINED
SUPPLY GND	14		32	FAULT
SELECT	13		31	INPUT PRIME
PAPER END	12		30	(R) INPUT PRIME
BUSY	11		29	(R) BUSY
ACKNOWLEDGE	10		28	(R) ACKNOWLEDGE
DATA BIT 8	9		27	(R) DATA BIT 8
DATA BIT 7	8		26	(R) DATA BIT 7
DATA BIT 6	7		25	(R) DATA BIT 6
DATA BIT 5	6		24	(R) DATA BIT 5
DATA BIT 4	5		23	(R) DATA BIT 4
DATA BIT 3	4		22	(R) DATA BIT 3
DATA BIT 2	3		21	(R) DATA BIT 2
DATA BIT 1	2		20	(R) DATA BIT 1
DATA STROBE	1		19	(R) DATA STROBE

(R) INDICATES SIGNAL GROUND RETURN

V.35 Interface

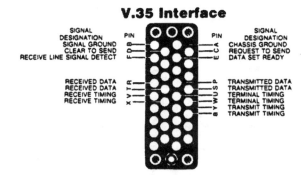

SIGNAL DESIGNATION	PIN		PIN	SIGNAL DESIGNATION
SIGNAL GROUND	B		A	CHASSIS GROUND
CLEAR TO SEND	D		C	REQUEST TO SEND
RECEIVE LINE SIGNAL DETECT	F		E	DATA SET READY
RECEIVED DATA	R		P	TRANSMITTED DATA
RECEIVED DATA	T		S	TRANSMITTED DATA
RECEIVE TIMING	V		U	TERMINAL TIMING
RECEIVE TIMING	X		W	TERMINAL TIMING
			Y	TRANSMIT TIMING
			a	TRANSMIT TIMING

The earliest UART chip for the personal computer market was the 8250, developed and manufactured by National Semiconductor Corporation in the early 1980s. At that time a standard modem operated at 300 bps, and a high-speed modem operated at 1200 bps. In these early days of PC development, it was well understood that the serial-port speed would never need to exceed 9600 bps—just as no one will ever need more than 640K of RAM, and all programs could be run from 5.25" diskettes.

The 8250 chip was subsequently upgraded to a slightly faster 16450. Both chips had only a 1-byte wide FIFO buffer. The next development in UART design was the 16550 and the 16550A UART, which fixed a bug in the original 16550. This new chip had a 16-byte wide FIFO buffer.

In older computers, the UART was a physically separate chip, but in newer computers the UART is embedded in a specialized chip set that performs many other functions associated with the serial port. When you use an external modem, you connect it to a serial port in the computer. If you use an internal modem, the UART is built into the modem chip set.

As modems execute sophisticated data compression and decompression algorithms, the data stream coming from the computer to the modem can be several times faster than the data stream leaving the modem towards the telephone line. With data compression ratios of up to 1:4, the computer can be sending data to the modem at rates four times higher than the rates at which the modem transmits data over the telephone line. Therefore a 28,800 bps modem can, under optimum conditions, require a UART operating at 115,200 bps. When selecting transmission speed in the setup screen of a communications program, you should always choose the computer-to-modem (higher) speed, while the modem decides—through negotiation with the remote terminal—what the actual transmission speed over the telephone line should be.

Because many UART chips cannot reliably handle transmission speeds of over 100,000 bps, a problem arises. Depending on the computer vintage and generosity of the computer manufacturer, one will find older and less-expensive 8250/16450 UART chips, or the newer 16550 chips that can handle transmission speeds of up to 115,200 bps.

Because the 8250/16450 chip has only 1 byte of buffer storage rather than 16 bytes in the 16550 chip, the data in the buffer can be easily overwritten before it is transferred to the computer memory or to the modem. When data is overwritten, the error-correction protocol—be it MNP 2-4 or ITU-T V.42—kicks in, detects the error, and requests re-transmission of a data block of 128 to 1024 bytes. This re-transmission will slow the data transfer.

An easy way to determine the type of UART in your computer and modem is to run the MSD or the WhatCom program on the enclosed disk. If your PC has Microsoft DOS version 3.1 or later, then it should also have a utility called Microsoft System Diagnostics (MSD). MSD provides useful information about a PC, including the type of UART chip found in each of the installed serial ports. The MSD and WhatCom programs must be run from a DOS prompt as described in Chapter 16. The programs will not function properly if Windows is running. If you are using Windows 95, click the START button and select "Restart the computer in MS-DOS mode?" from the popup menu.

To run MSD from a DOS prompt:

1. Type MSD, then press Enter. The MSD diagnostic screen appears.
2. Type C. The COM port screen appears.

The last line in the dialog box displays the type of UART that is present in each serial port.

To run Whatcom:

1. Follow the steps described in Chapter 16 to initialize the program.
2. Display the COM port assignments.

The type of installed UART can also be found by going to the Windows 95 Control Panel, selecting Modems, then Diagnostics folder, then selecting your modem and clicking on More Information. You will then see your current modem settings, the type of UART installed, and its maximum speed in Baud, which should really say "bps." If the serial port in your computer or the UART in the internal modem is not a 16550 chip, file transfer will not occur at the maximum possible speed with an up-to-date modem. If you need a faster file-transfer rate, the only solution is to install in your computer a new serial-port card that contains a 16550 chip, or install an internal modem with the same chip. With various 56,000 bps standards on the horizon, it is expected that a UART faster than the 16550 will soon become a new standard for serial communications.

Another solution to the slow UART problem is to use a proprietary UART from Hayes Corporation that operates at speeds of up to 230,400 bps. The UART, called the Hayes ESP Communications Accelerator, is included in an enhanced serial port that solves the problems of reduced throughput and data loss for high-speed communications.

Hayes ESP features dual 1,024-byte transmit and receive buffers. To reduce data loss, Hayes ESP features automatic flow control. As the 1,024-byte receive buffer becomes full, the Hayes ESP sends the appropriate flow-control signal to the modem. Received data remains buffered until the computer is ready to read it.

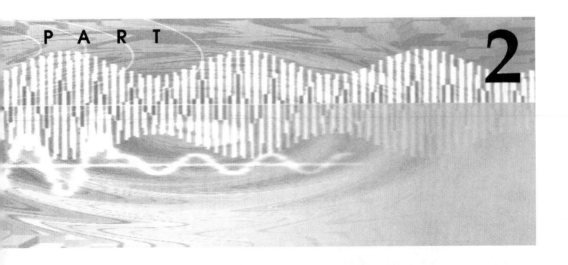

Modems for the Personal Computer

In Part 1 we explored the basics of data communications. Part 2 delivers the payoff—applying this knowledge to the wonderful gadget attached to your computer that can put you in touch with the rest of the world.

The personal computer, and the modem associated with it, provide a convenient and economical way to exchange digital information across the street or across the globe. In the early days of computing in the 1980s, business applications far overshadowed personal applications. Today the opposite is true. Because of the large personal computer market, measured in hundreds of millions of units, the inexpensive personal computer modem surpassed the capabilities of the commercial business modem from a few years ago, and became a standard for both the home and business markets. The only reason to use commercial modems today is for compatibility with older applications or to transfer a large amount of data over special transmission facilities.

In terms of sheer numbers, the personal computer market is the principal user of modems, and therefore drives the direction of modem technology today. The reasons for the popularity of modems, particularly in the United States, include: the nearly-complete lack of restrictions for connecting modems to the public switched telephone network; the availability of high quality communications software; and the standardization and low cost of personal computers and modems.

Thanks to standardization of hardware and software, users can exchange data with each other even if their modems and computers are made by different manufacturers or use different operating systems. All that is required is that the data communications parameters—such as bit rate, number of data and stop bits, and parity—as set by the communications software, are the same at both ends of a data connection.

In 1986, Hayes Corporation, the pioneer of modem design, had between 40% and 50% of the personal modem market. Today, personal computer modems are a commodity item, with modem-chip manufacturers selling their chip sets to thousands of small and large modem manufacturers. Currently, the two largest modem manufacturers in the U.S. are U.S. Robotics and Motorola Corporation.

Thanks to the ingenuity of designers, increased competition, and the availability of inexpensive modem chip sets, many modem features that were previously found only in the commercial market are now available at low prices in the personal modem market. For example, a number of years ago, I tried to establish—from my home phone—a data connection to a mainframe computer. Unfortunately, in order to gain access to the mainframe, a certain touch-tone code was required. But my home phone used a rotary dial, and the modem was not equipped with an automatic touch-tone dialer. Finally, with some ingenuity, the problem was solved with a tape recorder by recording touch-tone signals from another phone and playing the tape back to the mainframe computer after the connection was established using the rotary dial. The only serious problem remaining was that the connection established with so much effort would break down when someone would make a loud noise in the house. The sound would be picked up by the acoustic modem coupler, the only type of personal modem then available, and would break the connection.

Today the same connection can be established in seconds by means of an auto-dialing direct-connect modem. The data communications software resident in my home computer now uses a script file that sends the proper codes and passwords when it receives a prompt from the mainframe computer.

As discussed in Chapter 4, modems that are currently used with personal computers follow standards codified by international organizations, at first the CCITT and later the ITU-T.

The higher-speed standards, such as V.34 bis, are downward-compatible—meaning the 33,200 bps modem can also operate at lower speeds to be compatible with earlier modems, all the way down to the now-archaic 300 bps Bell 103 protocol. I still remember paying about $150 at Radio Shack for my first 300 bps modem. In all modern designs, the calling modem will recognize the maximum transmission speed of the called modem during the handshake sequence, which occurs at the beginning of a data call. If this speed is lower than the maximum speed of the calling modem, the calling modem automatically adjusts its transmission speed downward to match that of the called modem.

Today, all personal computer modems follow the so-called *intelligent* mode of operation, which includes auto-dialing and compatibility with the basic Hayes AT command set discussed in Chapter 9. All personal computer modems can communicate over the switched telephone network in full-duplex mode. With the continuing price erosion in the personal computer industry, modems that have speeds less than 28,800 bps are becoming obsolete, and 33,600 bps modems can now be purchased for less than $150. Currently, approximately 90% of the personal modem market consists of the 14,400 and 28,800 bps modems, but they are slowly being replaced by the 33,600 and 56,000 bps modems.

The following chapters will discuss all the important aspects of modems that are used with personal computers. Chapter 7 shows how to select a modem and what options to choose. Chapter 8 explains how to install a modem. Chapter 9 describes how a computer and modem communicate, and explains the modem language—referred to as the AT command set. Chapter 10 presents the all-important communications software. Chapter 11 shows what your modem can connect to—such as the Internet, bulletin boards, and other computers. Finally, Chapter 12 provides a look inside a modem by describing a typical modem chip set.

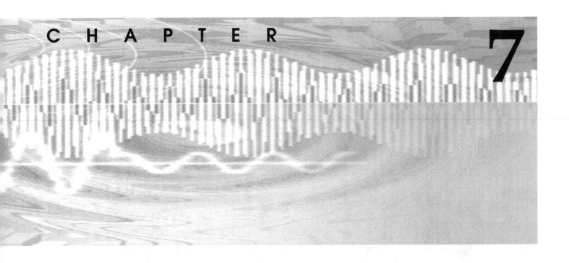

C H A P T E R 7

Selecting the Right Modem

The personal computer user is presented with many choices when selecting a modem. The first one is whether to choose an internal or external modem. Then comes the question of assorted features and how to balance their importance against the cost. This chapter will hopefully answer all these questions.

Internal or External Modem?

A personal computer modem can be purchased as either an internal card or as an external unit, which is then connected with a cable to the computer's serial port. Therefore the first choice is to decide if an internal or external modem is best for you.

Figure 7-1 shows a picture of an external modem, while Figure 7-2 shows a typical connection between the computer, an external modem, a telephone outlet, and a telephone set. Figure 7-3 shows a picture of an internal modem, while Figure 7-4 shows a similar connection for an internal modem. Figures 7-2 and 7-4 are further discussed in Chapter 8.

If the computer is used only to access bulletin boards, computerized information services, and other fully automated services that do not require human assistance, only one RJ11 jack is needed on the modem. However, if you plan to use the modem in conjunction with a contact manager or auto-dialer to make voice calls, it is usually desirable for the modem to have two RJ11 jacks—one for the telephone line, and one for a telephone set. This way, one can easily go back and forth between data and voice communication. Lack of a second jack will require a Y connector that would allow you to plug in a telephone set, as shown in Figure 7-5.

Figure 7-1 External modems (courtesy U.S. Robotics)

Figure 7-2 Hook up of an external modem to PC and phone outlet

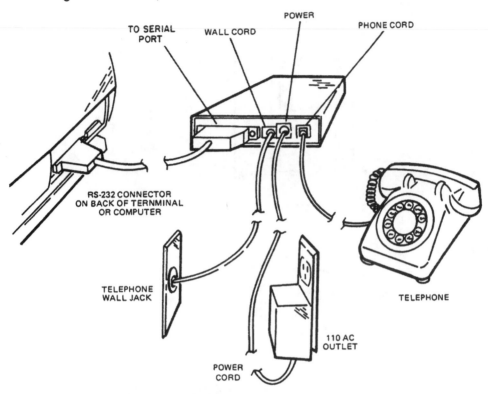

Figure 7-3 Internal modem (courtesy U.S. Robotics)

Figure 7-4 Hook up of an internal modem to a phone outlet

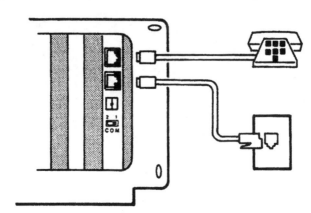

Figure 7-5 Modem connection with a "Y" plug

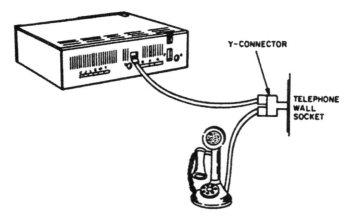

Similar components can be found in both internal and external modems, while some are specific to one or the other. The main structural difference between the two types of modems is that the internal modem has a built-in Universal Asynchronous Receiver and Transmitter (UART) chip, which translates the parallel signals coming from the computer bus (one byte or eight bits, at a time), to serial signals (one bit at a time) and vice versa.

An external modem uses the UART that is built into the computer's serial port. An external modem requires a separate power supply, which is a small transformer plugged into a power outlet. It also requires an enclosure, which is typically a 6" × 5" × 1" metal or plastic box. An external modem also has a set of indicator lights that are visible on the outside of the box (described in detail in Chapter 16), while the internal modem has no indicator lights. The cost of the power transformer, enclosure, and indicator lights makes the external modem about 20% to 30% more expensive than an equivalent internal modem.

Both internal and external modems have a small built-in speaker so you can hear the ring, busy and dial-tone signals on the telephone line during the modem's dial-up sequence. The volume of the speaker can be adjusted by software commands, though in some modems you may find a volume-control knob. Internal and external modems have one or two RJ11 telephone jacks. All external modems also have a DB-25 jack to accommodate the RS232-C cable to the computer's serial interface. A modem also usually has a set of jumpers or DIP switches for various parameters such as COM port and IRQ settings, although more recent modems are completely controlled by software and have no jumpers or DIP switches to set.

Figures 7-6 and 7-7 show individual components of two modem configurations.

Figure 7-6 Block diagram of an internal modem

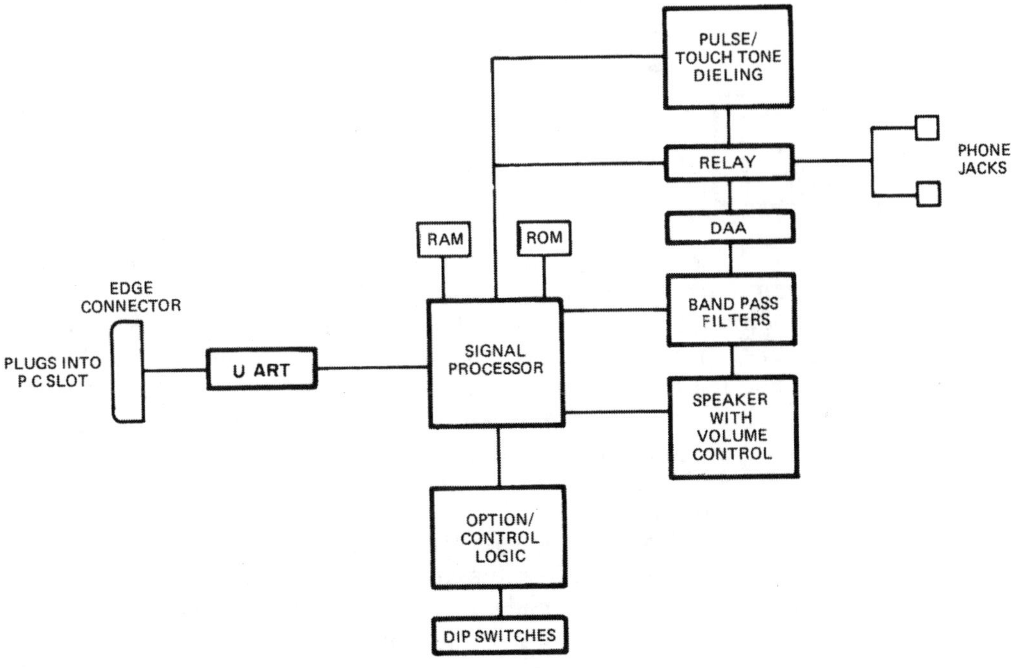

Figure 7-7 Block diagram of an external modem

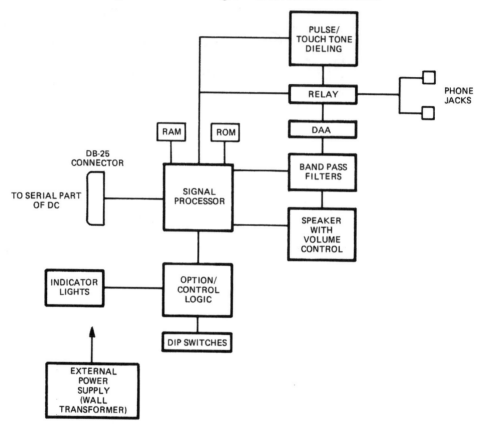

An internal modem and an equivalent external modem do not differ in terms of transmission quality. Still, each modem is built differently and each has its advantages and disadvantages from the consumer's point of view. Although differences in modem hardware should not be important to the user, there are functional differences that make choosing the proper type of modem quite important. Following are some pros and cons to consider.

1. Power Supply.
    The internal modem derives power from the computer's internal bus. This saves the cost of a power transformer compared to an external modem. The disadvantage is that the computer's power supply must be able to support the internal modem's power requirements. However, most computer power supplies can easily support the internal modem.

    Conclusion: Internal modem has the advantage.

2. Cable Clutter Next to the Computer.
   Using an internal modem obviates the need for a cable from the power outlet to the external modem, and for the thick serial cable from the computer to the external modem. This arrangement reduces clutter around the computer desk.

   Conclusion: Internal modem has the advantage.

3. Ease of Installation.
   To install an internal modem, you must open the computer case and insert the modem card—not an easy task for people who are "all thumbs" and are afraid of mechanical gadgets. You may also have to set some miniature jumpers on the card to set the correct COM port and IRQ settings. The Plug-and-Play feature of Windows 95—when it works, which unfortunately is not always the case—will make installation of an internal modem quite simple.

   Conclusion: External modem has the advantage for the "mechanically-challenged."

4. Serial Port Availability.
   On some computers, a serial port is a separate option in the form of a circuit card that plugs into the computer's bus on the motherboard. It may be a socket on a plug-in card, or it may be a socket on the computer's motherboard. Physically, the connector is a DB-25 jack with 25 pins, or a DB-9 jack with nine pins. The connector protrudes from the back or side of the computer's chassis. When connected to an external modem, the serial cable completes the connection between the computer and the external modem's serial port jack.

   An internal modem does not require a separate serial port connector in the computer. All serial port circuitry is already included in the modem plug-in card, which is mounted inside a card slot located on the computer's motherboard. The only external connections for an internal modem are one or two RJ11 jacks—one for the telephone cable that goes to the telephone outlet on the wall, and a telephone cable that goes to the telephone set if a second RJ11 jack is available on the internal modem.

   An internal modem is recommended if all serial-port jacks in the computer are already occupied by devices such as a mouse, serial printer, serial plotter, and other serial devices. If the two standard serial ports—COM1 and COM2—are in use, an internal modem can possibly be assigned to COM3 or COM4, which are supported by MS-DOS and Windows, provided that there is an associated IRQ available in the computer.

Another aspect to consider is that the serial port in the computer may not support a fast modem. To operate at 28,800 bps or higher, the UART on the serial card must be a 16550 rather than the older 8250 or 16450.

Conclusion: Internal modem has the advantage.

5. Card Slot Availability.
    An internal modem occupies a card slot in the computer chassis. Because the number of card slots is limited, a card slot may not be available if other cards occupy all available slots.

    Conclusion: External modem has the advantage if no card slots are available.

6. Indicator Lights.
    A major advantage of an external modem is its indicator lights, discussed in detail in Chapter 16. These lights can help pinpoint a problem in a data connection. They not only indicate if the modem is powered up, but also whether a carrier is detected and if data is being transmitted and received.

    Conclusion: External modem has the advantage if you wish to see an indication of the state of a data connection.

7. Portability.
    The same external modem can easily be attached to any computer that has a serial interface, such as an IBM-PC, an Apple MacIntosh, or even an early-vintage TRS-80 Model I. However, because it is small and portable, it can also "walk away" easily without anyone noticing.

    Conclusion: External modem has an advantage if you want to use one modem to connect to more than one computer at different times.

8. Power Switch.
    An external modem can be quickly turned off by flipping its power

switch without turning off the computer. If you hear an annoyed voice instead of the modem handshaking chirps on the modem speaker, you may want to quickly turn off the modem!

Conclusion: External modem has the advantage if you want to be able to turn a modem off without turning off the computer.

9. Cost.
Last but not least is the difference in cost between an internal and external modem. An internal modem will typically cost 20% to 30% less than an external modem of similar functionality. The reason for the difference, as explained earlier, is the additional cost of the enclosure, indicator lights, and power supply for an external modem.

Conclusion: Internal modem has the advantage.

Fax Modem

A major application for personal computer modems is facsimile transmission, or fax for short. Fax is the transmission of images or text such that the received image is a reasonable facsimile of the original image. Selecting this important modem feature today is simply a no-brainer. Because fax circuitry is included in practically all currently-available personal computer modems at no extra cost, this is certainly a feature to look for.

There is a considerable difference between transmitting the character A as in ASCII code (hex 41 or decimal 65), and transmitting the same character as a graphic picture of the letter A, which is a series of white and black dots that forms the letter. When the remote computer receives a sequence of ASCII codes for letters "m", "o", "d", "e" and "m," it can interpret this sequence as the word "modem." It can also do a spell-check and transfer it to any document in a word processor. If the computer receives a picture of the same five characters, it treats it as any other picture. Whether the picture represents Mona Lisa or the word "modem" is open to interpretation. A person looking at a picture of the word "modem" will recognize it immediately, but a computer requires a special optical character recognition (OCR) program to interpret the meaning of the picture with the word "modem" in it.

Historically, the stand-alone fax machine came first. It had a paper feeder and a built-in scanning device for documents that were to be transmitted. It also had a built-in phone or outlet for an external phone, and it had a thermal-paper printer for received documents. A scanning device in a fax machine scans each line of material as it is being sent. The stand-alone fax machine became popular for business applications in the 1970s when prices for such machines dropped to under $2,000. It certainly beat the Postal Service and even Federal Express, ensuring a nearly instant delivery of documents. As fax machine prices declined to $500 and less, their use spread to small businesses and personal use.

In the mid-1980s, more and more documents were generated in the form of computer files rather than pieces of typewritten paper. Often, pages combined text and pictures that were generated on a PC using desktop-publishing programs such as PageMaker or Ventura Publisher. There was a desire to be able to send those files to a remote fax machine directly, or to a remote computer capable of receiving a fax, without having to first print the document and feed it into a fax machine.

To answer this need, in 1988 Intel Corporation developed a computer fax board that would plug into a PC and would transmit the picture of a document in a graphical (PCX) format to a stand-alone fax machine or remote computer equipped with a similar board. The cost of the board was over $500, and it had a limited commercial success. The source of the transmission was no longer a printed document but a computer-generated file. Of course, a separate scanner attached to a computer would still be needed to transfer the paper image to the computer, which could then be transmitted using the fax board.

In early 1990s, as modem technology progressed, chip manufacturers began incorporating fax protocols and technology in standard modem chip sets. Today, practically every modem includes fax capabilities. Depending on the commands sent to the modem, you can choose either the data or fax mode of transmission. Although strictly speaking, a fax transmission is another mode of data transmission, one still has to distinguish between fax transmission—which is a message that can be interpreted by a fax machine or a computer as an image of a page—and data transmission—which is a message that can be interpreted by a computer as a sequence of ASCII or binary codes.

Other Features to Look For

Like a typical American automobile with dozens of options, the personal computer modem comes with dozens of optional features, each touted by the respective manufacturer as the most important feature of the decade. The features differentiate basically-similar models, following the same basic standards listed in previous chapters. Some of these optional features are of real importance, some are more like tail fins on cars in the 1950s. Of course, what one person finds to be of limited value can be very important to another.

Becoming familiar with the list of available features and comparing them with your real needs should help you make an informed buying decision.

Many modem features are discussed in detail throughout this book. What follows is a summary of the more important standard and optional modem features offered by various manufacturers for the personal modem market, and their relative importance.

The first set of features no longer represents options, as was the case in the first edition of this book 10 years ago. These features should now be standard in all commercially-available personal modems. Still, it may be worthwhile to make sure that they are implemented in the modem you purchase. If any of these features are missing, look for another modem. It just means that the modem manufacturer is not aware of the current technological trends and is probably on the verge of bankruptcy. Therefore these standard features do not have specific values assigned to them.

To help in making a quantitative comparison among similar units, a value of between 1 and 10 is assigned to each optional feature. Values 1 to 3 mean they are of limited importance; values 4 to 6 mean this feature is nice to have but not necessary; and values 7 to 10 mean they are essential—you do not want to be without them. However, these are subjective numbers and may have to be adjusted to suit your own circumstances.

Standard Modem Features

Auto-Dial with Rotary and Touch-Tone Option

This is one of the most important features, and it should be included in all modems. A preferable hardware implementation for the on-hook/off-hook switch and pulse dialer is an electromagnetic relay rather than a solid-state switch. A relay, though slightly more expensive and prone to failure than a solid-state switch, will provide complete electrical isolation between the telephone line and the computer, whether the modem is in use or not. This can be important in areas that are prone to lightning strikes. Inadequate isolation between a telephone line and computer could possibly interfere with phone conversations even when the modem is not being used.

Full-Duplex Operation

This feature means that data can flow in both directions at the same time. It is usually a software option included in appropriate transmission protocols.

Hayes Compatibility

All modern data communications software relies on modem compatibility with the basic Hayes AT command set. Therefore this feature is absolutely essential. These commands are discussed in detail in Chapter 9.

Automatic Answer Capability

This, and the next two features, Originate/Answer and Automatic Speed Select, are of importance if you exchange data with individual computer users and do not limit yourself to the Internet, bulletin boards, information services, and so on. The feature is somewhat similar to an automatic answering machine. When you are not available, others can still call and leave digital messages as long as your computer is on.

Originate/Answer Mode Capability

This is somewhat similar to the previous feature. All bulletin boards and information services operate in the Answer mode, with the subscriber's modem in the Originate mode. However, if you try to communicate with a remote modem and both modems can only operate in the Originate mode, then a connection cannot be established.

Automatic Speed Select

This is a feature normally embedded in V. series modem protocols. If you call another modem or another modem calls your modem, your modem should set itself to the highest mutually-acceptable transmission speed. This feature is of particular importance if you deal with many callers who use different transmission speeds.

Call-Progress Tone Recognition

This feature will recognize dial and busy tones. With the appropriate software, it is possible to avoid blind dialing and to get easier access to busy bulletin boards.

Speaker with Volume Control

A speaker should be included in an internal or external modem so you can hear the call-progress tones. It would help to know if an incorrect number was dialed and a person answers instead of another modem. Although early external modems often featured a knob to adjust the speaker volume, current modems provide only a software volume adjustment using the ATMn command.

Error Detection and Correction

Most modems include built-in error detecting and correcting protocols such as the V.42 and MNP 2 to 4 (described in Chapter 4). It is preferable for these protocols to be implemented in a modem chip set rather than in software that runs on a computer. The statement in the specifications to look for is V.42 "capable"—not good, versus V.42 "implemented"— good. The hardware implementation leads to faster data throughput, compared to a software implementation. Error correction and detection is always required when data compression is enabled.

Data Compression

Most modems also include built-in data compression protocols such as V.42 bis and MNP 5 (described in Chapter 4). Again, it is preferable for these protocols to be implemented in a modem chipset rather than in software that runs on a computer. The statement in the specifications to look for is again MNP 5 "capable" versus MNP 5 "implemented." The hardware implementation leads to faster data throughput, compared to a software implementation.

Front Panel Indicators

This feature is only available on external modems, although some software packages simulate a set of modem indicator lights on the computer screen. However, if you have an external modem, with its associated jungle of cables, this can be an important feature. An external modem should have at least 5 or 6 indicator lights. These indicator lights are described in detail in Chapter 16.

Storage of Telephone Numbers

This feature, although highly touted in many advertisements, is of very limited importance, except for security modems. In general, communications software takes care of storing at least four phone numbers, which can then be automatically dialed by the modem.

Number of RJ11 Telephone Jacks

Both internal and external modems should have two female RJ11 telephone jacks, or one RJ11 jack to accommodate a phone, and a telephone cable that has a male telephone plug on one end to connect to the wall jack. Modems equipped with a single jack and no cable should be avoided, because you must use a Y-type connector if you want to connect a phone as well, which is a major inconvenience.

FCC Approval

Each modem used in the U.S. must be approved by the FCC. Lack of FCC certification will prevent you from legally connecting the modem to the telephone network and may also result in fines, confiscation of equipment, and embarrassment.

Optional Modem Features

Modem Transmission Rate

The recommendation is to buy what you can afford, and buy what is the common and current standard. A few years ago, it cost $1000 or more in order to buy a 14,400 bps modem, the common modem then a 1200 bps unit. In 1997, you can buy a 33,600 bps modem for between $100 and $200. A 56 kbps modem sells for not much more.

If you buy a "high-speed" modem, make sure you deal with a reputable manufacturer. If the modulation/speed standard is not yet commonly accepted, it could be changed and an upgrade to the modem might be required. Otherwise, you could end up with a modem that might not be able to communicate with other modems. Value = 8.

COM1, COM2, COM3 and COM4 Capability

This feature applies only to internal modems. Access to COM3 and COM4 serial ports provides increased flexibility if ports COM1 and COM2 are already occupied by other devices. In any case, a modem should have at least the option of COM1/COM2 selection. Value = 0 if it has COM1 only. Value = 6 if it has COM1 and COM2. Value = 8 if it has COM1, COM2, COM3 and COM4.

Independent IRQ Selection Capability

In general, specific IRQs (Interrupt Requests) are assigned to specific COM ports. For instance, IRQ3 is assigned to COM2 and COM4; IRQ4 is assigned to COM1 and COM3. Still, if you have many peripherals attached to your computer, it may help in avoiding IRQ conflicts by being able to select a non-standard IRQ for a specific COM port. This applies only to internal modems. Value = 6.

Flash EPROM

This feature lets you update modem firmware by simply download-ing the new software from a bulletin board or Internet site. Being able to fix a bug, or upgrade a specification without having to send the mo-dem to the manufacturer for chip replacement, is a really nice feature.

It becomes even more important when many modems are released before the appropriate standards are approved. The stan-dard might change when it is finally approved by the ITU-T, which would require a firmware upgrade. Manufacturers are reluctant to issue expensive recalls. However, they do not mind putting a soft-ware patch on their bulletin board or Internet site for free download-ing. Value = 6.

Speaker-Phone Capability

Some modems have a built-in speaker-phone capability. A small speaker, which usually doubles as a microphone, lets you speak without holding the receiver. This feature is of marginal importance because most phones already provide this capability. Value = 3.

Testing Capabilities

Many modern modems allow for a whole range of transmission testing, as described in detail in Chapter 16. These tests are initiated by issuing AT&T commands to the modem. It is possible to test a single modem and end-to-end performance between two modems. In the event of poor-quality telephone lines, it is a nice feature to have. Value = 6.

Demon Dialer

This feature, which is related to the storage of telephone numbers feature, will repeatedly call one or more numbers stored in the modem's memory. The modem will detect the busy signal and will continue dialing, alternating between two or more telephone numbers, if so instructed. This feature can usually be duplicated by communications software. Value = 5.

Data/Voice Capability

This feature lets the modem behave like a sophisticated answering machine with individual mail boxes assigned to various recipients or for providing special messages to callers. If two modems support the Data/Voice feature, it is possible to conduct a voice conversation while data is being transmitted on the same line. Of course, if data is being exchanged between locations X and Y, then the voice communication must also be between X and Y. Value = 6.

Automatic Switching of Voice and Data Calls

If you use the same telephone line for voice, modem and fax, you may want to look for this feature. A modem equipped with this feature will recognize if the incoming call is a voice call, a data call to your modem, or a fax call, and route it to the computer, phone or fax, as appropriate. Value = 6.

Touch-Tone Decoder

Some modems have the ability to decode the incoming touch-tone signals. This feature, if recognized by the software, allows you to add security features to the modem by excluding callers who do not send a specific code with their touch-tone phone when they call you. Value = 5.

Size of Modem Card (Internal Only)

Older versions of internal modems were built on full-sized cards. Newer versions, which use more LSI circuits, usually fit on a half-size card. Because there is only a limited number of full-size slots inside a PC, and those slots may already be occupied by other cards, a smaller modem card is definitely preferable. Value = 5 for a half-size card.

Ease of Setting Internal Switches

Switches should be easily accessible without having to disassemble the complete modem. Software switches—whose settings are stored in the modem—that use low-power CMOS memory are preferable to those that use DIP switches, which in turn are preferable to pluggable jumpers. On the bottom of the list are soldered jumpers, often found in some old vintage modems.

Assigned values are as follows: Value = 10 for software switches. Value = 9 if it has easily-accessible DIP switches. Value = 6 if it has difficult-to-access DIP switches. Value = 5 if it has pluggable jumpers.

Caller Identification

Some modems decode caller-ID information from the signal sent between rings by the telephone company to those who subscribe to the Caller-ID service. A stand-alone Caller-ID detector costs around $40 and occupies space on your desk. If you subscribe to this service, it is a worthwhile feature. Value = 6.

Modems Working Only With a Specific Operating System

There are modems that will work only with computers that run a specific operating system. For example, WinModem, from U.S. Robotics, will only operate with Windows. The advantage of WinModem is that it makes full usage of the Plug-and-Play features when using it with the Windows 95 operating system, which should make installation a snap.

Such modems are slightly cheaper than general-purpose modems because they use, to a larger degree, the computer's own CPU for signal processing. I personally do not care to tie myself to a single operating system. In particular, for diagnostic purposes, it is good to have a modem that can work with both DOS and Windows programs. On the other hand, if you want to install the modem and forget about it, then WinModem may be for you. Value = 0.

Bundled Communications Software

Nearly all modems today come with several diskettes and possibly a CD-ROM in the box. Typically, you might find a simple terminal emulation program, a fax send/receive program, and initialization software to sign up with a large Internet service provider (ISP). The quality of the terminal emulation and fax software is slightly above of what you might find bundled with your operating system, but not as good as stand-alone commercial software and shareware programs. The ISP offers usually include some free time with the ISP before you will be charged. Try it and keep the service if you are satisfied. If you are not satisfied, talk to your friends about their experiences with a local ISP. They can sometimes provide better service than a regional or national ISP.

How Well Will My Modem Work?

Now that you have selected the correct assortment of features, you will probably find that a store or computer show will offer many units that will satisfy your requirements. There are brand-name modems and there are generic modems with unknown names. Generic modems will typically cost 20% to 30% less than their brand-name equivalents.

In my experience, if your Internet service provider (ISP) or bulletin board has a good local point-of-presence (POP)—which means they have local numbers to dial into wherever you might live—and you have a quiet and short telephone loop to the local telephone office, then any modem will do.

However, if you have to dial long distance to get to your ISP or bulletin board, or if you have a noisy telephone line, then some modems will do better than others. In general, the more expensive brand-name modems will do a better job. Always make sure that you deal with a reputable dealer who will honor the guarantees and exchange the modem if necessary.

Problems you might encounter on a poor connection will primarily be the difficulty in signing on to your ISP or bulletin board. You might have to dial repeatedly, and even after you establish a connection, you might encounter data loss and disconnects. Typical symptoms of data loss are:

1. Your user name or password are not immediately recognized.
2. Errors occur during file transfers, even with error control enabled.
3. Online text has missing characters or is not formatted properly.
4. Online menus appear jumbled or are missing some words.
5. You frequently get disconnected.

Of course it is possible that your connection is so poor or your service provider's equipment is so decrepit that no modem can operate successfully in that environment.

Remember that specifications listed on the modem box are often misleading. If a modem complies with a specific ITU-T standard, it only means that it does not violate that standard, but it does not mean that it satisfies every option listed in that standard. Some of these options may be of no importance in your particular situation, while others may be very important.

Furthermore, no personal computer modem I have worked with ever published performance specifications—such as the maximum throughput for specific signal-to-noise ratios under less-than-optimum conditions. In other words, as with any other consumer product, let the buyer beware.

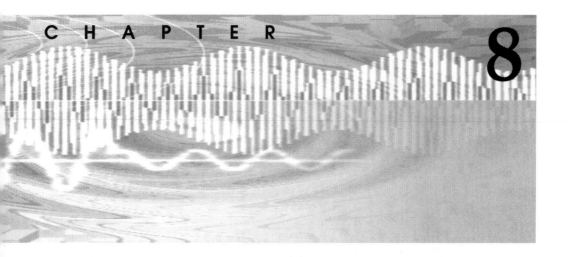

Modem Installation

You have now researched the market, selected the best modem to fit your needs, purchased and unpacked it. The next step is to install it. There are five steps in the installation process, all of them important to your final success. They are:

- Modem preparation—set the appropriate jumpers and switches, should it be necessary.
- Plugging it all together—install the internal modem inside the computer or connect the external modem with the proper cable to the serial port. In a laptop computer with a PC-CARD (formerly called PCMCIA) slot, insert the PC-CARD modem into that slot.
- Letting your computer know about the modem—configure your computer to recognize the modem.
- Letting the software know about the modem—configure your communications software to work with the modem.
- Monitoring the modem—monitor the operation of your modem with a single circuit.

Modem Preparation

Both the internal and the external modem require a few one-time hardware and/or software settings to make the modem compatible with the computer and the communications software. Depending on the modem manufacturer and the specific model, these settings are performed either with:

1. Software commands
2. Pluggable jumpers
3. DIP switches
4. Any combination of the above

In some modems, a jumper or DIP-switch setting becomes the default setting, which can still be changed with a software command. The settings control important modem parameters such as COM port and IRQ selection. An incorrect setting will make exchange of data with another computer impossible.

COM Port Setting

The first and most important setting is the selection of the COMmunications serial port, or COM port, the physical interface between the computer and modem. Computer architecture provides for a number of serial ports, not all of which are normally implemented. These ports are designated as COM1, COM2, and so on. An IBM PC and IBM compatibles are limited in most cases by hardware considerations to only two serial ports implemented as jacks—COM1 and COM2—although the operating system recognizes two additional ports—COM3 and COM4.

An external modem can only be assigned to a port that physically appears as a DB-9 or a DB-25 jack with male polarity, usually located at the rear of the computer chassis. An internal modem can be assigned to any serial port that is implemented by the computer, such as COM3 or COM4. These ports do not necessarily have to appear as a serial port jack.

To work properly, the COM port you select for your modem should not be used by any other device, such as a serial printer or a serial mouse. If, for example, a serial printer is assigned to COM1, then the modem should be assigned to COM2, if it is free. For an internal modem, the assignment of COM ports is made with jumpers or switches. For an external modem, the assignment of COM ports is done by plugging a serial cable from the modem into the DB-9 or DB-25 jack that is associated with the appropriate serial port on the computer.

Some internal modems provide the option of using serial ports COM1, COM2, COM3 or COM4, while others allow for the use of only COM1 and COM2. If the communications software supports all four COM ports, then a modem can operate on COM3, while a printer operates on COM1 and a serial mouse on COM2, for example. Such an arrangement will also require a specific IRQ setting for each port, because COM1 and COM3 normally share the same IRQ. This will be described next. Some specialty modems rely on the Plug-and-Play feature of Windows 95 and are completely software configurable.

Installing a PC-CARD modem in a laptop computer will require configuring the PC-CARD slot to behave like a COM port. This is normally accomplished by installing a modem driver, which is a program supplied with the modem on a diskette or included in the operating system of your laptop computer. As standardization has not yet reached the world of laptop computers, instructions for modem installation will vary from computer to computer. It is highly advisable to read the User Manual that comes with your modem.

IRQ Settings

Internal and external devices that are built into or attached to a computer, such as a keyboard, internal clock, or serial port, alert the CPU that they require service when they detect any activity on that device (such as when a key is pressed or released, or data is sensed on a serial port). This service request is sent to the CPU by means of an Interrupt ReQuest (IRQ) line. The IRQ lines are numbered consecutively IRQ1, IRQ2, and so on. The number of IRQ lines depends on the computer architecture. Sixteen IRQ lines are typical in current systems. The service request directs the computer to stop what it is doing and take care of the request, then return to what it was doing earlier. The computer knows which device issues the interrupt by sensing from which IRQ line it received the interrupt request.

In the event of a communications device such as a modem connected to a serial port, an IRQ request means that a byte has been received by the serial port from the modem and should be transferred to the computer's memory. Once the transfer occurs, the modem will allow the next byte to be received.

When the CPU receives an interrupt request, it halts what it is currently doing, saves its current status, performs the requested task, restores its previous status, and goes back to whatever it was doing. Interrupts are sent over one of the IRQ lines on the computer's bus. The IRQ lines run between all card slots on the motherboard. If two devices use the same IRQ, it is likely that either one or both devices may not operate as expected.

The COM port that a modem is connected to has a default IRQ assigned to it by the computer's BIOS. The default assignments are IRQ4 to COM1 and COM3, and IRQ3 to COM2 and COM4. If modems are automatically assigned to the same IRQs that are used by other devices, either the modem or other device must be changed to an alternate IRQ. Some devices do not have the ability to change their IRQ assignment, which means the modem must be able to select an alternate IRQ. Fortunately, many of today's modems are capable of addressing alternate IRQs. It is very important when assigning IRQs to make sure that no two devices will use the same IRQ at the same time.

The IRQ setting must be performed only on internal modems. It is done by choosing the proper jumper or DIP switch settings, or by software settings if your modem is software-controlled. You will find the jumper or DIP-switch setting that corresponds to a specific IRQ in your modem's User Manual. Although external modems do not have IRQ or COM port settings, you must still set the IRQ for the serial port that the modem attaches to.

If your PC operates under MS-DOS version 3.1 or later, then it should also have a utility called Microsoft Diagnostics (MSD), mentioned in Chapter 6. MSD provides useful information about a PC, including the COM port and IRQ assignments. Use either MSD or the enclosed program, WhatCom, to check the current assignment of serial ports and IRQs on your computer, and for advise on which IRQ to choose for the internal modem. Both MSD and WhatCom should be run in pure DOS mode, not in a DOS window under Win 3.1 or Win95. Although the programs are capable of being executed in an MS-DOS "window," they will not run properly and the computer may freeze, which will require rebooting.

There are two ways to get into the pure DOS mode from Windows 3.1 or Win95:

Method 1:   For Win95, click the Start button, then select the "Restart the computer in MS-DOS mode?" option from the menu. For Windows 3.1, press Ctrl-F4 to exit Windows.

Method 2:   When the computer first starts or reboots, hold the F8 key down. For Win95, a menu appears with the title "Microsoft Windows 95 Startup Menu." Select option 6, "Command prompt only." For Windows 3.1, a DOS prompt (C:\>) appears.

To run MSD:

1.   Switch to the directory or disk where the MSD program is located (usually the C:\WINDOWS directory), or make sure that it is in the search path.

2.   Type "MSD" and press Enter. The MSD opening screen appears, as shown in Figure 8-1. The opening screen shows the basic information about the computer, such as the type of BIOS, amount of memory, and DOS version.

3. To get additional information, click on any of the 13 choices.

Still, what MSD displays is not necessarily accurate. For example, MSD seems to think that some Pentium processors are really a 486DX. I guess not all software is perfect.

Figure 8-1 MSD opening screen

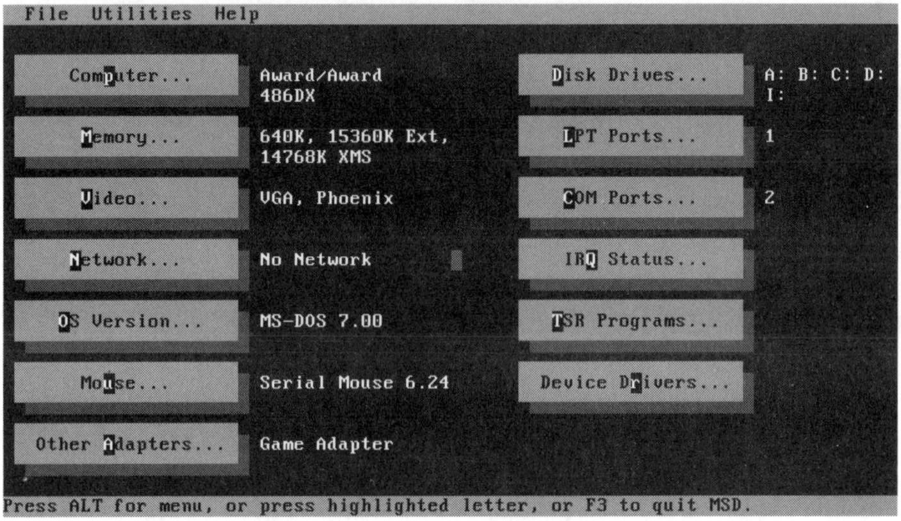

```
 File  Utilities  Help

    Com[p]uter...      Award/Award          [D]isk Drives...    A: B: C: D:
                       486DX                                    I:

    [M]emory...        640K, 15360K Ext,    [L]PT Ports...      1
                       14768K XMS

    [V]ideo...         VGA, Phoenix         [C]OM Ports...      2

    [N]etwork...       No Network           IR[Q] Status...

    [O]S Version...    MS-DOS 7.00          [T]SR Programs...

    Mo[u]se...         Serial Mouse 6.24    Device D[r]ivers...

  Other [A]dapters... Game Adapter

Press ALT for menu, or press highlighted letter, or F3 to quit MSD.
```

To run WhatCom:

1. Switch to the directory or disk where the WhatCom program is located, or make sure that it is in the search path.
2. Type "WhatCom" and press Enter. The first time you run the WhatCom program, it prompts you to type your name and company. Then the opening screen shows first the logo of Data Depot Inc., the program designers, followed by a menu of seven test and information screens, as shown in Figure 8-2. Tests performed by WhatCom on the modem and COM ports are discussed in more detail in Chapter 16.

Figure 8-2 WhatCom opening screen

Other Settings

Other software, jumper and switch settings on the modem card or on the external modem determine how the modem reacts to certain RS232-C interface signals. These settings are particularly important for auto dialing, and to some extent depend on the communications software. Frequently, the default settings—the settings the modem has when it leaves the factory—must be changed to operate with specific software packages. The default settings vary between modems from different manufacturers and even between different models from the same manufacturer. For specific information on your modem, check the User Manual.

Table 8-1 shows a list of selectable switch settings, which may be found in some external modems. Most modems of recent vintage will make these options software-selectable.

Table 8-1: Modem DIP Switch Settings in Some External Modems

Function	Switch Up	Switch Down	Default
DTR Response	Sense DTR signal	DTR forced ON	Down
Words/Numbers	Messages in words	In numbers	Up
Display	Messages not displayed	Displayed	Down

Table 8-1: Modem DIP Switch Settings in Some External Modems (continued)

Function	Switch Up	Switch Down	Default
Echo	Local Echo Enabled	Disabled	Up
Auto Answer	Auto Answer Enabled	Disabled	Down
Carrier Detect	Sense CD signal	CD forced ON	Down
Clear to Send	Sense CTS signal	CTS forced ON	Down
Configuration	User Defined	Factory Default	Down
AT Commands	Recognize	Ignore	Up

Some modems may also have a manual speaker volume adjustment in addition to the software control of the speaker volume. The modem speaker is bridged on the telephone line and plays the call-progress tones, such as dialing, busy, ring and answer. Listening to the call-progress tones provides the first indication that a connection is properly established.

Once the connection is established, the communications software will normally turn the speaker off, because it is no longer needed. A proper setting of the volume control may be required to avoid hearing those annoying loud tones coming from the modem while listening to the call-progress tones—on top of the ambient noise and whirring of the computer fan and disk drives.

Plugging It All Together

After setting correctly all the DIP switches and/or jumpers, you can take the next step—putting all the hardware together.

To install the internal modem, first open the computer case. Make sure that power to the computer is disconnected and that you ground yourself with a strap to a water pipe or other acceptable ground.

Opening the computer case requires removing a few screws and pulling the cover off. You should then see some empty card slots on the computer's motherboard. Notice that there are different types of slots. Typically you will see four slots for single inline memory module (SIMM) memory chips, perhaps four sockets for PCI cards, and three to four slots for standard ISA cards. In current vintage computers, most video cards plug into the narrow PCI slots, while most modems plug into the wider ISA slots. Plug your modem card into the correct slot, secure the modem card bracket with a screw, and replace the computer case.

The final two connections are a telephone cable that goes from the TELCO or LINE jack on the modem card to a wall telephone outlet, and an optional connection that goes from the other jack on the modem card marked PHONE to a telephone set. See also Figures 7-3 and 7-4 in Chapter 7.

Installing an external modem is even simpler. All you have to do is connect a serial cable from the modem's female DB-9 or DB-25 serial port jack to the computer's male DB-9 or DB-25 serial port jack. The cable must have a male polarity at one end and a female polarity on the other end. Next, plug the modem's power transformer into a wall AC outlet, and plug the low-voltage lead from the transformer into the power connector on the modem. The final two connections are the same as for an internal modem—a telephone cable goes from the TELCO or LINE jack on the modem card to a wall telephone outlet, and another cable goes from a telephone set to the other jack on the modem card marked PHONE. See also Figures 7-1 and 7-2 in Chapter 7.

Installing a PC-CARD/PCMCIA modem in a laptop computer that is equipped with a PC-CARD slot is even simpler. Before plugging the modem card into the PC-CARD slot, first make sure that your modem will fit in the PC-CARD slot in your laptop computer. There are currently 3 types of PC-CARD cards and slots: Type I, II and III. Most PC-CARD modems are mounted on Type II cards, although you may find other cards as well. They all use 68-pin connectors, but have different thicknesses. Type I is 3.3 mm thick, Type II is 5.0 mm thick, and Type III is 10.5 mm thick. A Type I card will fit all three slots; a Type II card can be used in Type II and Type III slots; a Type III card requires a Type III slot.

To connect a PC-CARD/PCMCIA modem to a phone outlet, you will need either a special connector to fit the slim profile of a card modem, or a special extendible built-in RJ11 jack similar to the one found on some Megahertz modems. The Megahertz Corporation, a subsidiary of U.S. Robotics, has a patent on such RJ11 jacks and markets them under the name RJ11 - XJACK.

If you use the same telephone line for voice, modem and fax, and your modem does not provide automatic switching capability, then you may want to add an automatic switch box to your system. That box will recognize if the incoming call is a voice call or a call to your modem or fax, and route it to the appropriate device. An automatic switch box provides four separate jacks for the following connections:

- Wall phone outlet
- Phone
- Answering machine
- Fax/modem jack in the computer

For a fax modem, you might also want to attach a scanner to your computer to be able to fax copies of paper documents.

Figures 8-3 and Figure 8-4 show all connections for the internal and external modems when used with an automatic switch box.

Figure 8-3 Automatic switch box with an internal modem

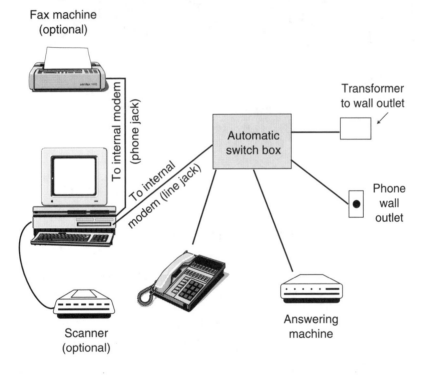

Figure 8-4 Automatic switch box with an external modem

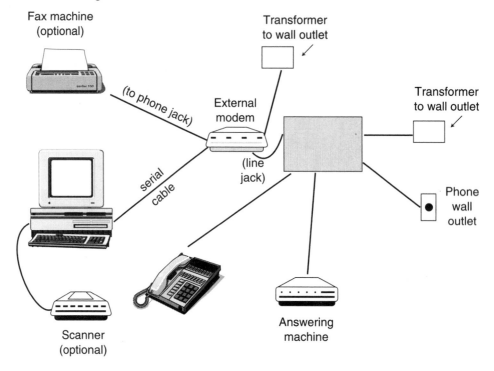

Letting the Computer Know About the Modem

The next step in modem installation is to notify your computer—or more precisely, its operating system—about the new arrival. The procedure depends on whether you have a DOS, Windows 3.1, or Windows 95 system.

In a DOS system, you do basically nothing. DOS does not care if there is a modem connected to your system. The communications software will take care of that. You just plug everything together as shown in Figures 7-2 or 7-4, run your communications software, and hope for the best.

In a Windows 3.1 system, you inform your computer about the modem as follows:

1. Choose Main -> Control Panel -> Ports.
2. Click on the COM Port that you will use for the modem.
3. Choose Settings, then choose a Baud Rate (bps) to suit your modem.
4. Choose Data Bits, Stop Bits, Parity, and Flow Control and make the appropriate setting for each. In general, you can

leave the settings at their default settings of 8 data bits, 1 stop bit, no parity, and RTS/CTS flow control. These defaults can be changed by your communications software.

5. Under *Advanced*, choose the correct Base I/O Port Address and correct IRQ. Again, you may leave them at their default settings unless you are aware of conflicts with other devices.

In a Windows 95 system, the *Plug-and-Play* feature, sometimes called *Plug-and-Pray* should take care of this part of the installation. In most cases, Windows 95 will even recognize the type of modem you have. In general, whenever you add, remove or replace a modem, follow this procedure:

1. Remove the modem and corresponding COM port from the system devices in the Control Panel. Then reboot the computer.
2. Click on the Add New Hardware icon in the Control Panel. The Add New Hardware Wizard panel appears.
3. At the next prompt, allow Windows 95 to detect the new hardware (modem) and select an available COM port. Follow the remaining prompts until the end of hopefully-successful installation.

This procedure should work with most modems. Windows 95 takes advantage of information about a modem's features and designation, which are stored in the modem's ROM. However, if your modem does not store this information—and some low-cost generic units do not—the following additional steps will be required.

4. If your modem is not detected in Step 2, go back to the Add New Hardware Wizard panel and answer No to the prompt "Do you want Windows to search for your new hardware?"
5. After clicking on Next, you will see a list of devices including Modem. Select Modem from the list.
6. Add a new modem by following the instructions on screen. Again, you can allow Windows 95 to detect the modem, or select it from the list on the screen.
7. The last choice is to select the driver. If you have the correct driver on the diskette provided with the modem, or a driver obtained from the modem manufacturer's BBS or Internet site, use that driver. If you do not have a driver, use the driver supplied with Windows 95.

The following article, by Joan Latchford, originally appeared in the October, 1996 issue of *read.me*, the newsletter of the Personal Computer Club of Toronto. It depicts the trials and tribulations of a modem installation. It is reprinted here with permission from the author and the newsletter publisher, Howard Solomon:

> First, I want to say that the Plug 'n Play U.S. Robotics Sportster Winmodem 28.8 is superb. Software updated to 33.6 kbits per second, Web pages appear quickly and the CNET Newscaster no longer sounds as if he has barely overcome a speech impediment.
>
> Installation is smooth: remove the cover from your computer, insert the modem firmly into an empty slot, connect the telephone jack and switch the computer back on. A message appears in Windows 95 identifying your Winmodem Sportster 28.8 and gives four choices for driver selection. 'Driver from disk provided by manufacturer' is pre-selected so all you have to do is insert the provided disk and press enter. But ...
>
> USR's Winmodem reminds me of the archetypal little girl who, 'when she was good, was very, very good but when she was bad, she was horrid.' (Of a normally sunny disposition, we just know some event must have occurred to set her off - but what?) Once installed USR's Winmodem performed flawlessly for a couple of weeks until one morning I rebooted my computer to receive a message that: 'The modem is being used by another dial-up networking connection or another program.' Not true ...
>
> Now Winmodem's 'Plug 'n Play' installation had connected it to COM 4. And there, I believed lay the problem, since my previous modem installation was on COM 2. However, neither the Winmodem Manual nor the installation software provided any instructions on how to change the installed COM port.
>
> I am a born 'tweaker' so it was natural for me to spend several days into the wee small hours in documented attempts to change my installation to COM 2. While I found a means to do this (In Win95 Open 'Control Panel', double click on the Winmodem icon, de-select 'Use Automatic Settings' and 'Force COM 1-4', double click on COM 2 and restart your computer) it did not resolve the refrain that: 'The modem is (still!) being used by another dial-up networking connection or another program.' I re-installed Win 95, Netscape Gold, removed and re-installed Winmodem and my dial-up network connection several times with equal lack of success. After several phone calls to Keating Technologies (Toronto distributors for U.S. Robotics) I was told to return the modem. While I waited for a replacement, my husband downloaded the software update file 'win336.exe' from U.S. Robotics' Web site. A new modem arrived a couple of days later. Steve, of Keating's Tech Support told me to unpack the update file (win336.exe) to a floppy and to insert it when Winmodem asked for the manufacturer's disk. It was rejected, so I used the original 28.8 install disk, run the 33.6.exe update file immediately after and rebooted my

system. The replacement Winmodem was again setup on COM 4 but after I had re-installed my dial-up network connection it was recognized and made its usual raucous connection to my Internet Service Provider. When I changed the installation to COM 2 (to see if I could replicate the original problem) everything continued to work. Since I have a system crammed with peripherals it seemed logical to test for installation conflicts without suspecting an individual defective modem. However, a great deal of time would have been saved if the Winmodem software came with documentation for changing the COM port. 'Plug n' Play'? I would have to say 'I guess so,' while recommending either an online 'help' feature or a specific trouble shooting 'Wizard'. Ironically, while I waited for a replacement modem, my USR Sportster 14.4 re-installed on COM 2 came up with the reliability of a Ford model T -- but Oh, how I missed the Jaguar swiftness of a fully functioning Winmodem 28.8!

Letting the Software Know About the Modem

You know it, the computer knows it, now comes the third step—the communications software also has to know it. In general, the operating system does not share its knowledge of the modem with the current crop of communications software. This may change in the future, which would save you this step. However, most of the current software requires an explicit indication from the user as to what specific modem is being installed.

Each communications program will display a setup menu similar to the one shown in Figure 8-5. The entries describe the basic parameters for the local computer, and for the BBS or ISP to which you are planning to connect. This includes COM-port selection, number of bits used per character during transmission, transmission rate, parity, and flow control.

Notice the Connection mode option. Normally it would show "Connect through modem." However, you can also select "Answer through modem" or "Direct to port" for a direct connection through a serial port to another computer that has a null modem, as described in Chapter 6. Other options, such as those shown in Figure 8-6, list hundreds of modems with associated initialization strings to choose from. If the vintage of your software is the same or more recent than the vintage of your modem, then there is a good chance that your specific modem will be listed. In this case, just select it and hope for the best.

Figure 8-5 Setup menu in a communications program

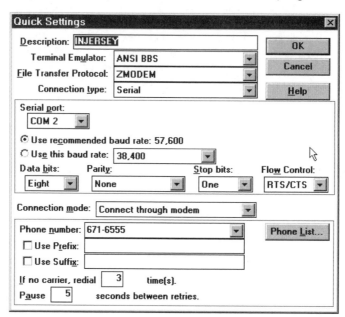

Figure 8-6 Modem selection in a communications program

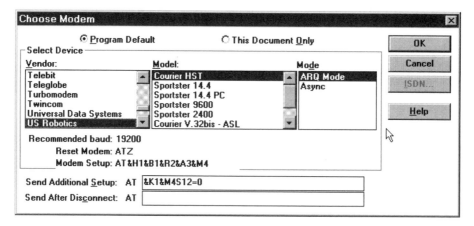

What the modem selection means to the communications software is simply to determine which initialization string it should send to the modem at the beginning of each communication session. These strings can get quite elaborate. The following shows a few examples of Hayes modem initialization strings:

1. Hayes OPTIMA 144 + FAX144:

   ```
   AT&FW2&C1&D2&K3&Q5S7=60S46=138S48=7S95=47
   ```

2. Hayes OPTIMA 288B V.34/V.FC + FAX:

   ```
   AT&FW2&C1&D2&K3&Q9S7=60S46=138S48=7S95=47
   ```

3. Hayes V-series ULTRA Smartmodem 14400:

   ```
   AT&FW2&C1&D2&K3&Q5S7=60S38=10S46=2S48=7S95=63
   ```

4. For Mac PPP:

   ```
   AT&Q5S36=7&C1&D0&K3
   ```

5. Hayes ACCURA 288 V.34/V.FC + FAX:

   ```
   AT&FW2&C1&D2&K3&Q5S7=60S36=7S46=138S48=7S95=47
   ```

The modem selection in the communications program may also determine how to interpret values stored in various S registers, which are discussed in Chapter 9. Though some of these registers are standardized by modem manufacturers, some differ from modem to modem. A general-purpose communications program is not written for a specific modem, but should satisfy thousands of users who own many types of modems that have many different features. Therefore the only chance for the program to customize its operation to a particular modem is to address it with a series of specific commands designed for it. These commands are included in the initialization strings.

An initialization string is composed of several commands that are written using the AT modem command-set language, which I will discuss in more detail in Chapter 9. The AT commands that comprise the modem language, when issued, enable or disable certain modem functions. The initialization string, which is the first series of AT commands that get sent to the modem, should make the most effective use of features of your specific modem.

Now what should you do if your modem is not on the menu of your favorite communications program? With the proliferation of generic modems, it is practically impossible for the software designer to keep track of all new arrivals. Another problem is that some manufacturers assign the same name, such as the Sportster modem from U.S. Robotics, to a whole family of modems. Modems from different

manufacturers and even different models from the same manufacturer use different subsets and extensions to the modem language.
Therefore a legitimate command issued to one modem may completely disable another modem.

Most communications programs, as illustrated in Figure 8-6,
show the initialization string for the selected modem and also let you
edit that string. If you find a different initialization string specified in
your User Manual from the one suggested by the software, or find an
initialization string by speaking with the technical support people of
the modem manufacturer, then replace the string suggested by the
software with the hopefully-better string.

If your modem is in the orphan category and you cannot find its
name in the software modem listing, and nobody is able to help you,
then choose the generic modem option from the menu. Otherwise,
choose any modem from the menu, but delete the initialization string
completely. You may find that leaving the initialization string blank is
often better than specifying a wrong initialization string.

This should start you going and allow you to connect to your
ISP or bulletin board. If you understand the basics of the AT language
as described in Chapter 9, and study your modem's User Manual,
you might come up with an optimum initialization string that will
take advantage of all important features of your modem, especially
its data compression and error-correction capabilities.

The communications software will allow you to choose many
other options, such as your terminal emulation and other options
shown in Figure 8-7. If you are connected to a mainframe computer,
you could possibly select a VT-100 terminal. For BBS communications, the popular selection is the ANSI BBS-type terminal emulation.
Other choices include display colors, default downloading protocols,
keyboard mapping, and much more.

Figure 8-7 Terminal settings in a communications program

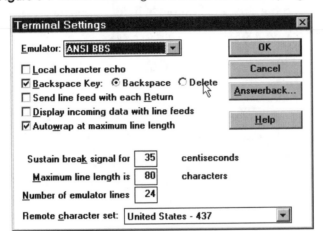

Monitoring the Modem

The modem has now been installed and hopefully works great. Another useful addition to the physical setup is a phone line switch/monitor with an amplified speaker, as shown in Figure 8-8. I realized the need for such a device after I unsuccessfully tried to send a fax message from my computer to a friend in Switzerland. Because of some problems with the automatic voice recognition in my friend's modem, my fax modem just kept trying to send, without me realizing what was happening. Later, I became aware of the problem after noticing 10 one-minute calls to Switzerland on my phone bill.

Communications software provides the choice, using the ATM command, to either monitor your call until the connection is established, monitor the call continuously, monitor the call continuously except during dialing, or to not monitor the call at all. Once the internal speaker volume is set to high, low or off using the ATL command, you cannot easily change the volume while the call is in progress. With my simple circuit, you can both monitor your call at any time you wish, and change the speaker volume any time you wish, not just at the beginning of the call.

Another great feature of this circuit is that you can, by flipping a double pole-double throw (DPDT) switch, quickly change the connection from modem/fax to a voice connection, or you can terminate the call made by the modem. This may come handy if your modem misdialed and you hear a voice instead of the modem or fax tones. You just flip the switch and you are in voice mode. The schematic and view of the monitoring circuit are shown in Figures 8-8 and 8-9. I found the circuit to be a big help for both data and fax calls.

Figure 8-8 Schematic of a monitoring circuit

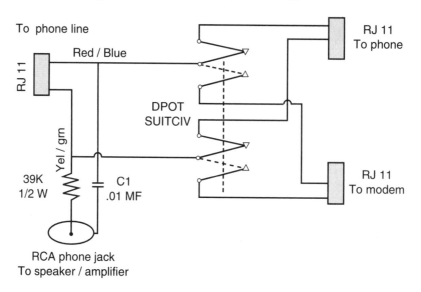

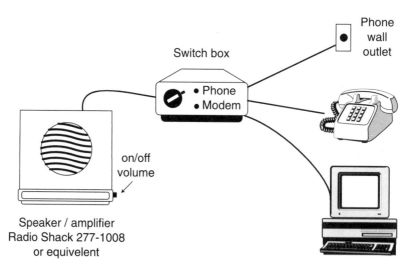

Figure 8-9 View of a monitoring circuit

The easiest way to construct the switch/monitor circuit is to combine some commercially-available inexpensive parts with a couple of electronic components. I used an A/B phone-line switch box and a small speaker/amplifier. The A/B phone-line switch box is carried by Radio Shack, or can be bought at computer fairs for under $10. The small speaker/amplifier can be bought from Radio Shack for around $10. I drilled a hole in the switch box and installed an RCA type jack. I then soldered the resistor and capacitor shown in Figure 8-8 between the RCA jack and the phone line in the A/B switch. I then connected the speaker/amplifier to the RCA jack with an audio cable.

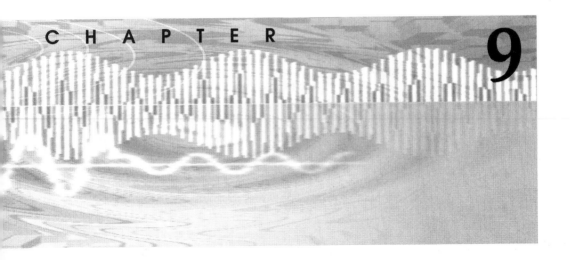

CHAPTER 9

Talking to the Modem

If you have followed the previous chapters, your modem is now installed and operational, and both the computer and the communications software know about its existence.

This chapter explains the principal modes in which a modem operates—the local, the online originating and answer mode, the offline, and the handshaking mode, as well as the concept of flow control—starting and stopping the data transmission to prevent a data buffer overflow and data loss. This chapter also describes the language in which the computer and the modem use to communicate with each other. This language consists of commands that change values of the so-called S registers, which are stored in the modem's own internal RAM (random access memory). You will learn how to send commands to the modem directly from the keyboard, and what kind of response you should expect. This type of direct conversation with the modem is great for exploring modem features and for troubleshooting.

Modes of Operation

Early modems, such as my first 300 bps external modem bought at Radio Shack for $150, were very simple. When you were feeding the modem a bit of data coming from the computer over the serial port at one end, the same bit of data, after being modulated, was spilled out towards the phone line at the other end.

Well, soon the price of my modem dropped down to its paper-weight value and a new breed of smart modems arrived on the scene. These modems could understand certain commands written in a language developed by Hayes Corporation, and could distinguish whether the data coming from the computer should be transmitted to the remote computer (the computer across the phone line), or whether it should be interpreted as a local modem command and not be transmitted over the phone line.

Transmission Modes

To establish a data connection, a computer that uses a current-vintage smart modem must first initiate the call. Unless otherwise agreed upon, that modem will remain, for the duration of the call, in the Originate mode, while the remote modem will remain in the Answer mode.

The two transmission modes—the Originate and the Answer modes—are physically different because they use different modulation schemes with different carrier frequencies. The modem that initiates the call does not have to remain in the Originate mode, although some modems can only operate in that mode. However, if your modem is calling a modem that can only operate in the Originate mode, the calling modem must be able to operate in the Answer mode as well. The selection of the transmission mode was accomplished in older modems by manually flipping a switch. In all current-vintage modems, it is accomplished by sending an appropriate AT command to the modem. For example, the command ATDT555-1212 sent from the computer to the modem puts the calling modem in the Originate mode and makes it dial 555-1212. However, using the command ATDT555-1212R will dial the same number and put the calling modem in the Answer mode.

Operational Modes

To better grasp the interaction between a modem, data communications software, and a computer, you must understand the four operational modes of the modem and their transitions. Depending on its current operational mode, the modem will interpret the data coming from the computer as either data to be transmitted to the remote computer, or as modem commands. The relations and transitions between the four operational modes are shown in Figure 9-1. The four modes are as follows:

1. Local Command Mode.
    In this mode, the modem is disconnected from the telephone line. It will accept dialing instructions and settings of data communications parameters, such as bit rate and parity.

When typing on the computer, you communicate with the modem's internal registers. This operational mode is changed to the Handshake mode when a call-dialing command, such as ATDT5551212, is received by the modem.

Figure 9-1 Modem operating mode transitions

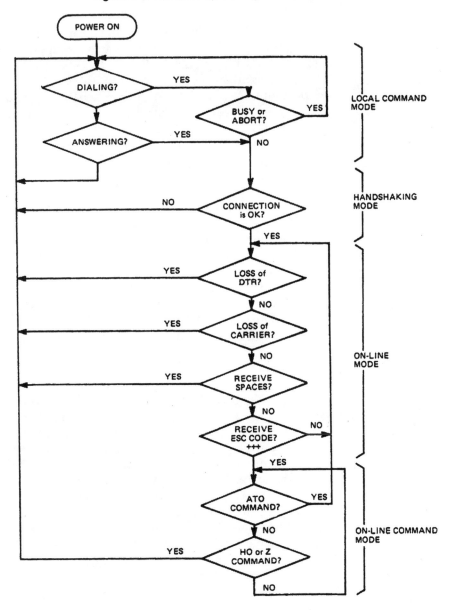

2. Handshake Mode.

    This mode is under control of the local and remote modems, and not under the control of the local computer. The local modem checks that the carrier from the remote modem has been received. Once the connection is established, the calling modem sends a training sequence to adjust the remote modem's adaptive equalizer. There is also a check for compatibility of the two modems. For example, if the remote modem can support 28,800 bps and the local modem can only support 14,400 bps, the remote modem must fall back to 14,400 bps.

    If the line conditions are poor due to excessive noise or other line problems, the transmission rate between the two modems would be set even lower. During the handshake sequence, the two modems negotiate error correction and data compression protocols that are acceptable to both modems. If the handshake is successfully concluded, the calling modem shifts to the Online mode and sends a CONNECTED message to the local computer, followed by the negotiated transmission rate and the mutually-agreeable protocols. Otherwise the NO CARRIER message is echoed to the computer, and the modem will make the phone line go on-hook (disconnected) and revert to the Local Command mode.

3. Online Mode.

    This is the data-transfer mode. All data sent to the modem from the computer, whether they are characters typed on the local computer or files being uploaded, is sent to the remote modem and its associated computer. Similarly, all data that is received by the modem from the remote computer is passed through to the local computer.

    While transmitting data to the remote modem, the local modem constantly listens for a specific sequence of bytes, for the so-called escape sequence. In order to leave the Online mode to again communicate with the modem directly, you must send it an escape sequence (by default, it is +++, but you can change it) with a 1-second guard pause before and after the escape sequence. Upon receiving the escape sequence, the modem changes to the Offline mode, sometimes referred to as the Online Command mode. In this mode the modem is still keeping the connection to the remote modem alive, but it can now detect modem commands. A loss of carrier will also change the operating mode from the Online to the Offline/Local Command mode.

4. Offline Mode.
   In this mode the connection remains open in the background, while the local computer can still send commands to the modem in order to perform certain functions before resuming the exchange of data with the remote modem. For example, you could instruct the modem to open or close a memory buffer for received data, enable local echo, read the local directory, or even terminate the call.

Figure 9-1 shows transitions between the four operating modes in terms of modem commands. A data communications program also provides the capability of shifting back and forth between operating modes. For example, Smartcom for Windows, described in Chapter 10, will switch from the Online mode to the Offline mode when an appropriate menu or button is activated. In some communications programs, pressing the Escape key automatically switches the modem to the Offline mode. Similarly, in some communications programs, pressing the Return key returns the modem to the Online mode. To enter the Local Command mode from the Online mode, you must terminate the call in progress by issuing a disconnect command while in the Offline mode.

Making the Modem Listen

While in the Local Command mode, you can speak and listen to your modem without any connection to the phone line or to the remote computer. You can find out about modem features and see how the modem responds to various commands. This type of direct conversation is helpful in exploring modem features as well as for troubleshooting.

This example uses the simple terminal-emulation program called Hyper Terminal provided free with Windows 95. It will put the modem into the Local Command mode. There is also a similar program in Windows 3.1 called Terminal. Although the procedure differs from one communications program to another, it should be fairly easy to translate the approach described here to your favorite communications software.

Configuring Hyper Terminal in Win95 in Local Command Mode

1. Click on the Hyper Terminal icon in the Accessories folder. You might also locate the Hyper Terminal program by clicking on the Start button in the Win95 opening screen and se-

lecting Programs/Accessories/Hyper Terminal from the Start menu.

2. A panel with several icons named AT&T, Compuserve, and so on, appears. Click on the icon named Hypertrm.exe.

3. The next panel is named Connection Description. Type a name for the direct connection, such as DIRECT COM 2. Also select an icon from the icon images provided. Click OK.

4. The next panel is named Phone Number. From the Connect Using dialog box, choose Direct to Com 2 rather than the default that lists the name of your modem. If your modem is connected to COM1 instead of COM2, choose Direct to Com 1 instead. Other entries, such as the phone number, are blanked out and should not be entered. Click OK.

5. The next panel is named COM 2 properties. You can leave the defaults as is. They are: Bits per second = 2400, Data bits = 8, Parity = None, Stop bits = 1, Flow control = Hardware. Do not change the advanced settings. Click OK.

6. At this point you will see a new panel named DIRECT COM 2, or whatever name you gave it in Step 3. Now you can communicate directly with the modem and watch its responses. Test it by typing AT followed by a carriage return. The modem should display OK on the computer screen immediately below the AT command you have typed. Remember that all modems accept AT commands written in all upper case, and most modems accept AT commands written in either upper or lower case, but not both. Therefore ATI7 or ati7 would work fine on most modems, but ATi7 may not work.

When you are finished, choose the Disconnect icon or pull down the Cali menu and choose Disconnect. If you forget to disconnect and try to close the panel, a dialog box appears, asking if you want to disconnect. Choose YES. A new dialog box appears, asking whether you want to save the session DIRECT COM 2. Choose YES again. From now on, the new icon you selected in Step 3 will appear in the Hyper Terminal panel. You will no longer have to repeat Steps 2 through 5, and will be taken directly to Step 6.

Information from the Modem

Here we will discuss in more detail what you will discover about your modem when you communicate directly with the modem. Tables 9-1 and 9-2 are only a preview of what a U.S. Robotics Sportster modem indicated when it was interrogated with some selected ATI commands. Your specific modem may or may not respond to the same commands, so consult your modem's User Manual.

Table 9-1: Modem Configuration From ATI4 Command

```
Command Sent: ATI4
Modem Response:
U.S. Robotics Sportster 33600 Fax Settings...
  B0   E1   F1   L1   M1   Q0   V1   X4   Y0
  BAUD=2400   PARITY=N   WORDLEN=8
  DIAL=HUNT   ON HOOK
  &A3  &B1  &C1  &D2  &G0  &H1  &I0  &K1  &M4  &N0
  &P0  &R2  &S0  &T5  &U0  &Y1
  S00=000 S01=000 S02=043 S03=013 S04=010 S05=008 S06=002
  S07=040 S08=002 S09=006 S10=014 S11=070 S12=050 S13=000
  S15=000 S16=000 S18=000 S19=000 S21=010 S22=017 S23=019
  S25=005 S27=000 S28=008 S29=020 S30=000 S31=128 S32=002
  S33=000 S34=000 S36=014 S38=000
  LAST DIALED #: T6716555
OK
```

Table 9-2: Modem Configuration From ATI7 Command

```
  Command Sent: ATI7
  Modem Response:
  Configuration Profile...
  Product type          US/Canada Internal
  Options               V32bis,V.FC,V.34+
  Fax Options           Class 1/Class 2.0
  Clock Freq            92.0Mhz
  EPROM                 256k
  RAM                   64k
  EPROM date            10/13/95
  DSP date              10/13/95
  EPROM rev             1.1
  DSP rev               1.1
  OK
```

As you can see from Tables 9-1 and 9-2, the ATI4 and ATI7 commands provide the current status of modem registers, assorted settings, and information about the EPROM version. When you contact the modem manufacturer about any problems related to your modem, you will have to provide some of this information.

Flow Control

The mechanism of the modem flow control starts and stops the data stream between the computer and the modem. If the computer sends data to the local modem, or to the remote terminal or computer at a speed higher than either can absorb, then the data stream from the local computer must be stopped. To prevent an overflow and subsequent loss of data, the remote modem either sends a specific character—called XOFF—to the local computer, or drops the CTS serial interface line to indicate to the local computer that it should stop sending data. To resume data flow when enough space is available in the modem's input buffer, the modem either sends the XON character back to the computer, or raises the CTS to indicate that the computer can resume sending data.

The flow control with XON and XOFF characters—normally Ctrl-Q (decimal 17) for XON, and Ctrl-S (decimal 19) for XOFF—is called *software flow control*, or more appropriately, the XON/XOFF control. Flow control with the CTS (Clear to Send) signal and its response—the RTS (Ready to Send) signal—is called hardware RTS/CTS flow control.

As mentioned in Chapter 6, the Data Terminal Equipment (DTE) and the Data Communications Equipment (DCE) are not always well defined in the serial interface protocol. Therefore another option for the hardware flow control is to use the DTR (Data Terminal Ready) and DSR (Data Set Ready) serial-interface signals. This type of control is referred to as hardware DTR/DSR flow control. Hewlett Packard uses still another version of the RTS/CTS flow control called ENQ/ACK in some terminals.

The flow-control selection is performed in the setup screen of all data communications programs, including Hyper Terminal. The hardware flow control should be selected over the software flow control for transmission of binary data files. The reason is that the modem cannot distinguish whether the XON and XOFF characters are meant to start and stop data flow, or are legitimate characters in a binary data file.

If your communications program provides an option of several types of hardware flow control, such as RTS/CTS, DTR/DSR or ENQ/ACK, you should try them all, if possible—one at a time, of course—to discover which flow control provides the fastest and most trouble-free operation.

AT Commands—Basic, Extended and Proprietary

Hayes Corporation, the pioneer and for many years a leader in the development of *smart* modems for personal computers, developed—in the early 1980s—a set of modem programming commands and modems that could understand those commands. The basic set of commands became the de-facto standard for the personal computer industry.

The commands are frequently referred to as AT commands because of the AT prefix required for most commands. The AT prefix asks the modem to start paying ATtention to whatever follows, and to interpret the character string following the AT as a modem command, rather than as data to be modulated and transmitted to the remote computer.

When a specific AT command is sent to a Hayes or Hayes-compatible modem, it will dial a remote computer with the number indicated in the command. You can tell the modem to use pulse or touch-tone dialing, and set the data transmission parameters such as the number of data bits, parity, number of stop bits, and transmission speed. After establishing a connection, the modem will start transmitting the data. AT commands, besides dialing, can also control speaker operation, initiate various diagnostic procedures, and read or write to various modem registers that control the modem's status.

In addition to the AT commands, various modem registers—the so-called S-Registers originally introduced by Hayes and described later in this chapter—also became a de-facto standard throughout the modem industry. For example, the S0 register stores a number that is equal to the number of rings the modem should wait for before it answers the call when it is in the auto-answer mode. For example, sending the command ATS0=3 tells the modem to wait for three rings before answering a call and sending a carrier signal.

The advantage to modem manufacturers in using the same commands and registers as the de-facto standard is that, if they follow the rules, most communications programs will work with their modems.

Over the years, many modem manufacturers, including Hayes, developed additional commands that are not included in the original Hayes command set. Some of these commands were adopted by many modem manufacturers and are called *extended commands*. Other commands apply only to specific modems and are therefore called *proprietary commands*.

The extended and proprietary commands follow the AT prefix with a special character such as $, &, \ or +. For example, the U.S. Robotics Sportster modems feature several help files. The command AT$ issued to the modem in Local Command mode sends back to the computer a list of AT commands that are applicable to that modem.

Modem manufacturers provide many features in their modems, such as error detection and correction, and data compression. These features can be enabled or disabled by issuing the appropriate AT commands from the extended command set. Similarly, most modems provide assorted self-testing features. These tests are activated by a set of extended AT&T commands, which are described in more detail in Chapter 16.

Other manufacturers, such as IBM, tried to completely bypass the Hayes command set by developing their own commands and appropriate data communications software to use those commands. These efforts have been largely abandoned, as most modem chip manufacturers decided to stay with the Hayes standard. Chapter 12 provides a complete set of extended and proprietary AT commands and S registers that are implemented in a specific modem chipset manufactured by Rockwell International Corporation.

AT commands can be organized into the following categories:

Basic Command Set

An upper-case character followed by a digit. For example, ATM1.

Extended Command Set

An & (ampersand) and an upper-case character, followed by a digit. This is an extension of the basic command set. For example, AT&M1. Note that AT&M1 is not the same as ATM1. The extended AT command set was adopted by many modem manufacturers.

Proprietary Command Set

Usually started by either a backslash (\), a percent sign (%), or other special character. These commands vary widely among modem manufacturers. For this reason, only a few of these commands are listed here.

Register Commands

ATSr=n, where r is the number of the register to be changed and n is the new value that is being assigned.

It should be remembered that most users hardly ever see AT commands. The user normally deals with communications software, which implicitly issues AT commands to the modem. Still, it is nice to know what the software is doing—in particular, if data transmission does not proceed smoothly. Also, an initialization string that consists of AT commands should be selected during the software installation procedure. Because different initialization strings are sometimes provided by the software and hardware manufacturers, knowledge of AT commands can help in making the right selection.

AT Command Structure

The AT prefix is required for all commands except for the A/ command. In most modems the characters AT and the remainder of the command must be in upper or lower case, but not both. All modems will accept upper case only. Therefore the command ATI7 or ati7 will work fine, but ATi7 will not. A command consists of a letter identifying it, such as H for the command designating the on-hook/off-hook condition, followed by an optional parameter value. If the parameter value is not specified, such as ATH, then the parameter value is assumed to be equal to 0. Therefore the ATH command is equivalent to ATH0 (hang-up and disconnect).

Several commands can be linked together into a character string, generally not exceeding 40 characters, including any spaces in between (which are not necessary but help in reading the command). The maximum length of the string should be given in the User Manual and is limited by the amount of memory in the modem. Each command string (except for A/) must be terminated by a carriage return. The backspace key can be used to edit a command before the carriage return is pressed. The important AT commands shared by most modems are listed in alphabetical order in Table 9-3.

Table 9-3: Basic AT Commands

A/   Repeat last command. A/ is the only command that is not preceded by AT nor followed by a carriage return. Each AT command is loaded into a buffer before it is executed. It remains in the buffer until the next command is loaded. When the modem encounters the A/ command, it reads the content of the modem buffer and executes it immediately.

A   Put the modem in Auto-Answer mode. When this command is executed, the modem will immediately go into the Answer mode regardless of the condition of the S0 register. Setting S0 to 0 would normally prevent the modem from answering a call.

D   Dial the following number. This command puts the modem in the Local Command mode, ready to dial a number. If followed by T, it uses tone dialing; if followed by P, it uses pulse dialing. An example of this command is ATDT 1-(213)-555-1212.
Dashes and parentheses are ignored when they are a part of the dialed number. A comma (,) instructs the modem to wait a specified number of seconds for a second dial tone. The wait time is determined by the contents of the S8 register. A semicolon (;) at the end of the dialing sequence puts the modem in Offline mode after dialing. A call from a PBX, where 9 must be dialed first, followed by a four-second wait, with touch-tone dialing, could be accomplished with the command ATDT 9,,1-(213)-555-1212.
The AT command set provides no capability to wait for a second dial tone after dialing 9. The wait time is fixed by the two commas to four seconds, because the default for S8 register is set to two seconds. If R is the last character in the dialing string (such as ATDT555-1212R), then the call will be dialed in the Answer mode.

Table 9-3: Basic AT Commands (continued)

The difference between the Answer and Originate mode is the choice of carrier frequencies. On a connection, one modem must always be in the Originate mode and the other modem in the Answer mode. Normally the calling modem is in the Originate mode. As explained earlier, this feature is required when placing a call to an Originate-only modem.

E   Echo ON/OFF. This command echoes (or sends) characters back to the local terminal (or computer). The setting of this parameter depends on the setting of the called modem. Normally, on a full- or half-duplex connection, the remote modem will echo each character back to the calling modem. However, if the screen is blank while you are typing a command to the local modem or to the remote computer, this parameter must be turned ON by typing ATE1. If, on the other hand, all characters appear twice on the screen, the echo command should be turned off by typing ATE0.

H   Onhook/Offhook command. Sending ATH or ATH0 command to the modem causes the modem to terminate the telephone connection (go on-hook). If all the normal log-off procedures for your bulletin board or service provider do not work, and you are still on line, then this is the command of last resort. A simple way to test this command is to type ATH1. This is the same as lifting the phone receiver off the hook, and the dial tone should be heard through the modem speaker. Then terminate the call by typing ATH0, or simply ATH.

I   Indicate product or version code of the modem. Typing ATI3 to most modems returns the modem designation.

M   The M command controls the modem's built-in speaker used for listening to call-progress tones (dial tone, ring, busy, and so on) In the default setting (ATM1), the speaker is turned on until the carrier is received from the remote modem, then it is turned off. ATM2 leaves the speaker on for the entire length of the connection. ATM0 leaves it off during the entire call, including the dialing period. ATM3 keeps the speaker off only during the dialing period, and leaves it on during the remainder of the call.

R   Reverse originate and receive frequencies to let the modem dial to an originate-only modem. This command modifier is discussed under dialing commands.

Table 9-3: Basic AT Commands (continued)

O   Switch the modem to Online mode after executing commands in the Offline mode. This command is not followed by a parameter value—the only correct use is ATO and not, for example, ATO1.

Sr=y Set the r register to value y (decimal). For example, sending ATS8=5 to the modem makes each comma in the dialing sequence correspond to five seconds instead of the default value of two seconds.

Sr?   Display status of the r register. If, after sending the command ATS8=5, the command ATS8? is issued, the modem will echo back "005" to the terminal.

Z   Restore all default settings based on the position of the DIP switches, if any. Z0: DIP switch # 7 determines reset. Z1: Reset to &W0 settings. Z2: Reset to &W1 settings. Z3: Reset to &F0 settings. Z4: Reset to &F1 settings Z5: Reset to &F2 settings. The Z command can be used like the H command to disconnect the modem. The difference is that the Z command will reinitialize all registers based on the various DIP-switch settings.

$ Help menu. It displays a list of all basic commands.

Table 9-4 lists some extended and proprietary AT commands. It is just a sampling of these commands, since they vary from one modem to another. You will see a large list of modem commands provided for a specific modem chipset in Chapter 12. For a list of commands applicable to your specific modem, check your modem's User Manual or contact the modem manufacturer.

Table 9-4: Extended AT Commands

&An	n=0	Disable /ARQ (Automatic Resend ReQuest) result codes
	n=1	Enable /ARQ result codes
	n=2	Enable /Modulation codes
&Bn	n=0	Floating DTE speed
	n=1	Fixed DTE speed
	n=2	DTE speed fixed when ARQ

Table 9-4: Extended AT Commands (continued)

&Cn	n=0	CD always on
	n=1	Modem controls CD
&Dn	n=0	Ignore DTR
	n=1	Online command mode
	n=2	DTE controls DTR
&Fn	n=0	Load factory 0, no flow control
	n=1	Load factory 1, HW flow control
	n=2	Load factory 2, SW flow control
&Gn	n=0	No guard tone
	n=1	550 Hz guard tone
	n=2	1800 Hz guard tone
&Hn	n=0	Disable TX flow control
	n=1	CTS
	n=2	XON/XOFF
	n=3	CTS and XON/XOFF
&In	n=0	Disable RX flow control
	n=1	XON/XOFF
	n=2	XON/XOFF characters filtered
	n=3	HP Enq/Ack Host mode
	n=4	HP Enq/Ack Terminal mode
	n=5	XON/XOFF for non-ARQ mode
&Kn	n=0	Disable data compression
	n=1	Auto data compression
	n=2	Enable data compression
	n=3	Selective data compression
&Mn	n=0	Normal mode

Table 9-4: Extended AT Commands (continued)

	n=4	ARQ/Normal mode
	n=5	ARQ mode
&Nn	n=0	Choose highest link speed
	n=1	300 bps
	n=2	1200 bps
	n=3	2400 bps
	n=4	4800 bps
	n=5	7200 bps
	n=6	9600 bps
	n=7	12,000 bps
	n=8	14,400 bps
	n=9	16,800 bps
	n=10	19,200 bps
	n=11	21,600 bps
	n=12	24,000 bps
	n=13	26,400 bps
	n=14	28,800 bps
	n=15	31,200 bps
	n=16	33,600 bps
&Pn	n=0	North American pulse dial
	n=1	United Kingdom pulse dial
&Rn	n=1	Ignore RTS
	n=2	RX to DTE/RTS high
&Sn	n=0	DSR always on
	n=1	Modem controls DSR
&Tn	n=0	End Test
	n=1	Analog loopback (ALB)

Table 9-4: Extended AT Commands (continued)

	n=3	Digital loopback (DLB)
	n=4	Grant remote DLB
	n=5	Deny remote DLB
	n=6	Remote digital loopback
	n=7	Remote DLB with self-test
	n=8	ALB with self-test
&Un	n=0	Variable link-rate floor
	n=1	Minimum link-rate 300 bps
	n=2	Minimum link-rate 1200 bps
	n=3	Minimum link-rate 2400 bps
	n=4	Minimum link-rate 4800 bps
	n=5	Minimum link-rate 7200 bps
	n=6	Minimum link-rate 9600 bps
	n=7	Minimum link-rate 12,000 bps
	n=8	Minimum link-rate 14,400 bps
	n=9	Minimum link-rate 16,800 bps
	n=10	Minimum link-rate 19,200 bps
	n=11	Minimum link-rate 21,600 bps
	n=12	Minimum link-rate 24,000 bps
	n=13	Minimum link-rate 26,400 bps
	n=14	Minimum link-rate 28,800 bps
	n=15	Minimum link-rate 31,200 bps
	n=16	Minimum link-rate 33,600 bps
&Wn	n=0	Store Configuration 0
	n=1	Store Configuration 1
&Yn	n=0	Destructive
	n=1	Destructive/Expedited

Table 9-4: Extended AT Commands (continued)

Command	Description	Comments
	n=2	Nondestructive/Expedited
	n=3	Nondestructive/Unexpedited
&Zn=s		Store phone number
&Zn?		Query phone number
&$		Help menu. Displays a list of all extended commands
	**Selected Proprietary Commands:**	
Command	**Description**	**Comments**
\A0 or \A	Maximum MNP block size	(64 character maximum)
\A1		128 character maximum
\A2		192 character maximum
\A3		256 character maximum
%C0 or %C	Data com-pression Enable/Disable	Disabled
%C1		MNP5 enabled
%C2		V.42bis (BTLZ) enabled
%C3		MNP5 & V.42bis (BTLZ) enabled
%D0 or %D	Data com-pression	512 BLTZ dictionary size
%D1		1024 BLTZ dictionary size
%D2		2048 BLTZ dictionary size
%D3		4096 BLTZ dictionary size
%E0 or %E	Escape method	ESCAPE DISABLED
%E1		+++AT method (default)
%E2		<BREAK>AT method
%E3		BOTH methods enabled
%E4		Disable "OK" to +++
%E5		Enable "OK" to +++
\J0 or \J	DTE Auto Rate Adjustment	Disabled
J1		DTE rate is adjusted to match carrier rate.

Table 9-4: Extended AT Commands (continued)

\N0 or \N		Normal connection (see below for definitions)
\N1		Direction connection
\N2		MNP Auto-reliable connection
\N3	Connec-	Auto-reliable connection
\N4	tion type	V.42bis reliable link with phase detection
\N5		V.42bis auto-reliable link with phase detection
\N6		V.42 reliable link with phase detection
\N7		V.42 auto-reliable link with phase detection

After executing a command, the modem will return to the computer the result code. Depending on jumper or switch settings, the code can be either numeric, e.g., "1" or it can be a word, e.g., "CONNECT." Table 9–5 lists the result codes, their numerical equivalents, and their meanings:

Table 9-5: AT Result Codes

Code	Equivalent	Explanation
0	OK	Command was executed without an error.
1	CONNECT	The carrier was detected, the handshake sequence was successfully completed, and the modem is ready to communicate.
2	RING	There is an incoming call. Unless the modem is set to answer mode, the call will go to the telephone set.
3	NO CARRIER	Either the modem timed out after dialing, the connection was terminated, or thecarrier was lost during a call.
4	ERROR	The command just sent to the modem contains an error. This could be a syntax error or the command exceeds 40 characters, the size of the buffer.
5	CONNECT	This code means that a connection was successfully established at 1,200 bps. Result codes higher than 9 will usually refer to other connection rates. Consult your modem's User Manual.
6	BUSY	Busy signal was detected from the remote modem.

Table 9-5: AT Result Codes (continued)

Code	Equivalent	Explanation
7	ABORT	Dialing or handshake was interrupted due to user's interaction.
8	DIS-CONNECT	For some reason the connection has been lost.

The AT commands can be sent to the modem through the data communications software when the modem is in the Local Command Mode or in the Offline (Online Command) Mode. For a quick test, the AT commands can also be sent directly from DOS, bypassing any data communications software. Sometimes this is a good test of the modem if there is an unidentifiable problem that may be in the communications software or in the Windows operating system. In the following example, we use a short DOS file called MYPHONE.BAT. Such a file could easily be written by means of any text editor, such as EDIT. Just type EDIT at the DOS prompt, type the following text, then save it by pulling down the File menu and selecting the Save As option, giving it the name MYPHONE.BAT. Then select the Exit option from the same menu.

```
File MYPHONE.BAT:

REM MYPHONE.BAT dials via modem connected to COM2 port
MODE COM2:2400,n,8,1,p
ECHO ATZ > COM2:
ECHO ATDT555-1212 > COM2:
PAUSE
ECHO ATH > COM2:
```

Typing MYPHONE at the DOS prompt will first close the relay in the modem and leave the phone line off-hook. Then it will dial the number 555-1212 and display a DOS message "press any key...." Pressing any key will send the ATH message to the modem, which will disconnect the phone and put it in the on-hook mode. When executing the MYPHONE.BAT batch file make sure that you are in the "pure" DOS mode and not in a DOS window, as explained in Chapter 8.

The modem can also be accessed directly via internal ports. These ports can be addressed with assembly programming or with certain commands in higher level languages. For example, the statement INP(123) in BASIC will return the status of port #123. The statement OUT 123,222 will send the decimal byte value of 222 to the port

#123. The port addresses assigned to a built-in modem or to a serial port connected to a modem vary from computer to computer and can usually be found in the technical manual or in the Windows Control Panel. For example, the IBM-PC assigns port addresses 3F8-3FF (hex) to COM1 serial port, while another computer may assign addresses 2F8-2FF to the same port. Setting of communication parameters could also be performed by reading and writing to those ports. However, direct control of the modem via ports requires considerable programming experience and should not be attempted by a novice.

S Registers

Modems are frequently described as "smart" or "dumb." The dumb modem has now practically disappeared. The smartness of the modem is associated with dialing features and its capability to respond to the AT commands. All "smart" features set by the data communications software, as a sequence of the AT commands described above, are stored in the modem's memory in so-called S registers. These registers are in addition to any internal RAM modem memory, which may be required for storing passwords and acceptable telephone numbers in security modems or buffering data in error correcting and other specialized applications.

The number of S registers varies from modem to modem. For example, the U.S. Robotics Sportster uses registers S0 through S38. Some of these 8-bit registers store a single number between 0 and 255. This number refers to a specific function; e.g., if S0 = 5 then the modem will wait for five rings before answering a call in the Auto Answer mode. Other so called bit-mapped registers store up to eight binary values. To interpret the meaning of data stored in those registers you have to translate the decimal number stored in them into a binary value and then check the Users Manual to find the meaning of bit 1, bit 2, and so on.

Modem manufacturers, in general, agree on modem functions controlled by the 13 lowest registers, S0 through S12, and in interpretation of the stored values. Therefore, most data communications programs use just those registers. The higher numbered registers vary in function assignments from modem to modem. For detailed assignments of these registers, consult the User Manual for your modem. These higher registers are controlled by customized software and by special initialization strings.

Table 9-6 shows the functions controlled by the 13 registers S0 through S12. The value stored in each register can be read with the ATSn? command, where *n* is the register number. Most of the registers can be changed or written to with the ATSn=y command, where *n* is the register number and *y* is the new assigned value.

Table 9-6: Assignments of S-Registers S0 through S12

Register	Default	Range	Description
S0	1	0–255	Sets the number of rings before the modem answers. If S0=0, then the modem will not answer automatically. The AA indicator on the front panel of an external modem will normally light up to show that the modem has been configured to the Auto Answer mode. On some modems, this register also stores a number related to the quality of the telephone line. Changing to the Local Command Mode and typing ATS0? will echo a number between 0 and 50. Any number below 40 indicates a poor telephone connection.
S1	0	0–255	Counts and stores the number of rings from an incoming call.
S2	43	0–255	Sets the ASCII value for the escape code. ASCII of 43 equals "+". If the register is set to a value larger than 127, then the escape sequence will be disabled.
S3	13	0–127	ASCII value used for the carriage return code. This character also serves as the command terminator. If the communications equipment is nonstandard, then a value different from 13 may have to be assigned.
S4	10	0–127	ASCII value for the line-feed code. The line-feed character is generated by the modem after result codes are returned to the computer.
S5	8	0–32, 127	ASCII value for the backspace code. It should not be set to a printable ASCII character (33-126) or to a value larger than 127, otherwise the modem will not recognize it.
S6	2	2–255	Dial-tone wait time in seconds (time between off-hook condition and start of dialing). Used in the "blind" dialing mode (the modem does not "listen" for the dial tone).
S7	60	1–60	Wait time in seconds before the end of dialing and carrier detection. If carrier is not detected, modem will disconnect.
S8	2	0–255	Pause time in seconds represented by the comma in the dialing sequence; e.g., ATDT9,5551212. The pause is used when dialing through a PBX or when using an alternate long distance service.

Table 9-6: Assignments of S-Registers S0 through S12 (continued)

Register	Default	Range	Description
S9	6	1–255	Time in tenths of a second between carrier recognition and its acknowledgment.
S10	7	1–255	Time in tenths of a second between carrier loss and disconnect. If set to 255, the modem will not disconnect due to carrier loss. The delay keeps the connection on, even if the carrier temporarily fails due to transmission problems or other reasons. On a noisy line the value should be increased from the default value.
S11	70	50–255	Controls the speed of the tone dialer (interval in milliseconds between tones). The default is 7.1 digits per second (tone of 70 ms + 70 ms interval between tones). Not used in some modems.
S12	50	20–255	Guard time before and after escape sequence (+++) required by the modem to recognize it. Increment is 20 ms, the default is 1 sec (50x20 ms). Care should be taken in choosing guard times shorter than the transmission time for a character. The time between the escape characters must be less than the guard time or the escape sequence will not be recognized by the modem.

Interpretation of registers higher than S12 varies from modem to modem. Some of these registers are bit mapped, some contain single values. Table 9-7 shows functions and assignments of these higher registers for the U.S. Robotics Sportster 33.6 modem.

Table 9-7: Registers S13 - S38 for U.S. Robotics Sportster 33.6 Modem

S13 Bitmapped
1 = Reset On DTR Loss
2 = Reduced Non-ARQ TX Buffer
4 = Set DEL=Backspace
8 = On DTR signal, autodial the number stored in NVRAM at position 0
16 = At power ON/RESET autodial the number stored in NVRAM at position 0
Strike a key when ready . . .
32 = Reserved
64 = Disable Quick Retrains

Table 9-7: Registers S13 - S38 for U.S. Robotics Sportster 33.6 Modem

128 = Escape Code Hang Up
S14 Reserved
S15 Bitmapped
1 = MNP/V.42 Disabled in V.22
2 = MNP/V.42 Disabled in V.22bis
4 = MNP/V.42 Disabled in V.32
8 = Disable MNP Handshake
16 = Disable MNP Level 4
32 = Disable MNP Level 3
64 = Unusual MNP-Incompatibility
128 = Disable V.42
S16 Test Modes
1 = Reserved
2 = Dial Test
4 = Reserved
8 = Reserved
16 = Reserved
32 = Reserved
64 = Reserved
128 = Reserved
S17 Reserved
S18 &Tn Test Timeout (sec)
S19 Inactivity Timeout (min)
S20 Reserved
S21 Break Length (1/100sec)
S22 Xon Char
S23 Xoff Char
S24 Reserved

Table 9-7: Registers S13 - S38 for U.S. Robotics Sportster 33.6 Modem

S25 DTR Recognition Time (1/100sec)
S26 Reserved
S27 Bitmapped
1 = V21 Mode
2 = Disable TCM
4 = Disable V32
8 = Disable 2100hz
16 = Enable V23 Fallback
32 = Disable V32bis
64 = Reserved
128 = Software Compatibility Mode
S28 V32 Handshake Time (1/10sec)
S29 V.21 Answer mode fallback timer
S30 Voice View deadman timer function
S31 TAD Audio level adjustment
S32 Connection bit mapped operations.
1 = V.8 Call Indicate enable
2 = Enable V.8 mode
4 = Disable V.FC modulation
8 = Disable V.34 modulation
16 = Disable V.34+ modulation
32 = Reserved
64 = Reserved
128 = Reserved
S33 V.34 & V.34+ Connection setup bit mapped control flags.
1 = Disable 2400 Symbol rate
2 = Disable 2743 Symbol rate
4 = Disable 2800 Symbol rate

Table 9-7: Registers S13 - S38 for U.S. Robotics Sportster 33.6 Modem

8 = Disable 3000 Symbol rate
16 = Disable 3200 Symbol rate
32 = Disable 3429 Symbol rate
64 = Reserved
128 = Disable Shaping
S34 V.34 & V.34+ Connection setup bit mapped control flags.
1 = Disable 8S-2D trellis encoding
2 = Disable 16S-4D trellis encoding
4 = Disable 32S-2D trellis encoding
8 = Disable 64S-4D trellis encoding
16 = Disable Non linear coding
32 = Disable TX level deviation
64 = Disable Pre-emphasis
128 = Disable Pre-coding
S35 Reserved
S36 Reserved
S37 Reserved
S38 Disconnect Wait Time (sec)

As mentioned earlier, S13–S38 register assignments will vary from modem to modem. In fact, considering the lack of standardization of many modem features and the dismal style and contents of most user and service manuals, it is often astounding that the majority of modem purchasers succeed in making them work.

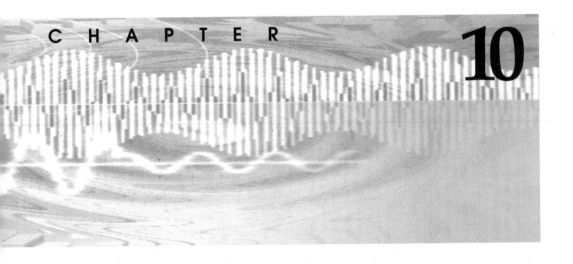

Data Communications Software

An average computer user does not care about the Hayes AT command set, the status of modem registers, or the details of file transfer protocols. All the user wants to do is to click on an icon and have the computer do the rest. The user may want to download some files from a bulletin board, browse the Internet, order a book from the library, get a stock quote, send a memo to a coworker, or transfer $500 from a checking to a money-market account. The job of the communications software is to perform all these functions unobtrusively and efficiently.

There are three ways in which communications software can be obtained—public domain software (freeware), shareware, and commercial software. Much of it can be obtained from various computer clubs, or downloaded from bulletin boards and from assorted Internet sites. If you have trouble in distinguishing between uploading and downloading, think of the remote computer as floating high in the heavens. You upload your files *to* it, and you download *from* it to your computer.

Public-domain software is free, but it is often not supported by its author or authors after it has been released. It often comes prepackaged with many modems. Public-domain software comes with some online documentation, which may or may not apply to the current release or to the specific computer/modem configuration of the user.

The next category of communications software, shareware, can be obtained from the same sources as public domain software, and from many shareware vendors who sell it at a nominal price of $10 or less. Shareware gives you an opportunity to test the software. If you decide to keep it, you should send the author the requested fee. Though shareware programs come with some form of online documentation, the author will usually promise to provide the user with additional online or hardcopy documentation after receiving payment.

Public domain and shareware programs will satisfy many users, but they seldom provide features and support required by the business community. Commercial communications software typically costs between $50 and $200, which may be high for a hobbyist, but is of little concern to the business community. Trial versions of commercial software with limited capabilities or having an expiration date, can often be downloaded from Internet sites. Commercial communications software provides a wide range of options and user support, with frequent program updates to keep up with new developments in modem hardware.

Many commercial software vendors provide limited telephone support. On more expensive packages, support is toll-free. After the support period is over, typically after three to six months, the software maker will often provide telephone support at a fixed per-minute or per-call charge. Many commercial software packages are sold as suites at prices just slightly higher than a single program. A communication suite would typically include a terminal emulation program to connect to bulletin boards, a fax program to send and receive faxes, and a set of Internet tools. Similarly, an office suite would contain a word processor, a graphics presentation program, and a spreadsheet program.

There are several types of communications software that will be discussed in this chapter. One is terminal emulation software such as the free Terminal program that comes with Windows 3.1, or the free Hyper Terminal program that comes with Windows 95. Another is the commercial package from Hayes, called Smartcom for Windows, to name a few.

Then there is the fax software, such as WinFax from Delrina Corporation with its built-in optical character recognition (OCR) software, as well as separate OCR software such as Omni Page from Caere Corporation. There is also Internet browser software such as Netscape's Navigator, or the proprietary software provided by America OnLine (AOL). This type of software uses a graphical user interface to the Internet. There are other types of software that use the Internet as the transmission medium. One type discussed here is the Internet Phone from Vocaltec for voice communication. Internet service providers (ISPs) use a different type of Unix-based character-oriented software on their servers, such as Lynx, which is very convenient to browse the Internet.

In addition to all those types of software, for efficient operation, you should also install file-transfer software, zip/unzip, and encoding/decoding software on your computer. You will find that the dividing line between the various types of software is sometimes blurred. For example, the proprietary AOL software might contain browser software, and terminal-emulation software always includes file-transfer software. Depending on your involvement in the communications field and your service provider or BBS, you will probably need one or more types of such software.

Terminal Emulation Software

Terminal emulation software is essential for connecting to bulletin board services (BBSs), and to Internet service providers (ISP) in the so-called shell or character mode. Historically, this was the first type of communications software in wide use. The early versions of the software, such as those written for my first Radio Shack TRS-80 Model I computer in early 1980s, did not even have commands to let the modem do its own dialing. The reason for this was that early modems did not have a dialing capability. The terminal software in those days took control of the telephone connection only after the connection was manually established.

Modern terminal software deals with a smart modem that understands a set of commands. The software first talks to the local modem in Local Command mode by sending to it the control codes (AT commands) to establish the connection to the remote computer. When the connection is established, the terminal-emulation software changes from the Local Command mode to the Online mode, and exchanges data between the local and the remote computer.

The early versions of terminal-emulation software were character and menu oriented, and operated in the DOS environment. Later versions of software, such as Smartcom for Windows from Hayes Corporation, which will be discussed later in this section, moved to the Windows environment and used a graphical user interface (GUI) instead of the text-based interface common to DOS programs. As the use of bulletin boards is slowly declining in favor of the graphic-oriented World Wide Web on the Internet, the terminal-emulation software is moving into the historic archives of computing and is being replaced by browser and proprietary software.

Basic Functions

The earliest form of data communication was a simple computer terminal with very limited storage and processing capabilities, connected to a local computer or a modem. The first and basic function of the communications software in the 1970s was therefore to make the general purpose personal computer, which was then becoming popular, emulate a computer terminal. Such terminals; e.g., the popular VT-100 from Digital Equipment Corporation, were frequently referred to as *dumb* terminals or *glass* teletypewriters.

One of the basic functions of a terminal is that if the user types "Hello," the word "Hello" should appear on the screen and be somehow transmitted to the remote computer. When someone at the remote computer types "I got your message," the same words should appear on your screen. If more than one line of text is being sent, then the previous line of text should scroll up the screen to make room for the next line. These are the absolute minimum requirements expected from communications software. In fact, this type of basic software is usually supplied free with any operating system.

As the next step, communications software should allow you to transfer ASCII and binary files between a local and remote computer. The transfer of binary files requires that the data transfer and error detection/correction protocols are part of the software. A single error in transmission of a binary file will usually render that file useless after it is received. It is also not enough to receive a file from a remote computer; you should also be able to save that file on disk and/or print it locally.

As modem technology progressed and modems gained new functions, terminal-emulation software was enhanced to handle these functions. Communications software should now take care of all dialing chores, retrieve the phone number from an address file, and set up proper transmission parameters associated with the call (such as transmission speed in bits per second, number of data and stop bits and parity). The communications program should then dial the remote computer, keep track of the call-progress tones by indicating Busy or No Answer, and should redial at prescribed intervals if no connection was established.

As soon as the carrier signal is received from the remote modem, indicating that a connection is established, the communications software should switch the local modem from the local command mode to the online mode. The modem will then become transparent to the computer—the output of the local computer will no longer be interpreted as modem commands, and will be sent to the remote computer.

The conversation will switch from the local computer talking to the local modem to the local computer talking to the remote computer. Other important parameters that should be set by the communications software are local echo, remote echo (also referred to as feedback), and the proper carriage return and line-feed combination. The local echo makes every character transmitted from the local terminal appear on the screen of the local terminal, and the feedback parameter sends each character received by the remote terminal back to the local terminal to appear on the display.

These options must be coordinated. Otherwise you will not see what was sent to the remote terminal. If both local and remote echo are active at the same time, then each character will appear twice on the screen. Instead of "Hello," you would see "HHeelllloo" appear on your screen.

Terminal Emulation

What one next expects from a terminal-emulation program are more esoteric but important features. For example, you should be able to emulate various popular terminals in addition to the generic glass teletypewriter. Different terminals have different keyboard layouts that the computer should be able to emulate. The need for emulation is caused by characteristics of remote computers that may be specifically designed to communicate with a certain class of terminals. These will typically be the plain teletypewriter with text just scrolling off, the DEC VT-100 series with screen cursor control, and emulations suited to bulletin boards such as the BBS-ANSI and RIP Graphics.

The last two emulation modes provide color, graphics and a Windows-type user interface. Of course, not all features of a specific terminal can be emulated. For example, the large-size character option on the VT-100, which is embedded in the VT-100 terminal firmware, could only be emulated by a Windows-type program.

The computer keyboard in the emulation mode should have certain keys reassigned to emulate the special keys present in the emulated terminals, such as the PF keys in the VT terminals. Even more important, the computer, in its emulation mode, should duplicate certain smart features of the emulated terminals. The main difference between a dumb terminal and a smart terminal obtained through emulation, is that a dumb terminal will let the user type at the keyboard and read from the screen, moving the cursor only left and right on a single line, while a smart terminal allows the cursor to move up and down on the screen and perform full-screen editing. In the emulation mode, the terminal also responds to special codes that are specified for a particular terminal, which might clear the screen, move the cursor, or define function keys. A smart terminal also has the ability to make input forms on the screen with highlighted fields for entering data, such as underline, reverse video fields, or a blinking field.

User Interface—Menus, Commands and GUI

The user friendliness in a terminal program can assume many facets. The program designer must weigh the simplicity of a user interface versus the program flexibility and consider the type of user community to which the program should appeal. In a commercial program, the designer also wants to maximize the number of potential purchasers, who might range from neophytes to sophisticated users.

The easiest way to interact with the user is by presenting a few simple choices, such as icons in a Windows-type environment, or an introductory menu in a DOS environment. Because the number of choices displayed on the screen is limited, the program might display additional choices in the form of icons, dialog boxes or sub-menus after the initial menu selection. A menu-driven program assumes only a minimal knowledge on the part of the user, so choices are always displayed on the screen. But it also has its limitations. A large number of choices leads to hierarchical menus that can make a menu-driven program very cumbersome to design and use.

An example of a menu-driven program is the once-popular commercial terminal emulation program called Crosstalk XVI for DOS, as shown in Figure 10-1. The opening menu displays a list of various BBS and service providers. All transmission parameters for these services are stored in program memory, and the user only needs to provide the access telephone number. Then the program displays a menu of assorted functions associated with the selected BBS. The program automatically dials the selected service and establishes a data connection. The user name, password, and so on, can either be entered manually or can be entered with a script file.

Figure 10-1 Crosstalk XVI for DOS, a terminal emulation program

```
                   CROSSTALK XVI Status Screen Online
Name  INJERSEY                        LOaded   C:INJERSEY.XTK
NUmber 671-6555                         CApture  Off
Communications parameters  Filter settings
SPeed 28800  PArity None  DUplex Full  DEbug Off  LFauto Off
DAta 8     STop 1     EMulate None   TAbex Off    BLankex Off
POrt 2     MOde Call   INfilter On    OUtfiltr On
Key settings           SEnd control settings
ATten  Esc             COmmand Esc          CWait    None
SWitch Home            BReak   End          LWait    None
Normal function key definitions
   F1 = atdt671-6555|F2 = 1|
   F3 = rlewart|F4 = Password|
   F5 = exit|F6 = @FK|
   F7 = @List|F8 = PS1='$PWD >'|
   F9 = @Bye|F10= @QUit|
Command? _
```

An alternate and more primitive method of user interface is the command-driven program, which is sometimes combined with a menu-driven program. In a command-driven program, the user—instead of choosing from a limited number of options displayed on the screen menu—issues short commands from a separate dictionary. The dictionary of any size can be a written document or an online help file. The user gains considerable flexibility at the expense of having to remember various commands. For example, Columbia University's Kermit is a powerful command-driven communications program.

Commands in a Command-Driven Communications Program

As an example, when using the MS-Kermit program, which is the Kermit version for MS-DOS, the commands available to the user are listed in alphabetical order. This list, shown in Table 10-1, should give you an idea of what you can expect from a modern communications program. Other communications packages will have a similar set of commands, although their names will most likely differ.

An MS-Kermit command is a line of words separated by spaces and ending with a carriage return (the Enter key). Example:

```
SET SPEED 2400<Enter>
```

Most words can be abbreviated and can be completed by pressing the Esc key. Example:

```
SET SPE<Esc> 24<Enter>  or even  SET SPE<Esc> 24<Esc><Enter>
```

For help, press the "?" key where a word would otherwise be expected.

Edit lines using the backspace key to delete characters, Control-W to delete words, and Control-U to delete a line. Control-C cancels the command.

Table 10-1: Frequently-Used MS-Kermit Commands

EXIT	Leave the Kermit program. QUIT does the same thing.
SET	PORT, PARITY, SPEED, TERMINAL and many other parameters.
SHOW	Display groups of important parameters. SHOW ? for categories.
CONNECT	Establish a terminal connection to a remote system or a modem.
Control-) C (Control-) followed by "C")	Return to MS-Kermit> prompt.
SEND filename	Send the file(s) to Kermit on the other computer.
RECEIVE	Receive file(s), SEND them from Kermit on the other computer.
GET filename	Ask the remote Kermit server to send the file(s) to us.
FINISH	Shut down remote Kermit but stay logged into remote system.
BYE	FINISH and logout of remote system and exit local Kermit.

Common startup sequence:

```
SET SPEED 9600
CONNECT
log in
start remote Kermit
put it into Server mode
escape back with Control-C
transfer files with SEND x.txt
GET b.txt
BYE
```

Table 10-2: MS-DOS Kermit commands, a functional summary

Local file management:	Kermit program management:
DIR (list files)	EXIT (from Kermit, return to DOS)
CD (change directory)	QUIT (same as EXIT)
DELETE (delete files)	TAKE (execute Kermit commands from file)
RUN (a DOS command)	CLS (clear screen)
TYPE (display a file)	PUSH (enter DOS, EXIT returns to Kermit)
SPACE (show disk space)	Ctrl-C (interrupt a command)
Communication settings:	**Terminal emulation:**
SET PORT, SET SPEED	CONNECT (begin terminal emulation)
SET PARITY	HANGUP (close connection)
SET FLOW-CONTROL	Alt-X (return to MS-Kermit> prompt)
SET LOCAL-ECHO	SET KEY (key mapping)
SET ? to see others	SET TERMINAL TYPE, BYTESIZE, other parameters
SHOW COMMUNICATIONS, MODEM	SHOW TERMINAL, SHOW KEY
File transfer settings:	
SET FILE CHARACTER-SET name	SET TRANSFER CHARACTER-SET
SET FILE TYPE TEXT, BINARY	SET SEND or RECEIVE parameters
SET FILE ? to see others	SET WINDOWS (sliding windows)
SHOW FILE	SHOW PROTOCOL, SHOW STATISTICS
Kermit file transfer:	**ASCII file transfer:**
SEND files (to RECEIVE)	LOG SESSION, CLOSE SESSION (download)
RECEIVE (from SEND)	TRANSMIT (upload)
MAIL files (to RECEIVE)	SET TRANSMIT parameters
Using a Kermit server:	**Being a Kermit server:**
GET files (from server)	SET SERVER TIMEOUT or LOGIN
SEND or MAIL (to server)	ENABLE or DISABLE features
REMOTE command (to server)	SERVER
FINISH, LOGOUT, BYE	SHOW SERVER

Table 10-2: MS-DOS Kermit commands, a functional summary (continued)

Script programming commands:	
INPUT, REINPUT secs text	:label, GOTO label
OUTPUT text	IF (NOT) condition command
DECREMENT or INCREMENT variable number	OPEN READ (or WRITE or APPEND) file
ASK or ASKQ variable prompt	READ variable-name
DEFINE variable or macro	WRITE file-designator text
ASSIGN variable or macro	CLOSE READ or WRITE file or logfile
(DO) macro arguments	END or POP from macro or file
ECHO text	STOP all macros and command files
PAUSE time	WRITE file-designator text
SLEEP time no comms sampling	SHOW VARIABLES, SHOW SCRIPTS, SHOW MACROS
WAIT time modem signals	

Command and Script Files

All but the simplest data communications programs will also support the so-called command and script files. A user can store settings of data communications parameters and responses to the prompts during the start of a session with the remote information service in these files. A command file contains all parameters required to establish a data connection, such as the telephone number, number of data and stop bits, parity information, full- or half-duplex setting, and so on.

A script file usually contains the user's name, and sometimes the password (often in encoded form) for a particular service, and standard responses to various prompts asked by the information service. A script file is written in a special language, called a script language, that varies from one data communications program to another. Script programs may contain many conditional statements that instruct the computer to provide certain answers such as the password, after a specific prompt such as "Your Password ? _".

The files are saved for each service and are then recalled as needed. At the push of a function key or a mouse click, the program will then dial the remote computer and, when the connection is established, will supply the user's identification or name and the password and will start an application session. Therefore the primary user who is familiar with the data communications software does not have to remember individual requirements of a specific service, as they are all stored in the script file. A casual user does not need to know any data communications parameters and only needs to know which button to press to start a data connection.

More advanced communications programs such as Smartcom for Windows, described next, will also learn by observing a user's keystrokes, translating them into a script and letting the user enhance the script with a text editor.

Terminal Emulation Programs for Windows

Following is a description of the various aspects of three modern terminal-emulation programs for Windows. The first one is Terminal, which is supplied with Windows 3.1; the second is Hyper Terminal, which is supplied as part of Windows 95; the third is a commercial program from Hayes Corporation called Smartcom for Windows.

They are each representative of the many communications programs available in the late 1990s. The Windows programs are not command driven like DOS-based programs are—they are menu-driven, with all user choices appearing in menus and activated by clicking on assorted menu choices and graphical icons.

Terminal (Windows 3.1)

This free program, which comes with Windows 3.1 and earlier versions of Windows, provides the basic facilities to set up a character-oriented session with your BBS or other character-oriented service provider. The program can be found in the Windows Accessories directory. After you click on the directory, select the Settings menu, which gives you a number of options. These are: Phone Number, Terminal Emulation (ANSI or VT-100), Function Keys (Macros), Text Transfers, Binary Transfers, Communications, and Modem Commands (initialization strings and a few basic AT commands). When you are satisfied with the settings, choose the Phone menu and select Dial.

Hyper Terminal (Windows 95)

Chapter 9 discussed how to set up the Hyper Terminal program to send commands to the modem and to read its responses. Now you will learn how to use Hyper Terminal to access a bulletin board operated by a computer club located in Eatontown, New Jersey.

First, set up Hyper Terminal to dial the BBS by going through the following steps:

1. Click on the Hyper Terminal icon in the Accessories folder. You can also locate the Hyper Terminal program by clicking

on the Start button in the Win95 opening screen and selecting Programs/Accessories/Hyper Terminal from the Start menu.

2. You should now see a panel as shown in Figure 10-2, with several icons named AT&T, Compuserve, and so on. Click on the icon named Hypertrm.exe (not hypertrm.dll). Depending on the size of your display fonts, you might not see the entire name of icons, so Hypertrm.exe may appear as Hypertrm... To make sure that you select the correct icon, click on it once with the left mouse button, then click once with the right mouse button and select Properties from the pop-up menu. The full name of the icon appears in the Hyper Terminal Properties panel.

Figure 10-2 Hyper Terminal opening screen

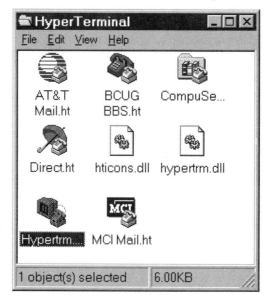

3. In the next panel shown in Figure 10-3, which is named Connection Description, type the name of your bulletin Board in the dialog box. For example, BCUG BBS.BCUG stands for the Brookdale Computer Users Group bulletin board. Also select an icon from the Icon dialog box. Click OK.

4. In the next panel, named Phone Number, type the phone number of the BBS, including any access codes and area codes. In the Connect Using dialog box, choose the default, which is the name of your modem. Click OK.

Figure 10-3 Hyper Terminal connection description panel

5. In the next panel, named COM2 Properties, choose the high-est speed for your non-compressed data transfer between the computer and your modem, which can be up to four times the transmission speed after data compression. For example, choose 115,200 bps for a 28,800 bps modem. Then choose the following parameters: Data bits = 8, Parity = None, Stop bits = 1, Flow control = Hardware. Do not change the advanced settings. Then click on DIAL and the modem will connect you to the BCUG BBS.

An example of a short session with a BBS using Hyper Terminal is shown in Chapter 11, *Connecting to the World*.

Smartcom for Windows

Smartcom for Windows is a commercial terminal-emulation program developed by Hayes Corporation. When compared to the free Terminal or Hyper Terminal programs, it provides more features and, what is most important, a powerful script or macro language to automate most communications functions. The easiest way to write a script is to let the program learn while you perform specific tasks such as signing on, typing your name and password, checking e-mail, checking usage statistics, and so on. A learned script can be taken without modification, or it can be modified with a text editor and script compiler that is included in the program. If you make a syntax error while modifying the script, the text editor will make the computer beep and will point to the error.

Each function or macro can be assigned to a button displayed at the bottom or side of the screen. Each set of buttons can be associated with a specific bulletin board or other service provider. When you connect to a different service, a different set of buttons will automatically appear. Some of the program features are shown in the next three figures. Figure 10-4 shows the opening screen with a display of the preset dial-up and sign-on scripts for each BBS and ISP. The program comes with a selection of standard sign-on scripts for major services. At this point you can add a new BBS to your list by clicking on Document and then on New.

Figure 10-4 Smartcom for Windows opening menu

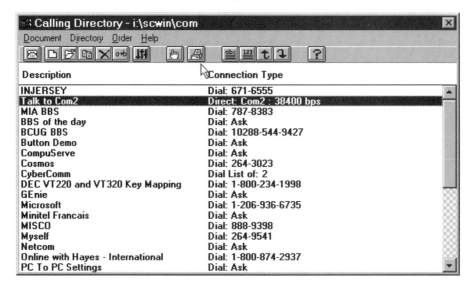

Figure 10-5 shows the screen appearance during a session with a BBS. Notice the row of buttons at the bottom of the screen, each corresponding to a specific function or script. This could be a simple sign-off such as "With best regards," or a complicated script that dials a BBS and responds to assorted prompts for User Name, Password, and so on. Although only 11 buttons appear at one time in a row, you can scroll through many rows.

Figure 10-6 shows a script that the program learned during a sign-on session to one of the ISP servers I subscribe to. After the program learned from my responses, I expanded the script by embedding several statements using the script language. For example, if the server displays a message saying I have new mail, my script will display a large box with words "New Mail Is Here." Because the server for my bulletin board service uses Unix, the final script statement lets the current directory path appear in each prompt. Notice that the Password is encrypted, so a person gaining unauthorized access to my computer cannot easily find out my password.

Figure 10-5 Smartcom for Windows script "buttons"

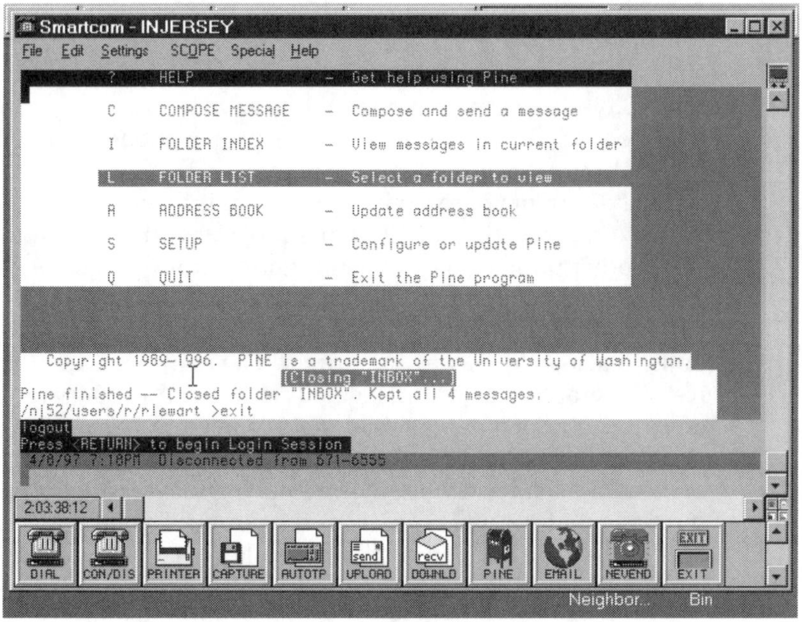

Figure 10-6 Smartcom for Windows sign-on script

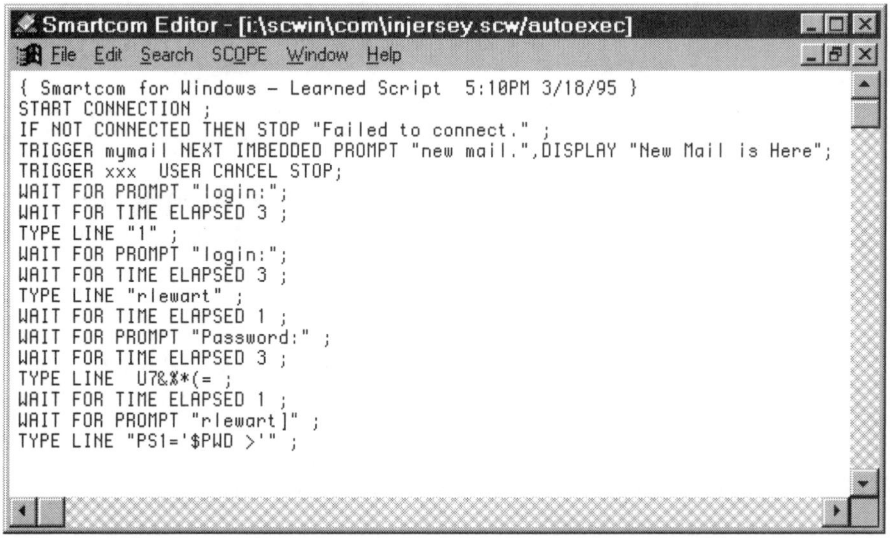

```
{ Smartcom for Windows - Learned Script  5:10PM 3/18/95 }
START CONNECTION ;
IF NOT CONNECTED THEN STOP "Failed to connect." ;
TRIGGER mymail NEXT IMBEDDED PROMPT "new mail.",DISPLAY "New Mail is Here";
TRIGGER xxx  USER CANCEL STOP;
WAIT FOR PROMPT "login:";
WAIT FOR TIME ELAPSED 3 ;
TYPE LINE "1" ;
WAIT FOR PROMPT "login:";
WAIT FOR TIME ELAPSED 3 ;
TYPE LINE "rlewart" ;
WAIT FOR TIME ELAPSED 1 ;
WAIT FOR PROMPT "Password:" ;
WAIT FOR TIME ELAPSED 3 ;
TYPE LINE  U7&%*(= ;
WAIT FOR TIME ELAPSED 1 ;
WAIT FOR PROMPT "rlewart]" ;
TYPE LINE "PS1='$PWD >'" ;
```

Facsimile and OCR Software

Modern communications software provided for PC-based modems makes sending or receiving a fax message a simple operation. Earlier versions of fax software required that a file that was originally created in a word processor or desktop publishing program be first converted into a graphic file (usually in the PCX format) by the original program or another program that could handle the conversion. This converted graphic file would be saved. A fax program would then be started and would send the proper modem commands to transmit the graphic file to its recipient.

The current approach in fax software is to integrate the software with word processing and graphics programs, and to treat the fax as just another printer. One of the most popular fax programs is WinFax from Delrina Corporation. The program makes sending a fax equivalent to printing a file from any Windows application. All one has to do is to select WinFax as the printer. The software takes care of the rest. If a graphics or text file already exists and is not part of a Windows application, one can read that file into any Windows program, such as Notepad or Paint, and print it from that application to the fax modem. To send a fax, the user simply prints the file from a word processor and specifies WinFax as the printer. The program then asks for the recipient and sends the file as a fax document. Received faxes are automatically displayed and stored.

The next two figures show the WinFax sending screen with assorted options, and the receiving screen. In the sending screen shown in Figure 10-7, the user selects the fax recipient from an existing list, or enters the name and phone number manually. The dialog boxes also allow for selection of a cover page and delivery schedule. The fax can be sent immediately or at a specified time.

The receiving screen shown in Figure 10-8 shows that a fax is being received. It also shows the transmission speed and the page currently received.

The program lets the user enter general settings that only need to be specified once, such as the option to display the call progress, prompting for keywords, and many more.

An important feature of WinFax is the built-in optical character recognition (OCR) software. It is possible to make the program recognize, or read, images of individual characters in the received fax message and translate them into text, which can then be handled by a word processor. The OCR program included with WinFax is quite powerful and makes very few mistakes. It is a subset of the Text Bridge program from Xerox Corporation and is limited to the fax formats supported by WinFax.

A more specialized commercial OCR program, Omni Page, was developed by Caere Corporation. It will also read a fax and provide many formatting options not included with WinFax. For example, the original formatting can be kept, including multi-column and single column text as well as tables. The Omni Page program also handles a wide array of graphical formats including various fax formats.

Figure 10-7 WinFax ready to send a fax

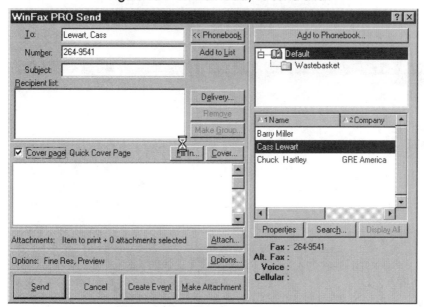

Figure 10-8 WinFax receiving a fax

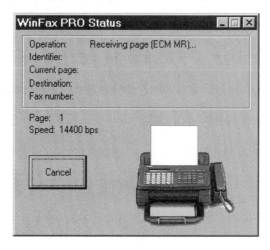

If documents generated by similar printers are scanned, the program will learn and improve the character recognition the second time around. Figure 10-9 shows the Omni Page program's opening screen with the picture of the page it is going to recognize. Figure 10-10 shows the program at work, recognizing a scanned or a faxed document. Notice the several panels in Figure 10-10. One of them shows a picture of the page that is automatically divided into regions, which are being recognized by the program, while the other panel shows how the program translates the contents of a page into ASCII text.

Figure 10-9 Omni Page opening screen

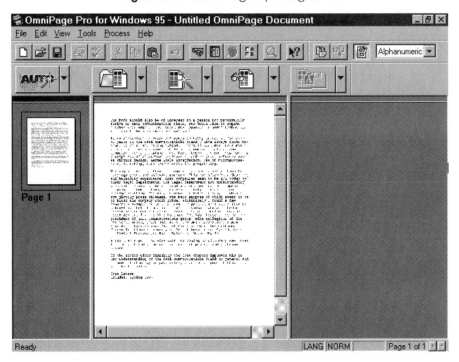

Figure 10-10 Omni Page recognizing a scanned document

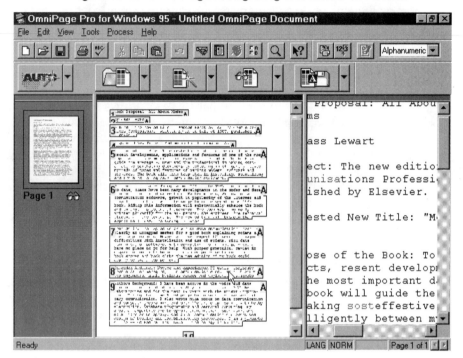

Internet Browsers

The most significant development in communications software in recent years is Web browser software. This type of software made the Internet accessible to a large segment of the people who were not previously familiar with computers.

Instead of cryptic commands and lines full of characters scrolling off the screen in a character-based environment (typical of terminal-emulation software), browsers introduced the user of communications software to the point-and-click environment of the graphical user interface as used by Windows and MacIntosh programs.

While terminal-emulation software sends and receives only the equivalent of ASCII characters, including some control characters such as carriage returns and line feeds, browser software also controls the cursor position and screen appearance such as color, background and fonts. Software control of the screen and cursor, and sharing of this information between the local and remote computer, made use of the GUI and mouse possible.

In browser software, similar to all Windows-based and MacIntosh-based software, the local or remote computer knows where the mouse-pointer is located on the screen, and when the mouse is clicked or double-clicked. This information is equivalent to issuing a specific command. For example, for the non-Windows based terminal-emulation software to enable or disable the saving of screen information, would require that the user type the command "CAPTURE OFF" or "CAPTURE ON," followed by a carriage return.

Browsers and other Windows-based software instead display the two commands on the screen. Pointing and clicking the mouse over CAPTURE OFF sends the computer a message that the mouse was clicked 2.67 cm from the top of the screen and 3.11 cm from the left side of the screen. This information is then interpreted as a request to turn the capture off. Similarly, clicking the mouse over CAPTURE ON sends the computer a message that the mouse was clicked 3.17 cm from the top of the screen and 3.55 cm from the left side of the screen. This information is interpreted as a request to turn the capture on.

Once the cursor and screen status are available to the local and remote computers, the screen can be *painted* by putting images at specific screen locations. Browser software can interpret the common picture encoding formats, such as JPEG and GIF, convert the data stream into actual pictures, and display them on the screen without further user intervention.

Similar to picture decoding, browser software recognizes and executes multimedia files such as sound and video clips in assorted formats.

Browser software opened data communication to anyone owning a computer, a modem, a telephone, and a connection to a bulletin board and/or Internet service provider. It started a computer revolution, where the world becomes smaller and smaller and communicating with a person half way around the globe is nearly as easy and inexpensive as communicating with a person next door.

Browser development started with the Mosaic program developed at the University of Illinois in the mid 1990s. The leading browser programs are currently Netscape Navigator from Netscape Corporation and Internet Explorer from Microsoft. New features are constantly being added by both companies. To develop a Web page requires writing a program in HyperText Markup Language (HTML). The syntax of the language is constantly expanding to accommodate new features. One can easily see the source code of a web page written in HTML by selecting Document Source from the View menu of Navigator, or saving an Internet page from the browser and viewing it using an ASCII text editor such as Notepad.

A separate program used with a browser is the so-called Windows socket, or Winsock for short. Until the introduction of Windows 95, the most popular program in this category was Winsock Trumpet developed by Trumpet Software International Pty Limited in Australia. Windows 95 includes a free Windows socket program that pushed out Winsock Trumpet from the leading market position. A Windows socket program makes your computer appear to the Internet like another Internet server. The program installs the so-called Transmission Control Protocol/Internet Protocol, or TCP/IP for short, which is the communications standard that allows all computers to send and receive information across the Internet network, regardless of the computer, modem, or program being used.

In addition to the TCP/IP protocol, another program is required for a seamless connection to the Internet from a home computer via a dial-up connection to the ISP.

That program is the SLIP (Serial Line Internet Protocol) or PPP (Point-to-Point Protocol). The choice between SLIP and PPP depends on what protocol is supported by the Internet Service Provider you deal with. The SLIP/PPP protocol provides the PC user with the ability to carry TCP/IP traffic over serial lines, such as dial-up telephone lines, provided that both computers—yours and the ISP's server computer—run the TCP/IP network software.

The best connection to the Internet is of course a direct connection to the network, which can cost several thousand dollars per month. A more affordable way, is a strictly character-oriented dial-up access to a shell account described later in this chapter. Many people find it less than satisfactory because it does not allow for GUI-based web browsers and for sending files via FTP directly to your own PC. By contrast, SLIP/PPP allows PC users to get quasi-direct Internet access from their own PC with just a simple modem, telephone line, and an account with an ISP. The user can then run GUI-based Web browsers, or become an FTP client and also run other programs as if he/she would be directly connected to the Internet.

SLIP/PPP is a quasi-direct Internet connection in the sense that the local computer has a direct communications link to the Internet, even if it is through an ISP. The requirements are that the local computer has the TCP/IP networking software installed to communicate with other computers on the Internet, and that the local computer has an identifying IP address so that it can be contacted by other computers on the Internet. The IP address on the local computer is most often dynamic; i.e., it changes with each Internet session.

Both SLIP/PPP and normal dial-up access involve dialing into a remote computer system (which is usually directly connected to the Internet) and logging in to the ISP. Some people may have difficulty understanding the difference. The key distinction is that with SLIP or PPP, your own PC is communicating using native IP with other computers on the Internet, while with normal host dial-up, your PC simply acts as a dumb terminal to the remote computer (server) at the ISP location, which then communicates with other computers on the Internet using the server's IP. When you run an FTP client program with a host dial-up, the files you receive are stored on the ISP's server, not on your own computer. With SLIP/PPP, however, the files are received directly on your own PC. You can run GUI-based clients (Telnet, Web browsers, and so on) on your own PC and make direct use of any Internet service.

The connection to your ISP will become transparent and you will be able to establish several simultaneous sessions going out over the Internet. You can, at the same time, download files via FTP, send e-mail, and browse through various Internet sites.

A Windows socket program also includes a dialing program, called Dialer, to establish a telephone connection with the ISP.

Setting Up an Internet Connection

Setting up an Internet connection from a computer operating under Microsoft Windows 95 to the Microsoft-operated ISP—the Microsoft Network (MSN)—is very easy. You just click on The Microsoft Network icon on the Windows desktop and follow a few simple prompts. Similarly, setting up a connection to a large commercial ISP such as AT&T or an online service provider such as AOL is relatively easy. You load the diskette or CD supplied by the service provider and follow a few prompts, which will include a request for your credit card number. However, setting up the initial Internet connection to one of the smaller ISPs can sometimes be a challenge.

Assuming that you have already selected an ISP based on their reputation, access and reliability, here are the three steps required:

1. Install TCP/IP on your computer.
   Click the Start button, point to Settings, and select the Control Panel. In the Control Panel, double-click on Network.
   In the Network panel, click on the Configuration tab, then

from the Select Network Component panel, click on Adapter and click Add.

You should now see the Select Network Adapters panel. Click on Microsoft in the Manufacturer List, and click on Dial-Up Adapter in the Network Adapters list. Click OK.

In the Network panel that reappears, click Add. The Network Component Type panel appears.

In the Network Component Type panel, click on Protocol, and click on Add.

In the Select Network Protocol panel, click on Microsoft in the Manufacturer List, and click on TCP/IP in the Networks Protocols list. Click OK.

When the Network panel reappears, click OK. This completes the first part of your installation.

2. Install Windows 95 Dialer.

Go to the Control Panel and double-click on Add/Remove Programs.

Click on Windows Setup tab, then double-click on Communications and select Dial-Up Networking from the Communications panel. Click OK. Now there is only one final step remaining.

3. Create and configure a dial-up connection to your ISP.

Click the Start button, point to Programs and Accessories, then double-click on the Dial-Up Networking program. You can also locate this program in the My Computer folder on the desktop.

Double-click on Make New Connection. In the Make New Connection panel, type the name of your ISP, then click Next and type the area code and phone number of your ISP. Click Finish.

Now go back to the Dial-Up Networking folder, click once on your new icon that has the name of your ISP. Now click with the right mouse button and select Properties.

In the My Connection panel, click on Server Type. In the Server Types panel, select PPP... Internet. Ask your ISP whether to check the Enable Software Compression box, then click on TCP/IP Settings.

You must find out from your ISP how to fill out the TCP/IP Settings panel. You will be presented with several four-digit entries to fill in. If your ISP has the so-called dynamic addressing, then you do not have to enter any of the numbers—they will be filled in automatically.

There are other options you may want to set at this time. Go back to the My Connection panel, and click on Configure instead of Server Type. Then click on the Connection tab in the Internal Properties panel. Click on Advanced. You can now choose the initialization string for your modem. Also check the boxes Use Error Control, Compress Data, and Use Flow Control. Under Modulation Type, choose Standard.

Figure 10-11 shows the nine configuration screens associated with the Internet dialer on my computer. Those on your computer may be slightly different, in particular the numerical values in the TCP/IP Settings screen.

Figure 10-11 Windows 95 Internet Dialer configuration screens

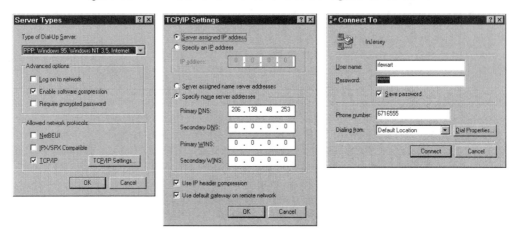

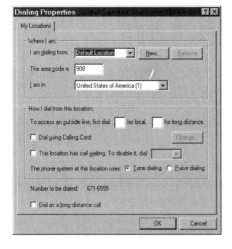

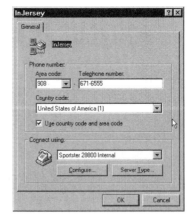

Figure 10-11 Windows 95 Internet Dialer configuration screens (continued)

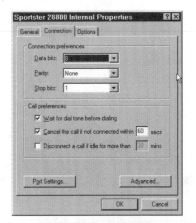

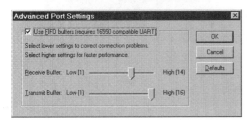

You have now successfully completed the "simple" procedure and are ready to use the Internet forever after, with no more difficulties.

After connecting to your ISP by using the Internet dialer, you can launch your browser software. Figure 10-12 shows the main screen of the Netscape Navigator browser with the Option menu displayed. This menu lets you choose colors, fonts, set the general screen appearance, enter your e-mail address and Usenet newsgroup server. Your service provider should supply you with this information.

Figure 10-13 shows bookmarks in the Netscape Navigator browser. Bookmarks are Internet addresses of Web sites that you frequently visit. You store a bookmark by choosing Add Bookmark from the Bookmarks menu.

Other examples of Internet sessions are shown in Chapter 11.

Figure 10-12 Internet browser—Netscape option screen

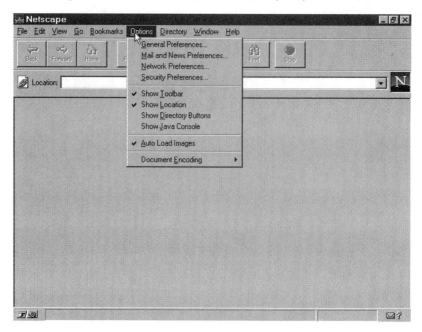

Figure 10-13 Internet browser—Netscape bookmarks

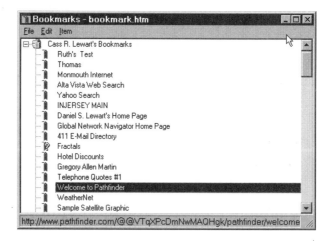

Voice Communication via Internet

An interesting type of software are Internet phone programs. The idea behind such programs is to communicate by voice over the Internet in a mode that resembles a phone conversation. You speak into a microphone that is plugged into a sound card in your computer, and you listen to the other party through the speakers connected to the sound card. The main operational difference between the Internet phone and a regular phone is that both participants must be online at the same time, and should be using the same proprietary Internet phone software. The call is free of any extra long-distance charges other than the cost of the local call to the ISP, even if the call is between two continents, but the quality is still far below what you would expect on a regular long-distance call. The process is somewhat similar to a ham radio exchange, except that you do not need a license and do not annoy your neighbors with TV interference.

There is a free program called CoolTalk that comes with the Netscape browser. When I tried it I could hear call-progress tones such as ringing and busy, but never succeeded in hearing a human voice. Later I downloaded a trial version of another program called I-Phone from the Internet. This time I could faintly hear someone answering my call, but again could not establish a two-way conversation. I was considerably more successful with a commercial program called Internet Phone from Vocaltec Corporation.

To run the program, I had to first start the Windows 95 dialer, which connected me to my ISP. Then I selected Chat Room from a menu. A list of online participants appeared on the screen. I clicked on a random name from Australia (country of caller appears next to each nickname in a chat room), and heard the phone-ring sound in my speakers. A few seconds later a fellow with a definite Australian accent answered. We had a short conversation—he could hear and understand me and I could hear and understand him. The quality in terms of a regular phone call was between poor and adequate—we had to repeat questions and there was a few seconds delay between questions and answers. Still, it was a thrilling experience. We talked for a couple of minutes. Then I tried the same with a fellow from Taiwan. Same good result, except that I had some difficulty in understanding him because of the accent.

The following two figures show a session in the Internet Phone Chat Room. The participant can click on a name of a person currently online and start a conversation. Figure 10-14 shows the main panel, which displays the name of the person connected to, and call statistics. Figure 10-15 shows the list of current chat-room participants. As you can see, the whole world is ready and willing to talk, even if they have nothing to say. Each of the chat-room participants can be called by clicking on their name.

Figure 10-14 Internet Phone—calling panel

Figure 10-15 Internet Phone—chat room

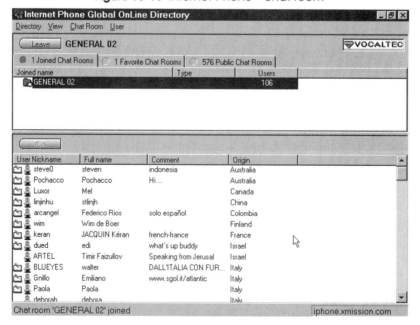

The Internet Phone program as well as other programs of this kind have many other interesting features. For example, you can set up the program as an answering machine and send or receive voice-mail messages while you or the called party are away from the computer. Besides speaking to random people in Chat Rooms, you can set up the program for a private conversation between you and a friend. Another feature that is possible is white-boarding, or exchanging and editing documents and photographs in real time.

System requirements for the Internet Phone, as well as other similar programs, are as follows (preferred values are shown in parentheses):

- Windows 95
- 486/66 (Pentium 120)
- 8 (12) Mbytes of RAM
- Half-duplex (full-duplex) sound card
- 14.4 kbps (28.8 kbps) modem
- Connection to an Internet Service Provider.

Similar programs allow for video conferencing and fall into the *See You - See Me* category. They require that both participants have special video cameras connected to their computers, such as the inexpensive QuickCam from Connectix.

Software on the ISP Server

Many ISPs offer an alternate access to the Internet other than via GUI-based Web browsers. This purely character- and command-oriented access makes the Internet appear as a BBS, but the environment is considerably less user-friendly. You may then ask "Why bother?" if one can click on a few icons in Netscape Navigator or MS Internet Explorer and easily browse the Internet. The answer is simple—speed and power. The back and forth conversation between your computer and Internet servers around the world no longer has to go through filters such as your computer, modem and transmission line between you and your ISP. If you issue an Internet-related command, it leaves the ISP server's computer at 50 to 100 kbps, several times faster than you could do from your computer directly. There are only two disadvantages to this process:

1. You cannot work with graphics, and
2. You have to learn a few new commands.

The operating system used by most Internet providers is Unix. Even if you know nothing about Unix, you must learn the important Unix commands.

First find out if your ISP provides this service, often called a Shell account. Most ISPs do not charge an extra fee for this once you subscribe to the regular SLIP/PPP service used by the GUI-based Web browsers. Then find out what you have to do to get to that account.

My ISP is INJERSEY in Asbury Park, New Jersey. Their access method to get to the Shell Account is to answer with a "1" to the initial login prompt, then answer with the user name and password to the succeeding prompts. After signing on, you get a new prompt and you are then on your own. You are then at the heart of your Internet account and in full control.

If you use a wrong Unix command and delete an important file, you might even lock yourself out. In that case, a sheepish call to your ISP support line will be necessary to restore your account.

Here are a few important Unix commands that might help you to browse around the Internet:

- ls - list files (like DIR in DOS)
- cp <file 1> <file 2> copy file 2 to file 1
- mv <file 1> <file 2> move or rename file 2 to file 1
- rm <file 1> delete file 1, use with extreme caution!
- cat <file 1> list file 1
- cd <new directory> change to new directory
- pico <file 1> open a simple text editor called "pico" with file 1
- vi <file 1> open a powerful text editor called "vi" with file 1

Here are a few Unix commands that perform Internet functions:

- lynx <URL> - go to URL location on the Internet
- ftp <URL> - start a file transfer session with the URL location
- ping <URL> - see how long it takes for a round trip to the specified URL
- pine - start program to send and receive e-mail
- tin/rtin - start program to access Usenet groups
- rz/sz - start file transfer program between server and your computer

One thing to remember is that files downloaded from a remote server to the Shell account are not in your local computer. To download them to your local computer requires an additional step. It can be done by either initiating an FTP connection between your computer and the server, or using the sz command on the server. The same applies to uploading files from your computer to the server, except that you would use the rz command instead. You will have to upload files to the server if, for example, you want to set up your home page for the whole world to see.

To upload a file from your PC to the server using the rz command, do the following:

1. At the prompt in the shell account, type "rz <filename.ext>". Be sure to substitute the actual filename for <filename.ext>, and do not type the quotes or brackets.
2. Press Return. You get a message that your server is ready to receive the file.
3. Your terminal-emulation software will have a command that lets you upload a file. Execute that command.

Note

If you have a large file, break it into a number of smaller files before uploading it to the server.

You should get a message that the server is receiving the file.

To download a file from your server to the PC, do the following:

1. At the prompt, type "sz <filename.ext>".
2. Press Return.
3. If you see some codes, ignore them.
4. Your terminal emulation software will have a command that lets your computer download a file. Execute the command to receive (download) the file. You should get a message that the server is sending the file, or your PC is receiving a file.

File Transfer Protocols

Every telephone connection will experience occasional bursts of noise that may result in data errors. If you transfer a long file of plain ASCII text, an occasional error may not be important However, error control of binary files is essential. A single undetected error during transmission of a binary file will, in general, make it unusable.

File transfer protocols are methods by which computers exchange binary files via a modem. The two primary functions of a file transfer protocol is to provide error detection and correction, and to make sure that the binary codes transmitted between modems do not interfere with control functions such as setting up flow control.

The file transfer protocols came into use before hardware error detection and correction became widely available. With hardware error controls such as those embedded in V.42-capable modems, the file transfer protocols provide additional insurance against error.

The file transfer protocols break the transmitted data into blocks and check parity and the cyclical redundancy check (CRC) in each block. The difference between a simple parity check, sometimes referred to as longitudinal redundancy check (LRC) and the more advanced CRC, is that the parity check derives only a single bit (0 or 1) from each byte, where in a CRC check, all bits in a block of typically 128 bytes are examined and a checksum value of two bytes in length—the so-called block check characters (BCC)—is computed and transmitted following each block of data.

The detailed algorithm for the U.S. (CRC 16) and the European (CRC ITU-T) implementations of the CRC algorithm is shown in Figure 10-16. The CRC algorithm is more sensitive to errors than a simple parity check. Two errors in a single byte will make a parity check appear correct, but will be detected by the CRC algorithm. If an error is detected by the parity check or the CRC algorithm within a block, then the whole block will be re-transmitted.

Some protocols adjust the block size to the frequency of errors. On a noisy line with frequent errors, the blocks should be small. If the transmission is fairly error-free, then the block size should be larger. As each block adds some overhead for acknowledgment, the transmission on a quiet line will be faster when using a protocol that uses variable block sizes. Following is a brief description of a number of file transfer protocols, which are normally included in most data communications programs.

Figure 10-16 Cyclical redundancy check (CRC) algorithm

Data Block = 11011011...

$$\frac{\text{Data Block}}{\text{CRC Constant (16 bit)}} = \text{quotient} + \text{remainder (16 bit)}$$

quotient is discarded

remainder = 16 bit or 2 BCC characters = $x_{16}x_{15}x_{14} \cdots x_1$

CRC 16 = $2^{x_{16}} + 2^{x_{15}} + 2^{x_2} + 1$

CRC CCITT (ITU-T) = $2^{x_{16}} + 2^{x_{12}} + 2^{x_5} + 1$

Xmodem Protocol

The Xmodem protocol was originally developed in 1979 by Ward Christensen, and although it lacks certain modern features such as variable block length, it is still very popular for personal computer data exchange. A more recent version of the protocol, Xmodem-CRC, adds a cyclic redundancy check (CRC) algorithm to further improve error correction. The regular Xmodem protocol catches 98% of errors, while Xmodem-CRC catches about 99.99% of errors.

Once the protocol is initiated, the receiving modem starts sending a NAK signal (Not Acknowledged, or ASCII 21) every 10 seconds. When the transmitting modem detects the signal, it starts sending the file to be transmitted in blocks of 128 bytes. At the beginning of each block is the SOH (Start of Header, or ASCII 01), followed by the ASCII character representing the block number, followed by the ASCII character representing the one's complement of the block number. This is followed by 128 data bytes and a checksum byte derived from the data.

The receiving modem makes sure that the block started with SOH, and that the block number and the checksum number are the same. If it finds that all three are OK, it then sends ACK (Acknowledged, or ASCII 06) and the transmission continues. If one of the three numbers is incorrect, NAK is returned, which results in retransmission of the entire block of 128 bytes. At the end of the file, the transmitting modem sends an EOT (End of Transmission, or ASCII 04); the receiving modem acknowledges with ACK, and the transmission is finished.

Because Xmodem is a relatively slow protocol, it should only be used when a faster protocol is not available.

X.PC Protocol

This public domain protocol, which was quite popular in the 1980s, can still be found in some communications packages.

The protocol provides the following features:

Multiplexing	The ability to support multiple data streams.
Flow control	The ability to control, for each data stream, the flow of data between transmitting and receiving DTEs and DCEs.
Error control	The ability to detect errors at the packet layer and to correct errors indicated by the link layer.
Reset and restart	The ability to reinitialize the communications paths at the packet layer if serious errors occur.

Ymodem-CRC Protocol

This protocol uses 1024 bytes per block, versus 128 bytes for Xmodem, to reduce overhead. If more than five errors are encountered in a block, the protocol reduces the block size to 128.

Kermit Protocol

In addition to the complete command-driven communications program discussed earlier in this chapter, there is also a Kermit data transfer protocol. This very popular protocol is supported by Columbia University in New York City. The following description is derived from the official Kermit documentation, which can be obtained from http://www.columbia.edu/kermit.

Like FTP, but unlike Xmodem, Ymodem, and Zmodem (with which Kermit is most often compared), most Kermit programs transfer files in text mode unless you specify otherwise. This means that record formats and character sets are converted appropriately when necessary. For example, between DOS and Unix, or between a MacIntosh and an IBM mainframe.

Unlike Xmodem, Ymodem, and Zmodem, Kermit is usually tuned for robustness to maximize the chances that file transfers will work, even under the worst conditions. It does *not* assume or require:

- a connection that is transparent to all (or any) control characters
- an 8-bit connection
- a clean connection
- a full-duplex connection
- large buffers along the communications path
- physical-link-layer flow control.

In short, it assumes that "everything that can go wrong will go wrong." Kermit is a pessimist, and behaves cautiously unless given instructions to the contrary.

As a result, Kermit transfers work almost every time. This is in contrast to protocols like Ymodem and Zmodem, which are tuned for speed, and therefore do assume and require all of the above to operate effectively. They fail to work when any of those preconditions are not met.

In fairness, non-Kermit protocols provide high performance right out of the box when all the preconditions hold, but in many cases they require extensive tuning to operate effectively. In others (for example, on mainframes), there is no way to make them work at all.

Kermit's robust default tuning is at the expense of speed. Therefore there is a widespread misconception that Kermit is a slow protocol. In fact, Kermit can be just as fast or faster than other popular protocols on any given connection. This requires adjusting only two or three parameters—packet length, window size, and control-character un-prefixing—to the highest levels your connection allows.

Zmodem Protocol

Zmodem, developed by Chuck Forsberg in the 1980s, is the preferred file transfer protocol for PCs. It ends each data block with CRC-32 error checking. It does not wait, however, for an ACK message from the receiving computer. It simply assumes that everything is OK, but still allows for re-transmission of bad blocks. This *streaming* feature makes Zmodem faster than many other protocols on moderately-good lines. Zmodem is also capable of multiple file transfers and crash recovery, in which an interrupted transfer can resume after it has been aborted.

Microcom Network Protocols (MNP)

Microcom Corporation developed a series of hardware and software error-correction schemes that are known by the collective name MNP. They were discussed in more detail in Chapter 4.

There are several levels of MNP protocols, with each higher level being more elaborate than the lower one. The first three levels are in the public domain and are based on CRC algorithms and block re-transmission. The three higher proprietary levels of the MNP protocols—Levels 4, 5 and 6—feature a combination of software and hardware embedded in special modems.

There are two aspects to this approach: error detection and correction, and improvement in data transmission efficiency by data compression. Using specialized modems provides the manufacturer with the advantage of being able to implement non-standard and possibly better technologies. The disadvantage to the user is that modems at both ends of a communication link must come from the same manufacturer. In order to exercise special features of these modems, they may require changes to the standard data communications software.

The MNP modems improve data throughput by using proprietary data compression. MNP data compression consists of buffering and analyzing the data, and then—depending on frequency of appearance of various ASCII characters—coding each character into four to 12 bits. For example, the most common letter of the English alphabet—e—would be coded in four bits, while the uncommon X would be coded in 12 bits. The assignments are dynamic and may change during a call.

To further maximize the data throughput, the block length of data is assigned dynamically and varies with the quality of the transmission facility. A quiet line would have long blocks, while a noisy line would have short blocks. These special modems operate in the synchronous mode by stripping start and stop bits.

Zip and Unzip

As mentioned in Chapter 4, there are several utility programs that compress offline files before transmission, and decompress them after they are received. The best-known programs of this type are PKZIP and PKUNZIP from PKWARE Inc. in Brown Deer, Wisconsin. A file is first zipped (or compressed) on a local computer by means of the PKZIP program. The new compressed file, which is in general substantially smaller than the original file, is then transmitted. At the receiving computer, the zipped file is decompressed by means of the PKUNZIP program and is restored to its original form.

The compression/decompression programs perform a function in software similar to the V.42 bis and MNP5 hardware protocols at the hardware level. If the file is already compressed by PKZIP or a similar program, then the hardware protocols will hardly affect the overall compression rate.

Encode and Decode

Certain operating systems can only handle ASCII characters with byte values between decimal 32 and 127. To translate binary files with byte values between 0 and 255 requires translation. Many programs, such as Netscape Navigator, perform this translation internally with programs such as BINHEX and MIME without the user noticing it. When, for example, Netscape Navigator receives an ASCII file corresponding to a translation of a graphic file in a known format, such as GIF or JPEG, it will automatically translate it into a binary file and display it. Similarly, e-mail programs automatically translate binary attachments into ASCII for transmission, and vice versa.

In this context, translation from binary to ASCII is called decoding, while the opposite process is called encoding. Sometimes an explicit encoding or decoding by the user is required if the browsing software did not perform proper translation. Such programs exist and are in the public domain or are available as shareware. An example is the UUDECODE.EXE and UUENCODE.EXE.

Data Encryption

Data communication links and e-mail are secure only to a limited degree. If a determined individual or a government organization tries to pry information from a computer, it can usually be done. Passwords of users are stored in files that can often be accessed by back doors in operating systems. Telephone lines can be tapped, and Internet traffic can be intercepted. Many companies and individuals therefore encode their data so that it cannot be decoded even if it is intercepted. There have been many encryption methods over the ages, from a simple character-substitution used by Ceasar, to code books used by German submarines, cyclical-character substitution used in Enigma machines during World War II, and the theoretically-unbreakable one-time pads used by the Russian spy Rudolf Abel in the 1950s. All these methods required a secure transmission of the coding password.

A popular encryption scheme based on recent advancements in cryptography is the two-key (public and private key) encryption method. It is based on the computational difficulty of factoring large numbers. The big advantage to this method is that there is no need for a secure password transmission. Although the method is not foolproof, it can be shown that, depending on the length of the private and public keys, it would take hundreds of years for the fastest computers with present technology to decode a message.

A company very active in this field is Pretty Good Privacy Inc. in San Mateo, California. The company markets several two-key data encryption programs that are integrated with popular e-mail programs. Its founder, Phil Zimmermann, gives frequent talks about the necessity for privacy in public and private life.

Person A using a Pretty Good Privacy (PGP) program first generates a pseudo-random private key, which is known only to Person A. A PGP algorithm then generates a public key based on the private key. The public key can be safely and freely distributed to all current and future recipients, such as Person B. Knowing Person A's public key does not help anyone discover the corresponding private key, which should be known only to Person A. A person only needs one key pair, but can have more than one.

Each key pair includes a User ID (such as the owner's name or e-mail address) so you and your colleagues can identify the owners of keys. Each private key also includes a pass phrase that protects it, like a password.

Keys are used to digitally sign a message or file and to authenticate a signature. When you sign a message, the program uses your private key to create a digital signature that is unique to both the contents of the message and your private key. Anyone can use your public key to verify your signature.

Verifying a signature proves that the message was actually sent by the signer, and that the message has not been subsequently altered by anyone else. The signer, alone, possesses the private key that created the signature.

When Person A sends a message to Person B, Person A encrypts the message before sending it with Person B's public key. Person A signs the message by encrypting the signature with Person A's private key.

When Person B receives the message, he or she decrypts the body of the message with B's private key, and the signature of the message with Person A's public key. This procedure not only assures that only Person B can read the message, but also authenticates that the message was signed by Person A.

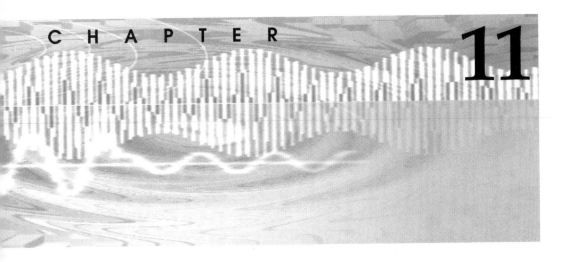

Connecting to the World

The computer is all fired up, the modem lights (if you have an external modem) are flashing, the hard disk is spinning full of communications software, and you have read previous chapters in this book. What to do next?

A few years ago the answer would be to connect to a local bulletin board, to connect to your workplace computer, or to connect to a large and frequently quite expensive commercial time-sharing computer service. Today, these three types of services have been joined by a fourth service, a new but healthy and fast-growing arrival, as more and more computer users explore the marvels of the Internet.

The dividing lines between the four types of services often blur—a bulletin board or commercial provider may also provide Internet access. All these services—the bulletin board, workplace computer, commercial services, and Internet service providers (ISPs), satisfy a range of needs. This chapter discusses these four types of services and what you can expect from each one.

Bulletin Boards

When connected to a remote computer, you can download and upload files. In case you have difficulty in distinguishing between downloading and uploading, think of the remote system—whether it is a bulletin board or a company mainframe computer—as something floating high up in the sky while your computer is planted firmly on the ground. You upload something *to* the sky, and you download *from* the sky to your local computer.

There are two types of bulletin board services (BBS for short)—a *single location* and a *global access* bulletin board. The first type is often a shoestring type of operation that is run as a hobby for the benefit of a few hundred affinity-group subscribers. The second type is a commercial service with hundreds of thousands or even millions of paying subscribers.

Single Location Bulletin Boards

This type of bulletin board is one of the earliest applications of data communications for the personal computer user. The resources to establish a local single-location bulletin board are minimal. All it takes is a full- or a part-time system operator (Sysop), a computer, one or more modems, appropriate software, and a telephone line for each modem.

Using 1997 prices, it is possible to set up a bulletin board for less than $1500 with a monthly expenditure of around $20. Many ambitious individuals, corporations and affinity groups have set up their own bulletin boards to exchange information with friends, customers or subscribers. The cost of running the board may be covered by a small and often voluntary contribution, or it can be borne by the organization sponsoring the bulletin board.

In a typical setup, a bulletin board is self-contained with all information residing on the hard disk on a single dedicated computer. The number of subscribers would typically be under 1000. Subscribers use their modems to dial the designated telephone number of the bulletin board, then identify themselves with a user name and password. For first-time users, the bulletin board software will ask for a full name and address, and will ask the new user to enter a password, which will then be stored on the BBS's computer. A typical dialog for an established user is shown in Figure 11-1.

Figure 11-1 Session With a BBS

```
WILDCAT! Copyright (c) 87,96 Mustang Software, Inc.  All Rights Reserved.
Registration Number: 95-7539.  v4.20 SL(Single Line).  Node: 1.
Connected at 24000 bps. Reliable connection. ANSI detected.
Comment: Connection was established at 24,000 bps
You have connected to node 1 on BCUG BBS
This system is operating on Wildcat! v4
Please make use of your real name on this BBS
What is your first name? cass
What is your  last name? lewart
```

```
Looking up your name. Please wait...
```

Comment: Wildcat program is checking the list of current BBS subscribers

```
Welcome CASS LEWART from Holmdel, NJ.
Password? [******          ]
```

Comment: Password is not echoed back for security reasons

```
Good morning, Cass, you are caller number 8,070.
Welcome To The Brookdale Computer User Group, Inc. Wildcat! v4.2 BBS System
 9:25am - 02/22/97
No bulletins have been updated since your last call.
Would you like to view the bulletin menu [y/N]? N
Checking for personal mail....No new personal mail found.
Press [ENTER] to continue?
Main Menu          Wildcat! v4
M  Message Menu     J  Join ConferenceN  Newsletter
F  File Menu        Y  Your SettingsP  Page Sysop
B  Bulletin Menu    C  Comment To SysopH  Help Level
?  Command Help     S  System StatsG  Goodbye & Logoff
I  Initial Welcome
Conference : For Everyone
Time Left  : 54  Time On : 0 Main Menu Command >>F
```

Comment: I am planning to download file RL_ART.ZIP

```
File Menu          Wildcat! v4
Q  Quit To Main     J  Join A Conf       L  List Files
M  Message Menu     H  Help Level        N  New Files Since
P  Personal Stats   ?  Command Help      D  Download Files
I  Info On A File   G  Goodbye/logoff    U  Upload Files
V  View A Zip File  S  Search Files      E  Edit Mark List
R  Read Text File   F  Transfer Info
Conference : For Everyone
Time Left  : 54  Time On : 0
File Menu Command >>D
Enter up to 9999 files. Press [ENTER] alone to stop.


                        Bytes   Time    Total BytesTotal Time
File # 1? RL_ART.ZIP    69,180  0.5     69,180             0.5
File # 2?
```

Comment: I selected one file to download.

```
Automatically logoff after last download is completed [y/N]? Y
Ready to send RL_ART.ZIP.
Please begin your Zmodem download now, <CTRL> X to abort...
```

Comment: My terminal emulation software starts the download.

```
B00000000000000
File Received: i:\scwin\rcv\rl_art.zip (67.5K bytes, 24.6 seconds, 2816
CPS, 1 retries)
```

Comment: The file has been downloaded to local directory i:\scwin\rcv.

```
RL_ART.ZIP    - SUCCESSFUL!   CPS = 2,823
Disconnecting in 10 seconds, press [H]ang-up or [ENTER] to remain online...
Seconds until disconnect:   1
TIMEOUT - Disconnecting...
Total time logged was 1 minute(s), with 53 minutes remaining for 02/22/97.
```

Comment: This BBS has a time limit for each user of 60 minutes/day.

```
Thank you for calling, Cass.
NO CARRIER
 2/22/97 9:18AM  Disconnected from 10288-544-9427
```

As seen from the recorded session, after logging in, a user is presented with a menu of options that are available for this particular bulletin board. The general appearance of the menus is determined by the software installed on the bulletin board and not by your terminal emulation software. One of the more popular BBS software packages is Wildcat from Mustang Software Inc. It is used by the bulletin board belonging to the computer club I belong to.

Depending on the arrangements with the Sysop, the BBS subscribers identified by their user name and password, can access some or all bulletin board options, often called bulletin board areas after signing on. The most important areas are Mail and File. Entering the Mail area allows a subscriber to send messages to the Sysop, or to other individual subscribers or to all subscribers. Bulletin board subscribers can also read messages addressed to them. This early form of electronic mail was very popular, but it was limited to subscribers of a single bulletin board, and had no facilities for exchanging messages with other systems.

The other popular use of bulletin boards is file exchange. Files stored in the File area of the bulletin board can be downloaded by subscribers. Similarly, a file can be uploaded from the subscriber's computer to the bulletin board. The file is usually checked for viruses by the Sysop, and if none are found, it is made available to other BBS subscribers.

Many bulletin boards also provide connection to some world-wide networks such as Fido. Fido is a message-based network that provides hundreds of special interest groups (SIGs), which can be selectively downloaded by Sysops. Table 11-1 shows a sample of Fido groups available through one of the bulletin boards to which I belong. A BBS connected to the Fido network will typically carry several hundred groups.

Table 11-1: Sample of Fido Groups

4. FIDO Private Netmail ONLY	98. FIDO - Zone 1 General Chat
101. FIDO-10th Ammendment	102. FIDO-12 Steps
103. FIDO-4DOS Echo	104. FIDO-4DOS/4OS2 and Take Command
105. FIDO-4X4 Echo	106. FIDO-Assembly Language Program
107. FIDO-African American Genealog	108. FIDO-American Atheist Online S
109. FIDO-disAbled User Information	110. FIDO-Abled Athletes
111. FIDO-Abortion Discussion	112. FIDO-Advanced Dungeons & Drago
113. FIDO-ADAM Internation Computer	114. FIDO-Adaptive Technology for t
115. FIDO-Adept XBBS Support Echo	116. FIDO-ADEPT SysOp to SysOp supp
117. FIDO-Attention Deficit Hyperac	118. FIDO-Sound Cards Echo
119. FIDO-Adoptees Information Exch	120. FIDO-ADS FILE ANNOUNCEMENTS
121. FIDO-AIDS & HIV	122. FIDO-AIDS.DATA
123. FIDO-AIDS & ARC	124. FIDO-Airgunners' Info Exchange
125. FIDO-Alaska Off Topic Chatter	126. FIDO-AllFix Support Conference
127. FIDO-Alternative Medicine	128. FIDO-Amateur Radio Echo
129. FIDO-Amiga International Echo	130. FIDO-Amiga Games
131. FIDO-Amiga Hard/Software ForSa	132. FIDO-Intl. Amiga & CDROM/CDTV/

The Sysop downloads on a daily basis—either by phone or satellite dish—all new messages in Fido groups that the Sysop wishes to publish on the BBS. If a message is addressed to you by another user of the Fido network—which can be located anywhere in the world—it is treated as though it were a message from another subscriber of your local bulletin board. You are then alerted that there is mail in your mailbox. Fido addresses you only by your first and last name. Therefore if your name is John Smith, you will find many messages from all over the world in your mailbox on subjects you were not even aware existed. So try to use a unique name for your user name to avoid this potential problem.

Commercial Global Services

Parallel with the development of single-location bulletin boards, the development of commercial ventures, referred to as *commercial global services*, or online service providers, started in the 1970s. Examples of such services are the big three: CompuServe, America OnLine (AOL), and Prodigy. These services provide local telephone access at hundreds of locations throughout the United States and around the world. Therefore a subscriber located just about anywhere should be able to place a local call to connect to the service provider, who may be located anywhere, even as much as thousands of miles away. This avoids having to pay for long-distance calls to a service provider.

The price structure of such services can be a monthly fee of from $10 to $20, or an hourly access fee of typically $3, and occasionally other fees to pay for special information access and for memory storage on the server. What these services offer beyond what is available from a single location bulletin board is no or low long distance telephone charges, specialized proprietary information, special services such as chat rooms that are available only from that particular service, and a large number of users, typically several million.

Before the advent of the Internet, these were very important factors for many users and led to a rapid expansion of large commercial services. Through advertising and special promotions of free initial access, these services signed up millions of subscribers. Each of these services emphasizes access to large amounts of proprietary information that are available only to subscribers—the so-called value-added information—along with access to many manufacturer's help lines and special interest discussion groups. Commercial services provide good customer support and a user-friendly interface for less-technical subscribers. Under pressure from competition, commercial services also lowered their fees, and in some cases even switched to a flat fee for unlimited use. This last change caused lots of grief to AOL subscribers for a while because it generated a sudden surge in use without a corresponding expansion of facilities. The result was a wave of busy signals and subscriber complaints.

Though the commercial global services were originally self-contained, they now offer access to many Internet services as well. They also opened their systems to electronic mail that originates outside their systems. An AOL or Internet user can therefore send and receive e-mail to or from CompuServe or Prodigy subscribers. Early implementations of commercial services provided a character- and menu-based interface similar to those used by single location bulletin boards. The general acceptance of the graphical user interface (GUI) by both Apple and the IBM PC community gave a new look to these services. A Windows-type point-and-click interface now provides easier access to all services. As new versions of display software become available, they can be automatically downloaded and installed on the subscriber's computer.

An example of a GUI interface that is provided by AOL is shown in Figure 11-2. Clicking on an icon leads to AOL's Internet access, or to chat rooms and other proprietary areas.

Figure 11-2 America Online interface

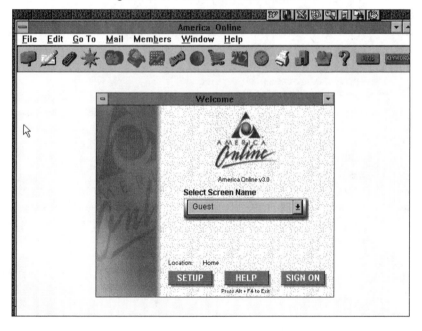

In a historical perspective, commercial global services satisfied the needs of many computer users until the mid 1990s. Their offerings were great when compared to single location bulletin boards, but became rather questionable once Internet access became universal and inexpensive, at least in the United States and many other countries. Due to the competition from the flat-fee unlimited-access ISPs, some of these commercial services are now on the decline. It is also due to the migration of manufacturer's help lines from these services to the Internet. If straight Internet access satisfies a user, then the additional layer of software and hardware from the commercial global services only slows down the exchange of information with an Internet server, and becomes more of a hindrance than a help.

Specialized Services

Another type of commercial global service is not on the decline at all—it is thriving and appears to be in no danger from competition from the Internet. These specialized services provide very narrow but detailed information for certain professions. Search services, such as Lexis-Nexis, deal mostly with the legal profession and provide personalized and often proprietary information. Although these services are quite expensive for personal use, their cost is often insignificant in a business environment in terms of the value-added information they provide. Similar to other commercial global services, these specialized services provide local point-of-presence (POP) dial-in access in many locations throughout the world. They also provide telephone help with minimal or no waiting, individual training, piles of manuals, and software to access the service.

Figure 11-3 is an example of a search session with Lexis-Nexis. The first screen shown provides a list of available libraries. Once a library of interest is identified, a Boolean search can be performed on its contents. Libraries fall into legal, medical and news categories. Most libraries provide only text information, legal briefs, articles, and legal opinions.

An example of an article dealing with the Ebola virus was found through a simple search in the MEDLNE library, as shown in Figure 11-4. The search produced 18 hits, and could be successively displayed and saved to the local disk.

Figure 11-3 Lexis-Nexis opening screen

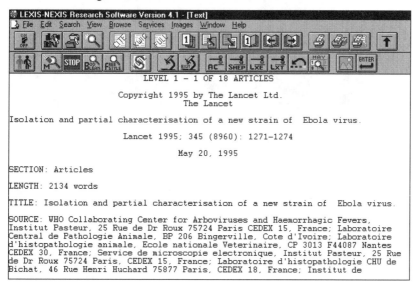

```
LEXIS-NEXIS Research Software Version 4.1 - [Text]
File  Edit  Search  View  Browse  Services  Images  Window  Help

                    LIBRARIES -- PAGE 1 of 2
Please ENTER the NAME (only one) of the library you want to search.
- For more information about a library, ENTER its page (PG) number.
- To see a list of additional libraries, press the NEXT PAGE key.
NAME   PG NAME     PG NAME   PG NAME   PG NAME   PG NAME    PG NAME   PG

-------General Legal-------- --Public Records--- --Helps-- Financial --News--
MEGA    1 2NDARY   2 LAWREV  3 ALLREC  4 INSOLV  5 EASY    6 COMPNY  7 NEWS    22
GENFED  1 ALR      2 MARHUB  3 ASSETS  4 LEXDOC  5 GUIDE   6 INVEST  7 REGNWS  22
STATES  1 BNA      2 LEXREF  3 DOCKET  4 LIENS   5 PRACT   6 NAARS   7 TOPNWS  22
CODES   1 ABA      2 HOTTOP  3 FINDER  4 VERDCT  5 TERMS   6 QUOTE   7 LEGNEW  22
CITES   1 CAREER   2          INCORP  4                    CATLOG  6 D&B     7 CMPGN   22
LEGIS   1 CLE      2                                       CUSTOM  6           WORLD   22

-------------------------------Area of Law---------------------------- Medical
ACCTG   8 CORP     9 ETHICS  10 HEALTH 11 LEXPAT  12 PUBHW  13 TORTS   14 GENMED 15
ADMRTY  8 CRIME    9 FAMILY  10 IMMIG  11 M&A     12 REALTY 13 TRADE   14 EMBASE 15
ADR     8 EMPLOY   9 FEDCOM  10 INSURE 11 MILTRY  12 STSEC  13 TRANS   14 MEDLNE 15
BANKNG  8 ENERGY   9 FEDSEC  10 INTLAW 11 PATENT  12 STTAX  13 TRDMRK  14
BKRTCY  8 ENVIRN   9 FEDTAX  10 ITRADE 11 PENBEN  12 TAXANA 13 UCC     14
COPYRT  8 ESTATE   9            LABOR  11 PUBCON  12 TAXRIA 13

Enter .NP for Individual States, International Law and more News information
```

Figure 11-4 Lexis-Nexis article on ebola virus

```
LEXIS-NEXIS Research Software Version 4.1 - [Text]
File  Edit  Search  View  Browse  Services  Images  Window  Help

                    LEVEL 1 - 1 OF 18 ARTICLES

                Copyright 1995 by The Lancet Ltd.
                        The Lancet

Isolation and partial characterisation of a new strain of  Ebola virus.

              Lancet 1995; 345 (8960): 1271-1274

                      May 20, 1995

SECTION: Articles

LENGTH: 2134 words

TITLE: Isolation and partial characterisation of a new strain of  Ebola virus.

SOURCE: WHO Collaborating Center for Arboviruses and Haemorrhagic Fevers,
Institut Pasteur, 25 Rue de Dr Roux 75724 Paris CEDEX 15, France; Laboratoire
Central de Pathologie Animale, BP 206 Bingerville, Cote d'Ivoire; Laboratoire
d'histopathologie animale, Ecole nationale Veterinaire, CP 3013 F44087 Nantes
CEDEX 30, France; Service de microscopie electronique, Institut Pasteur, 25 Rue
de Dr Roux 75724 Paris, CEDEX 15, France; Laboratoire d'histopathologie CHU de
Bichat, 46 Rue Henri Huchard 75877 Paris, CEDEX 18, France; Institut de
```

Certain libraries, such as the patent and trademark libraries, also contain image files. It is possible to view a complete patent application, then download and print all documents and drawings associated with it. Examples of such searches are shown in Figures 11-5, 11-6, and 11-7. These figures show pictures from two patents in the PATENT library, and a trademark from the TRDMRK library. Other libraries accessible through the service are huge databases containing property records for each state, for example. A text search on mainframe computers took only a few seconds to locate property assessments of all houses on our street in the Holmdel township.

Figure 11-5 Lexis-Nexis patent search

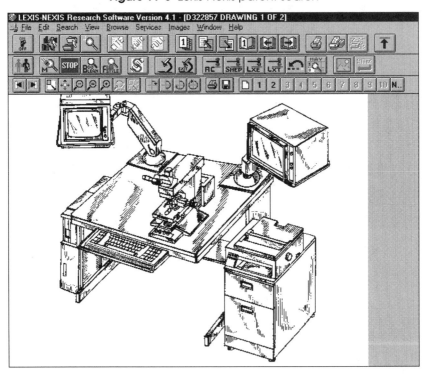

Figure 11-6 Lexis-Nexis patent search

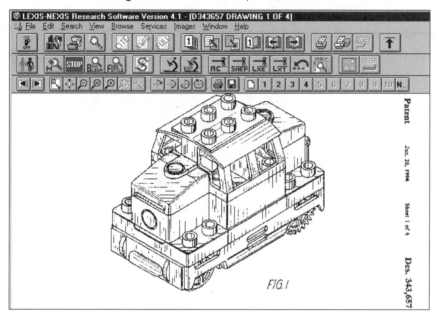

Figure 11-7 Lexis-Nexis trademark search

Sales and Telecommuting

The term *virtual office* is becoming the way of life for many workers in industrialized countries. People whose business leads them all around the country and around the world must be able to connect to their home office. In the days of the traveling salesman, it was the phone; today it is the computer modem connected to a wired or wireless phone. The other side of the connection—the office computer—provides an interface similar to that of a BBS. Information can be downloaded or uploaded from any location in the world, accessible through the wired or wireless telephone network.

With traffic congestion on highways getting worse and worse all around the world, many people reduce commuting between home and office. They *telecommute* by setting up an office with a computer and a modem at home. Some companies even eliminate fixed work locations and assign employees to different work areas, depending on the space available on a particular day. All information—e-mail, voice-mail and a telephone extension number, can move with the employee.

Another type of office connection is access to a mainframe computer. Before the advent of powerful personal computers, this was one of the main applications for data communication. A worker needing computer resources no longer had to bring his or her data or programs on punched cards or on magnetic tape and hand it to people feeding the big mainframe computer. Jobs could be submitted remotely to a time-shared mainframe computer, and the results could be seen and printed on remote terminals.

Access to a host such as an IBM AS400 or other mainframe computer may require setting the modem to synchronous transmission in order to satisfy the computer's protocol. The following are instructions on how to configure a Hayes or a Hayes-compatible modem for synchronous operation.

Configuring the Modem to Synchronous Tranmission

The originating modem should be configured for *synchronous originate* mode. It will dial a number stored in the modem's memory when a connection is started. You will need a dumb terminal or terminal-emulation software to configure the modem. To configure the originating modem:

1.   Attach the modem to a serial port on a PC or dumb terminal using a standard serial cable.

2. Configure the port-speed setting in the dumb terminal or the terminal-emulation software to match the speed that will be used on the synchronous port.

3. Configure the software for direct-connect or terminal mode and open the connection to the port.

4. Type AT&F&W and press Enter. The modem should respond with OK. If double characters appear, type ATE0 and press Enter to disable local character echo.

5. Type AT&Q2&S2&W and press Enter. The modem should respond with OK.

6. Type AT&Z0=T<phone number to store> and press Enter. The modem should respond with OK.

7. Type AT&D2&W and press Enter. The modem should respond with OK.

8. Type AT&C1E0Q1&W and press Enter. The modem should *not* respond with OK because character echo and result-code reporting have been disabled.

The answering modem should also be configured for synchronous operation in the answer mode. Although the answering modem is usually attached to the mainframe host system, you will first need to connect it to a dumb terminal for configuration. To configure the answering modem:

1. Attach the modem to a serial port on a PC or dumb terminal using a standard serial cable.

2. Configure the port-speed setting in the dumb terminal or the terminal-emulation software to match the speed that will be used on the synchronous port.

3. Type AT and press Enter. The modem should respond with OK. If double characters appear, type ATE0 and press Enter to disable local character echo.

4. Type AT&F&W and press Enter. The modem should respond with OK.

5. Type AT&Q1&S2&W and press Enter. The modem should respond with OK.

6. Type ATS0=1 (or the number of rings you want the modem to answer on) and press Enter. The modem should respond with OK.

7. Type AT&D2&W and press Enter. The modem should respond with OK.

8. Type AT&C1E0Q1&W and press Enter. The modem should *not* respond with OK because character echo and result-code reporting have been disabled.

Establishing a Synchronous Connection

Attach the originating modem to the serial port and turn the power on. When a connection is started, the modem automatically dials the stored number and attempts to connect to the other modem.

Attach the answering modem to the synchronous port on the host system. The modem should now answer incoming calls in &Q1 synchronous mode.

Internet—The World Listens and Talks to You

When Al Gore was running for the office of Vice President of the United States in 1992, he mentioned in one of his election speeches that we should all be united by the Information Highway. At that time he was just talking about a concept with no specific meaning attached to it. But a few months later the Information Highway suddenly appeared on the horizon, and its name was the *Internet*. From its humble beginnings as a file-and message-exchange network started by the government's Defense Advanced Research Project Agency (DARPA) and used by military establishments and universities, it grew into a universal access medium slowly approaching the spread of the telephone network.

The main differences between bulletin boards and the Internet are size, accessibility and type of transmission facilities. A local bulletin board could typically connect 1,000 subscribers; a large commercial service a few million. By contrast, the reach of the Internet is now hundreds of millions people. Local bulletin boards and commercial services are limited primarily to their own subscribers. By contrast, the Internet became a utility open to all.

A bulletin board operates through the switched telephone network provided by telephone companies. On this network, a call is established between two individual parties and is kept open exclusively for those two parties until one of them hangs up and disconnects. The Internet operates over a different type of shared network, similar in concept to a giant party line, where individual packets of data are switched between destinations. The so-called routers guide each packet of data to their destination, depending on the header contained in each packet.

In general, an Internet service provider (ISP) supplies an intermediate link between the Internet network pipeline and an individual user. An ISP typically connects to the Internet with a high-speed digital line, such as T1, and connects to subscribers via a bank of modems attached to telephone lines. Individual users then dial the ISP modems. In general, an ISP must have at least 1,000 customers to be cost-effective. Some ISPs can be reached on only one telephone number, while users in a different area may have to pay for a long-distance call. Larger online service providers, who also provide access to the Internet, such as commercial global service providers, have local point-of-presence (POP) access points in dozens or even hundreds of localities. As telephone charges to a distant location can be substantial, a large number of POPs attracts many new customers who might live in areas where those POPs provide local telephone access.

The most important services currently provided by the Internet include the graphics-oriented World Wide Web (WWW), e-mail, file transfer protocol (FTP) access, and access to Usenet groups and chat rooms. Before the introduction of the WWW (the Web) in the mid 1990s, a character- and menu-oriented program called Gopher was often used to access information from remote sites. The remainder of this chapter discusses these services and their importance to the development of the Internet's popularity..

World Wide Web

What Alexander Graham Bell did for the exchange of information, the World Wide Web (WWW, or the Web as it's now commonly called) did for the expansion of the Internet. There were probably many railroad telegraphers who claimed that the Morse code was great and that it should be used by everyone. In fact, Morse code was in many respects even better than speech, because the dots and dashes could not be easily misinterpreted even when transmission was poor.

The Web made the Internet into an easily-accessible medium that contained text, graphics and even sound and video. For Internet users, the Web is an asymmetric way of information exchange, somewhat like a library. You go through a library catalog, enter a few keywords, and select a title. In return you get back a book with thousands of words. Similarly, in a Web session, you enter a few keywords or click on a title, and get back many pages of information. These pages are stored on a server somewhere around the world, and are transmitted to your screen in a series of data packets.

It is comparatively easy even for individuals to create a so-called Web page, or Home page, then store it on an ISP's server and make it available to any Internet user around the world who knows about its existence. You can create the material either by means of Web-authoring software or directly by entering HyperText Markup Language (HTML) statements in a text editor. You can also create your document in an ordinary word processor such as Word or Word-Perfect, and use the HTML extensions to translate the word processor document into an HTML document.

HTML extensions are currently available for all major word processors. What you put on the home page is strictly up to you. You can show pictures of your grandchildren, as my wife does, or you can show a picture of the lunch you ate yesterday, as one student's home page showed. You can also put some *important* information like schedules of organizations with which you are associated.

It is very easy to create links to other locations. The Internet address is officially called the Uniform Resource Locator (URL). A URL starts with the character string "http://" and is followed by the address of the server and the user. Special characters in the URL, such as forward slash (/) followed by the tilde (~), point to a user directory and to specific documents.

Besides Web pages showing grandchildren and other trivia, you will also find pages that provide lots of useful information. While working on this book, I spent much time gathering information from Web pages prepared by small and large corporations. I found modem specifications, frequently asked questions (FAQ) files about modems, installation hints, AT commands, and much more.

To find something specific, you will normally start by gaining access to one of the Internet search engines, such as Yahoo or Alta Vista, and type a few keywords such as "modem" or "AT commands." You are then presented with a list of Web pages that includes those keywords. The list may be very large, with thousands of entries, but an intelligent search engine sorts the results in order of relevance. Listed at the top are those Web pages that have the best match to your keywords, followed by those that are less likely to be of use. Each listing may indicate "x% relevant" to give you an idea of just how relevant the search engine thinks it is. You then click on any URL displayed in the list and start collecting information. You can save each document to your local computer's hard disk or to a floppy. An example of a Yahoo search screen, ready to search on the word "modem" is shown in Figure 11-8, while a simple home page with links to many sites is shown in Figure 11-9.

There are many extensions to Web browsers, called *plugins*, which can be downloaded from various Internet sites. As mentioned in Chapter 10, these plugins turn your computer into a radio (with the RealAudio plugin), a telephone (with the Internet Phone plugin), or even a two-way picture phone (with the See U See Me plugin). Unfortunately the current status of Internet technology does not provide a quality that is comparable to that offered by commercial services. The telephone plugins give you voice quality similar to early amateur radio, and images are changing at a slow refresh rate, making the display a bit jerky. However, wide-band modems and expansion of the Internet may improve this capability in the near future.

Figure 11-8 Example of an Internet session—Yahoo search engine

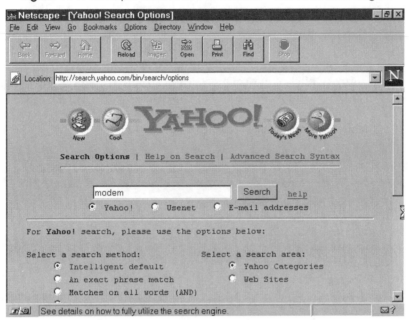

Figure 11-9 Example of an Internet session—my son's home page

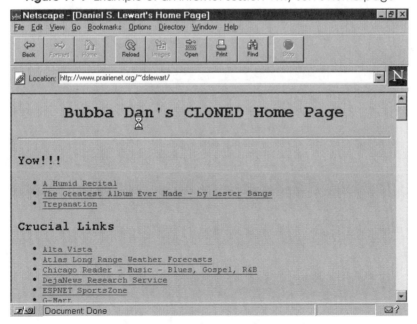

Gopher

Gopher is an older version of the Web before it became popular. The name Gopher was given to it by the University of Minnesota, the original proponent and developer of Gopher. A gopher is the university's mascot. It also points to the program's characteristic as it "burrows" through piles of information stored around the world. The program is strictly menu-driven and character-oriented. It is more of historical significance, since it is used less and less today. Figure 11-10 shows an opening screen of a typical Gopher session.

Figure 11-10 Gopher session

```
                Internet Gopher Information Client v1.12S
        Root gopher server: wiscinfo.wisc.edu
  -->   1.  About WiscINFO and WiscWorld Services at UW-Madison/
        2.  Search all Titles in WiscINFO <?>
        3.  General Campus and Community Information/
        4.  Directories of People, Organizations and Services/
        5.  Calendars and Schedules/
        6.  News Releases/Announcements, Newsletters, and Newspapers/
        7.  Employment, Financial Aid, Scholarships and Grants/
        8.  Academic Programs and Other Learning Opportunities/
        9.  Library Catalogs and Services (The Electronic Library)/
        10. Course Materials and Other Educational Resources/
        11. Computing Information and Services/
        12. Information for Faculty and Staff/
        13. Access to WiscINFO Resources by Information Provider/
        14. Other Information Sources and Gopher Servers/
        15. National Weather Service Forecasts.
  Press ? for Help, q to Quit, u to go up a menu, g to go to a location Page
```

As seen from Figure 11-10, the Gopher choices provide information about the starting site, in this case the University of Wisconsin. However, option 14 or 15 leads to other Gopher servers. And if a Gopher server address is already known, it can be reached directly using the g command.

E-Mail

Electronic mail, or e-mail for short, is probably the most important Internet application of all time. A short address, such as "rlewart@injersey.com," uniquely identifies anyone around the world. There are no more international boundaries with different postal rates, no long ZIP codes, and no licking of stamps and envelopes. The delivery is nearly instantaneous, and in most cases there is no additional cost for this service beyond the monthly payment to your ISP. Some companies in the United States even offer e-mail service for free, in exchange for putting up with short advertising blurbs appearing in the e-mail messages.

E-mail service was originally limited to plain-text ASCII messages. Now, most e-mail service providers allow a sender to include so-called *attachments* with the text message. These attachments can be binary files. Therefore it is now possible to enclose an attachment representing a scanned photograph or a formatted file from a word processor with your message. If you generated a formatted file with your word processor, you can now send it to a recipient who will hopefully have the same word processor. Your document can then be displayed in all its glory with headers, indentations and embedded graphics and images on the recipient's computer. It can then be further processed by the recipients in their word processor or graphics program.

The mechanics of e-mail is quite simple. An e-mail program can be accessed from a browser such as Netscape Navigator, which has it own e-mail facility built in, or it can be accessed from a separate program such as Eudora from Qualcomm Inc., which works with the Windows Socket program that is required by your browser's software. If you have a shell account with your ISP, you can use one of the e-mail programs installed on the server. My favorite is the program called Pine. Whatever program you use, it should provide the following capabilities:

1. Store and retrieve names from an address book.
   The address book stores e-mail addresses and nicknames. If you type "john1" in the address field of your message, it will be replaced by *John Smith <john.smith@injersey.com>*, which is the complete e-mail address as stored in your address book. An e-mail program also lets you display, print, delete and edit an e-mail address book.

2. Reply to a message.
   If you read a message addressed to you, and would like to respond, you can simply choose the Reply option in your e-mail program. The Reply option opens a *new message* that automatically contains the return address in the *Send to* field. For your convenience, the original message addressed to you can be included in the return message, if you choose this as an option. The beginning of each *quoted* text line usually has a ">" to indicate that it is quoted text. You can then respond to each point while retaining the original text in your message. When you are finished, just click on the Send button, or press Ctrl-x when using Pine in a shell account, and your message is on its way.

3. A signature block.
    You can create a short file with your address, telephone num-
    ber, and a clever statement if you wish, and it will be automat-
    ically attached to each message you send. This saves having to
    type a signature for each message, and also ensures that each
    signature is identical. Of course, you can change the signature
    at any time.
4. Offline use, save and delete.
    You can prepare messages before connecting to your ISP. This
    is called offline use. You can also move, save, and delete mes-
    sages if you want to reorganize them or clean out your files.
    The program will also give you statistics of messages current-
    ly saved, and of new messages that have been read and an-
    swered.

The following two figures show typical e-mail sessions using a
browser, and using a standalone e-mail program (Pine) in a character-
based shell account. Figure 11-11 shows the GUI-based Netscape
Navigator mail screen with messages stored in the local computer,
and new messages recently downloaded from the server. Figure 11-12
shows the character-oriented Pine opening screen.

Figure 11-11 E-mail with Netscape

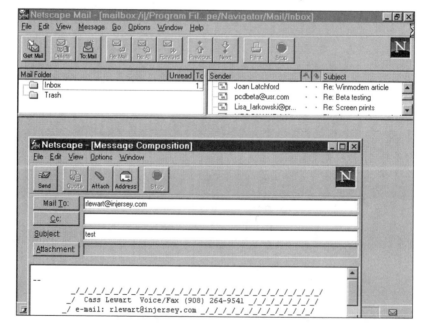

Figure 11-12 E-mail with Pine

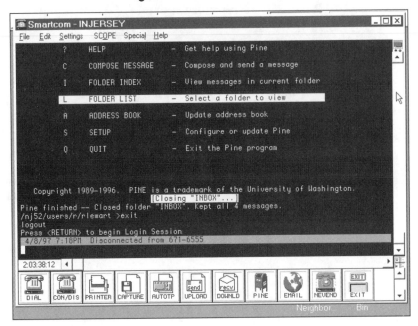

File Transfer Protocol (FTP)

File transfer using FTP is one of the earliest Internet applications. It is file transfer pure and simple. Literally millions of files are stored at large servers, many operated by universities all around the world. To use FTP, you need to know the directory structure of the server you are downloading from or uploading to. These structures can be quite complicated. For example, a utility I recently downloaded from a server at the Massachusetts Institute of Technology was a file explaining how to create a new Usenet newsgroup. The name of the server is:

```
rtfm.mit.edu
```

and the complete address of the file, which is embedded deep in the directory structure of the server, is:

```
rtfm.mit.edu/pub/usenet/news.answers/usenet/creating-newsgroups
```

When you log into an FTP site, the remote server asks for your user name and password. If the server is accessible to the general public, you can type "anonymous" for your user name, and type your e-mail address for your password. This is called an *anonymous* connection.

Figure 11-13 shows a session with an FTP program called WS_FTP. It shows the directory structure of your computer on the left side, and the directory structure of the remote server on the right side. You can move files back and forth by marking the file(s) and clicking on the appropriate arrow.

Figure 11-13 FTP session with WS_FTP

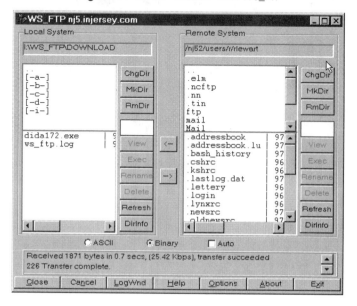

Usenet

For a typical computer user, a connection to the Internet means having access to e-mail and to the World Wide Web. E-mail lets you communicate with individuals or groups, while the Web lets you access information prepared by individuals and corporations. A third way to communicate with the world over the Internet, which is less popular but equally important, is the use of Usenet newsgroups. It is somewhat akin to the Citizen Band radio or a bulletin board, except that your audience is worldwide and can be targeted to narrow interest groups.

If you need information on a very specific subject, one of the easiest ways to get it is to send a message to a specific Usenet group. Your Internet provider will usually store Usenet messages from selected newsgroups on a computer that is accessible to their subscribers. Smaller Internet providers may contract with a remote Usenet server. Usenet servers often have the word "news" as part of their URL address, such as *news.injersey.com* for users of the *injersey.com* ISP. Some large organizations such as Microsoft and Netscape provide their own news servers, which carry only newsgroups related to their business. Some of these newsgroups are only accessible to employees of those companies, while others are freely accessible to all Internet users.

What is There and How to Find it.

My Internet news server carries over 18,000 newsgroups. Universities typically carry 2000 to 5000 newsgroups. Subjects range from the ridiculous to the sublime. You will find a newsgroup dedicated to the TV program NYPD Blue (alt.tv.nypd-blue) and another dealing with numerical analysis (sci.math.num-analysis). There is also a discussion group dedicated to prostitutes in Tijuana and plenty of newsgroups carrying articles for any particular group of common interest.

You can search for specific groups with the tin/rtin Unix program if you have a shell account, or with the Windows freeware program Free Agent. You can also access specific newsgroups with any Internet browser such as Microsoft's Internet Explorer or Netscape's Navigator (for Navigator, select Window, then Netscape News from the main menu bar). When connected, you can read messages posted in each newsgroup, or you can answer a message to a newsgroup using a follow-up, or you can reply to a message via e-mail, or you can post a new message. Starting with Netscape Navigator version 2.0, you can also decode (convert) and display GIF and JPG graphic files included in Usenet messages.

You can also search through names of newsgroups offline by means of a program called USENET, which I developed and which can be found on the enclosed CD-ROM.

To run the program, first copy the contents of the USENET directory on the CD-ROM to a newly created directory on your hard drive. Then get into DOS mode by either booting in the DOS mode or by opening a DOS window from Windows. After getting the C:\> prompt, change to the USENET directory on your hard disk. Optionally, you can run the program directly from the CD-ROM without copying it to your hard drive.

One of the two files in the USENET directory should be USENET.EXE; the other should be USENET.DBF, which is a database file that contains all information about newsgroups. Execute the program by typing this at the C:\> prompt:

```
usenet
```

You will be prompted for a keyword describing the newsgroups you are interested in. For example, typing "amiga" will find all newsgroups with the word "amiga" in their name or description.

Table 11-2 shows a partial list of Usenet groups that have the word "modem" in their name. The number after the group description refers to the number of new messages over a period of a week, which gives you an indication of the level of activity of that group. Some groups carry only occasional traffic.

Table 11-2: Result of USENET Keyword Search on "modem" Using USENET.EXE

alt.net.and.modem.games	3
biz.comp.telebit Support of the Telebit modem.	4
comp.dcom.modems Data communications hardware and software.	740
comp.os.os2.comm Modem/Fax hardware/drivers/apps/utils under OS/2.	105
comp.sys.ibm.pc.hardware.comm Modems & communication cards for the PC.	92
de.comm.isdn.computer ISDN-Karten/-Modems fuer Computer.	14
de.comm.modem Modems: Grundlagen, Konfiguration, Probleme.	0
fido.ger.comms About Terminalprogram's, Modem's, DFUE general.	0
fido.ger.modem Modems: Configuration, Installations etc.	0
fido.ger.zyxel Disk. ueber die Familie der Multi-Dienst-Modems Zyxel.	0
fido7.game.modem videogames over modem	0
fido7.hardw.hsmodem computer hardware talks	0
fido7.hsmodems high speed modems talks	0
fido7.modem modems	0
fido7.usr US Robotics modems	0
fj.net.modems Modem hardware, software, and protocols.	0
fj.net.modems.fax Fax-Modem hardware, software, and protocols.	0
fr.network.modems Discussions concernant les modems.	0
iijnet.dcom.modem Discussion of Modems.	1
ka.uni.dial-in Dial-In Zugang zur Uni, Modem, SLIP & Co.	0
mcmaster.services.network.modempool For discussions about the McMaster	0
misc.forsale.computers.modems Modems for sale and wanted.	261
relcom.fido.ru.modem Inter-network discussion on modems.	186
sanet.modems Anything concerning modems and telecom hardware.	0

My Recent Experiences with Usenet

I gave my old Kenwood shortwave receiver to my daughter, who did not have much use for it. I advertised it in the *rec.radio.shortwave* newsgroup and got several offers. The radio sold for over $200 within a week. By reading messages in that group, I also discovered that WNYE-FM in New York City carries my favorite BBC World Service programs in full stereo every night. Similarly, after my wife bought a copy of a very popular (and expensive) photo image editor program called Photoshop, from Adobe Inc., I won a copy of the same program at the PC Expo in New York. I was able to sell one copy to a fellow in England by advertising it in the Usenet group *comp.graphics.apps.photoshop.*

I was planning a short trip to Poland recently, so I put up a message to the *soc.culture.polish* Usenet news group, asking for travel tips. Within a few days I received a dozen answers from Poland, Canada and the USA. I was told about ATM machines (just a few and geared to local banks), taxi cabs (if you hail one on the street you will be ripped off—call 919 for an honest driver), and the cost of gasoline for our rented car (1.50 zloty per liter). Where else could I get this type of information?

Chat Groups (IRC)

Similar to Usenet newsgroups, where information is exchanged in the form of pre-composed messages, a chat group lets you chat online with members of a group who are also currently online. Many people enjoy this form of interaction—there are even reports of marriages that started from an online conversation. There is quite a number of Internet Relay Chat (IRC) servers on the Internet. When you issue the command "irc" in a shell account, the irc program on your ISP's server will first try to find an available IRC server and provide statistics about the selected server. As an example, the following is a recorded irc session:

```
 .... >irc
*** Connecting to port 6667 of server irc.sdsc.edu
*** Unable to connect to port 6667 of server irc.sdsc.edu: unknown host
*** Connecting to port 6667 of server irc.eskimo.com
Looking up your hostname...
Checking Ident
Found your hostname
No Ident response
*** Welcome to Newnet rlewart
*** If you have not already done so, please read the new user information
with /HELP NEWUSER *** Your host is irc.eskimo.com, running version nn-1.1
```

```
*** Your host is irc.eskimo.com, running version nn-1.1
*** I've been with the group since Thu Feb 13 1997 at 22: 26:29 PST
*** umodes available oilfuckdws, channel modes available mvspitonblk
*** There are 1342 users and 531 invisible on 58 servers
*** There are 67 operators online
*** 1034 channels have been formed
*** This server has 11 servers connected
*** Current local users:  639  Max: 708
*** Current global users:  1874  Max: 2233
```

At this point you can start conversing with the 1874 users in various groups. You can also ask for help by typing /HELP. All commands are preceded by the slash (/). For instance, typing /HELP will give you a list of available commands, as shown in Table 11-3.

Table 11-3: List of IRC Commands

/HELP				
!	:	ABORT	ADMIN	ALIAS
ASSIGN	AWAY	BEEP	BIND	BYE
CD	CHANNEL	CLEAR	COMMENT	CONNECT
CTCP	DATE	DCC	DEOP	DESCRIBE
DIE	DIGRAPH	DISCONNECT	DMSG	DQUERY
ECHO	ENCRYPT	EVAL	EXEC	EXIT
FLUSH	FOREACH	HELP	HISTORY	HOOK
IF	IGNORE	INFO	INPUT	INVITE
JOIN	KICK	KILL	LASTLOG	LEAVE
LINKS	LIST	LOAD	LUSERS	ME
MLOAD	MODE	MOTD	MSG	NAMES
NICK	NOTE	NOTICE	NOTIFY	ON
OPER	PARSEKEY	PART	PING	QUERY
QUIT	QUOTE	RBIND	REDIRECT	REHASH
RESTART	SAVE	SAY	SEND	SENDLINE
SERVER	SET	SIGNOFF	SLEEP	SQUIT
STATS	SUMMON	TIME	TIMER	TOPIC

Table 11-3: List of IRC Commands (continued)

TRACE	TYPE	USERHOST	USERS	VERSION
WAIT	WALLOPS	WHICH	WHILE	WHO
WHOIS	WHOWAS	WINDOW	XECHO	XTYPE
basics	commands	etiquette	expressions	intro
ircll	menus	news	newuser	rules

You can find more about a command by typing:

```
/HELP <command>
```

You sign on, select a group of interest by typing /LIST, then type /JOIN to join that group and start exchanging short messages with the group participants. Each participant is identified by a nickname. You can then address this particular person or the entire group.

Of course, seeing is believing. If you think you're chatting with a 20-year-old woman, there is no guarantee that it isn't a 60-year-old man. So a word of caution—try to stay anonymous unless you are sure that people you converse with are who you think they are.

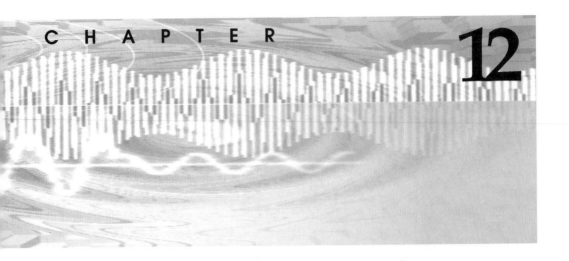

C H A P T E R **12**

A Look Inside a Modem— Modem on a Chip

Until now, we have discussed the modem as a separate subsystem—a black box located between the computer/terminal and the telephone line or some other transmission medium. In this chapter we will look in more detail at the individual circuits and their functions inside that black box.

The current trend in miniaturization, by means of very large scale integration (VLSI), led to the development of single and multiple chip sets of integrated circuits, which execute most modem functions. Adding a few discrete components results in an inexpensive single-board modem. Modems based on VLSI circuits are not only smaller and less expensive than earlier modems using discrete components, but they perform more functions and provide better transmission performance. This is due to the implementation—onto VLSI chips—of complex circuits such as adaptive equalizers and error correction and data compression, which, because of economic considerations, were omitted in earlier modem designs.

Rather than develop separate chip sets for various protocols and features, the trend in today's chip designs is to produce a more or less universal modem chip set. The modem designers then start with this universal chip set and customize it by choosing the set of features for each modem implementation.

There are three basic approaches to the design of modem chips—the first two are only of historical significance. The first and simplest approach, pursued in the early 1980s—was to integrate the earlier analog designs, which were implemented with discrete components, by means of analog ICs, such as operational amplifiers and comparators, and retain analog filters for frequency discrimination. The disadvantage of this approach was that all problems inherent in analog design, such as sample variations, were retained. This approach has therefore been abandoned, except for some low-volume proprietary circuits.

The second approach used in the mid 1980s was to perform digital signal processing (DSP) with a general-purpose microprocessor and with programs stored in ROM. This approach, because of its high cost-per-unit, is used only for low-volume proprietary circuits.

The third and current approach, started in the early 1990s, is to design a specialized digital chip set and provide it with all appropriate modem functions. This is the standard approach used today by major modem chip manufacturers such as Rockwell, Motorola and Lucent. Large production runs and competition drove the price of a 33,600 bps modem that was equipped with a specialized modem chip set down to less than $150.

The question that arises is, how are the analog signals, which are transmitted over the telephone lines or other analog facilities, handled by the digital modem circuits? What the modem chip designers do is to provide both analog-to-digital (A/D) and digital-to-analog (D/A) converters as interfaces between the digital signal processors and the analog world. Many analog functions, such as frequency filtering and modulation, can be performed more easily and with full repeatability in the digital domain.

Multiplication of a transfer function in the digital domain is equivalent to analog frequency filtering. Analog amplitude or frequency modulation is equivalent to digital convolution—which is a special type of multiplication for digital signals. Similarly, passing an analog signal through a low-pass filter having a cutoff frequency of F_c Hz is equivalent to the digital function of sampling an analog signal at the sampling rate of $2 \times F_c$.

Most of this chapter describes a popular modem chip set from Rockwell International Corporation. The chip set is representative of modem chip set designs from other modem chip manufacturers. First we look at the basic characteristics of the chip set and break down its functions in terms of individual chips. We will also look at how the chip set interfaces to the rest of the modem. For a better understanding of its operation, we will break down the chip set into functional building blocks and see how all those blocks are combined into a working modem. The following section lists the AT Commands that are accepted by this chip set. Finally, we will look at how a modem can be designed and tested with a mathematical simulation.

Basic Characteristics of the RC Chip Set

We will use the RC-series modem chip set, which is a two-chip or three-chip implementation of a 28,800, 33,600 and 56,000 bps modem from a major modem chip manufacturer, Rockwell International Corporation. The three digit number following the RC designation, such as RC288, indicates the maximum transmission speed in hundreds (RC288) or in thousands (RC56) of bits per second. This particular chip set is a good example of modern trends in VLSI-based modem technology.

Many modem manufacturers use the RC chip set as their building blocks, while other modem manufacturers use similar designs from other chip manufacturers. A modem chip set provides the processing core of the modem. The original equipment manufacturers (OEM) add a crystal, a few discrete components, and a digital access arrangement (DAA) interface circuit to complete the modem design. The RC chip set is a mixture of digital and analog functions apportioned to the digital and analog sections. The set supports and implements the protocols shown in Table 12-1 in either synchronous or asynchronous transmission mode:

Table 12-1: Protocols Supported by the RC chip set

Protocol	Transmission Speed (bps)
Bell 103	0 - 300
Bell 212A	1200
V.21	0 - 300
V.22	1200
V.22 bis	1200, 2400
V.23	1200Tx/75Rx, 75Tx/1200Rx
V.32	4800, 9600
V.32 bis	4800, 7200, 9600, 12,000, 14,400
V.FC	14,400, 16,800, 19,200, 21,600, 24,000, 26,400, 28,800
V.34	2400, 4800, 7200, 9600, 12,000, 14,400, 16,800, 19,200, 21,600, 24,000, 26,400, 28,800
V.34+	33,600 and all V.34 speeds
56K	56000Rx, 28800Rx and all V.34 speeds

Other basic characteristics of the RC chip set are as follows:

- 2-wire full-duplex operation
- Adaptive and fixed compromise equalization
- Auto/manual answer mode
- Auto/manual dial mode
- Tone or pulse dialing
- Call-progress tone detector
- Multiple test modes
- Variable character length in asynchronous mode, 8-11 bits
- Interface: RS-232-C functional, TTL electrical
- Supply voltages: +5 V. +12 V, -12 V
- Power consumption: 2 - 3 Watts

Functional Description of the RC chip set

The Rockwell full-featured RC chip set partitions the modem functions into three VLSI chips. There are also limited-feature implementations of the chip set that have only one or two separate chips. They then share some of the functions of the three-chip set. The three chips used in the full implementation are as follows:

1.   Modem Data Pump (MDP)
2.   Microcontroller (MCU)
3.   Compression Expansion Processor (CEP).

Figure 12-1 shows how the three chips interact. There is also a low-cost version of the modem configuration that only requires the first two chips. In the low-cost version, some of the modem functions are transferred to the host computer.

Modem Data Pump (MDP)

The modem data pump (MDP) can be thought as a versatile analog-to-digital (A/D) and digital-to-analog (D/A) converter. It generates analog data, fax, and voice signals required for transmission over analog telephone facilities. The MDP also converts the received analog data into digital signals for further processing by the other two chips and by the host computer.

The MDP performs the complex modulation algorithms described in the CCITT and ITU-T protocols. The MDP in this specific chip uses a basic clock frequency of 40.32 MHz, which is then subdivided to provide clocks for various modem functions. As a data modem, the MDP can operate in full-duplex synchronous and asynchronous modes at line rates of up to 28,800, 33,600, and 56,000 bps, depending on the specific chip set (RC288, RC336 or RC56). Using

Figure 12-1 Modem chip set—MDP, MCU, and CEPt

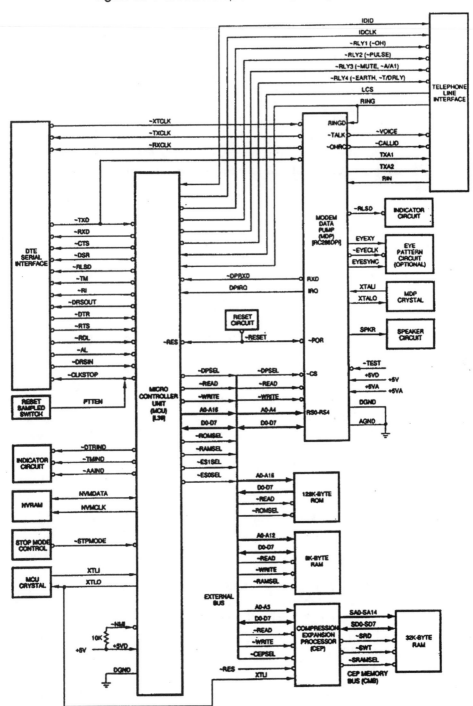

V.34 modulation, the MDP can adjust the internal adaptive equalizer to compensate for attenuation and delay distortion of the telephone line. The MDP can connect at the highest data rate that the transmission channel can support, with automatic fallback if the line conditions worsen.

As a fax modem, the MDP supports Group 3 fax at send-and-receive speeds of 14,400, 12,000, 9600, 7200, 4800, and 2400 bps. The optimum speed for both data and fax transmission is determined during the initial handshaking sequence and is adjusted up or down as the line conditions change.

The MDP also supports AT commands for modems that function in voice mode for answering machines and voice mail boxes.

Microcontroller (MCU)

The MCU performs error detection and correction, MNP 10 control, fax class 1 and class 2.0, DTE, and host interface functions. If you send a specific AT command to the modem, the MCU interprets it and sends appropriate instructions to the MDP. The MCU connects to the host computer via a serial interface or parallel microcomputer bus, depending on the installed MCU firmware. In the parallel interface operation, the MCU can connect to a PC-CARD (formerly called a PCMCIA card) connector directly, or with a Rockwell PCMCIA Interface Control Adapter (PICA) device and a PC-CARD plug.

The MCU connects to the MDP via dedicated lines and the external bus. The MCU external bus also connects to OEM-supplied ROM and RAM and, in the high performance configuration, to the CEP. The external ROM size is typically 128 to 256 kbytes, and the RAM size is 32 to 64 kbytes. In addition, non-volatile memory (NVRAM) can optionally be connected to the MCU over a dedicated serial interface. NVRAM is used for storing data such as user-selected modem configurations and telephone numbers.

Compression Expansion Processor (CEP)

The CEP included in the higher-priced modem implementations performs the dedicated data compression and expansion functions in V.42 bis/MNP 5 modes, providing maximum bi-directional throughput. The CEP host interface connects to the MCU external bus, and the CEP external memory bus connects to 32 kbytes of RAM. In modem implementations where the CEP is not provided, the compression/expansion functions are performed by software that runs on the host computer.

Modem chip set Interfaces

The RC modem chip set interfaces with the host computer and a number of external circuits. Each of these circuits requires one of the following interfaces.

Parallel Interface

A parallel 16550A UART-compatible host bus interface is provided for an internal modem implementation, but would be omitted in an external modem implementation. Eight data lines, three address lines, four DMA request/acknowledge lines, four control status lines, and a reset line are supported.

V.24 Serial Interface

A V.24/RS-232 logic-compatible DTE serial interface is supported. A clock stop signal is provided, which can be used to turn off the transmitter and receiver clocks to the DTE when in asynchronous mode.

Indicator Output Interface

Four indicator outputs are supported.

NVRAM Interface

A serial interface to an optional OEM-supplied non-volatile RAM (NVRAM) is provided. Data stored in NVRAM can have precedence over the factory-default settings. A 256 byte NVRAM can store up to two user-selectable configurations and up to four 35-digit dialing strings.

Speaker Interface

A speaker output, controlled by AT commands, is provided for an optional OEM-supplied speaker circuit.

External Bus Interface

The MCU external bus connects to the MDP and to OEM-supplied ROM and RAM. In the high-performance configuration, the external bus also connects to the CEP. The non-multiplexed bus supports eight bi-directional data lines and 17 address output lines. Read Enable, Write Enable, and Chip Select outputs are also supported.

Telephone Line Interface

MCU provides four relay control outputs to the line interface. These outputs may be used to control relays such as off-hook, pulse, mute, and talk/data. The relay outputs can also be used to drive Caller ID and voice relays. The MCU accepts the ring signal and the line-current sense signal from the phone line interface. The telephone line interface is shown in Figure 12-2.

Figure 12-2 Telephone line interface

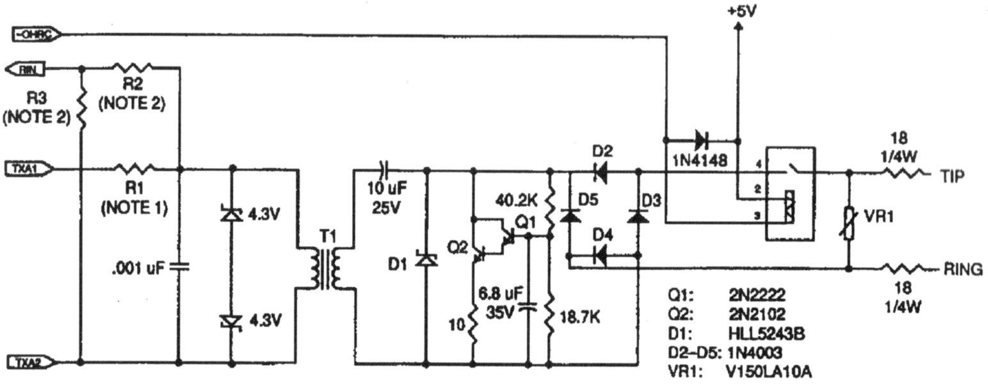

NOTES:
1. CHOOSE R1 VALUE TO OBTAIN 600 OHM INTERNAL IMPEDANCE.
2. CHOOSE R2 AND R3 VALUE TO OBTAIN A 6 DB LOSS FROM TIP AND RING TO RIN.

Business Audio Interface

For a voice-modem implementation, the MCU provides three outputs—one to select the volume control, one to control the volume up and down, and one to set the volume increment.

Eye-Pattern Generator Interface

Eye-pattern data, clock, and synchronization interface signals are provided. An external eye-pattern generator circuit can be added in order to observe modem performance as a function of line impairments. The purpose of an eye pattern is to qualitatively evaluate a modem by sending a random sequence of bits. The eye pattern, ideally a rectangle, shows interference between adjacent bits. If the eye pattern shows overlapping, then error-free detection is not possible. Eye-pattern testing of transmission facilities is discussed in Chapter 17.

Modem Building Blocks

Depending on the configuration selected by the user, the modem can assume different personalities. Figures 12-3 and 12-4 show detailed block diagrams of the transmitting and receiving sections of a modem that is configured for asynchronous operation. The individual blocks are discussed next.

Figure 12-3 Transmitting section of a modem

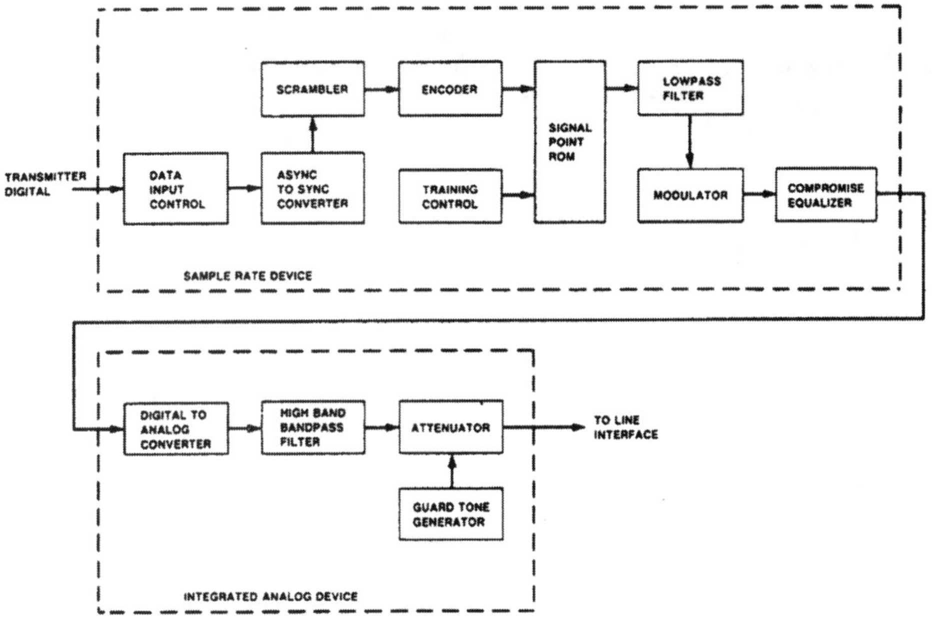

Modem building blocks perform separate well-defined functions. In the prehistoric days of telecommunications (the 1970s), before the development of VLSI chips, each of the following building blocks performed a specific function and was associated with a chip or even a circuit board. Today many of these blocks are combined into a single chip.

A modem has basically two sections: a digital signal processor (DSP) section, and an integrated analog device section. The high-speed DSP section is part of the MDP, MCU and CEP chips, while the integrated analog device section would be implemented on the MDP chip.

Figure 12-4 Receiving section of a modem

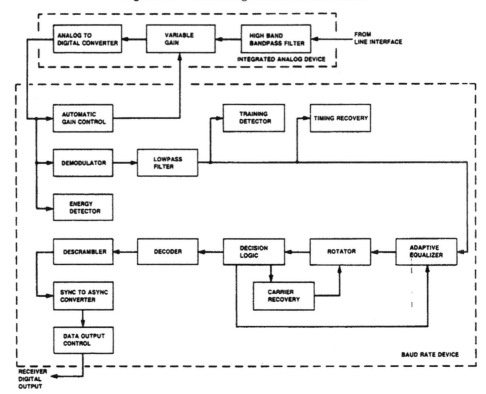

Digital Signal Processor

The digital signal processor (DSP) is a specialized 16-bit microprocessor designed for intensive numeric high-speed applications such as those required by A/D and D/A conversion, adaptive equalization, scrambling, and data conversion. One of its subsections is the *sample rate device*, which performs tasks at the rate at which the data signal is sampled. The other subsection is the *Baud rate device*, which performs tasks associated with groups of bits contained in each transmitted or received signal element.

The assorted modem characteristics, such as those associated with a 1,200 bps 212A-type modem, or with a 28,800 bps V.34-type modem, are stored in the internal ROM; the specific modem configuration, selected by the user, is stored in the NVRAM. The signal processor uses 16-bit words and a 32-bit accumulator to perform internal computations.

The basic building blocks of the signal processor are shown in Figure 12-5.

Figure 12-5 Signal processor building blocks

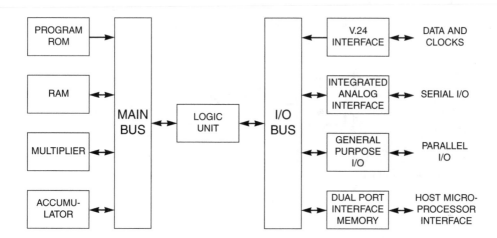

Integrated Analog Device

This block performs a multitude of functions that were earlier associated with discrete analog components. In discrete designs, such circuitry would occupy up to 80% of the board size; the use of VLSI circuits has contributed here most to modem miniaturization.

One of the main functions of this block—frequency filtering—can be performed in two ways. It can be done either by means of a switched-capacitor technology, which allows for the substitution of small integrated circuit capacitors for resistors, or by digital filtering.

The integrated analog device also performs many functions related to interfacing the modem to the transmission facility and the serial interface in the computer. The interface to a telephone line includes the data access arrangement (DAA) circuitry required by the FCC to protect the telephone network from electric shocks caused malfunctioning or poorly-designed equipment. Figure 12-6 shows the assorted functions located within the integrated analog device.

Data Input Control

This circuit controls the data flow from the computer via the RTS and DTR serial interface leads. In the test mode, the internal test generator would feed a test pattern at this point without regard to the status of the RTS and DTR leads.

Figure 12-6 Analog functions performed by the modem chip set

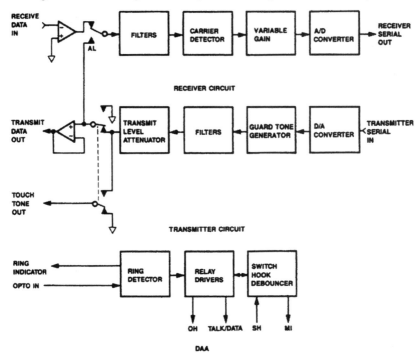

Asynchronous to Synchronous Converter

This converter is required only in the synchronous mode of modem operation. When the modem is in asynchronous mode, this circuit is bypassed. The function of the circuit is to fit the gaps in asynchronous data flow so that the remaining modem blocks can operate under clock control, as if the data would be flowing continuously.

Scrambler

The purpose of this circuit is to assure a pseudo-randomness of the data stream, even if the data consists of a long string of 1s and 0s. The receiving terminal is namely dependent on the randomness of received data to extract the timing information. The scrambling algorithm is part of a specific CCITT/ITU-T or Bell protocol, and must be the same for the receiver and transmitter. The scrambler has nothing to do with data security—the algorithms are public, and are a part of appropriate protocols.

The scrambling algorithm and the scrambling and de-scrambling circuits are shown in Figure 12-7. They would differ depending on the protocol used. The square boxes are shift registers, each introducing a one-bit delay. By inspection, or by writing a short program, one can see that a string of 1s will be translated into a pseudo-random sequence, but a string of 0s will remain a string of 0s after passing through the scrambler. The solution is to add a *kick* circuit that throws in a 1 after a sequence of 0s of predetermined length. The scrambler circuit is used in all but the 300 bps operation, because FSK modulation does not require timing recovery from the data stream.

Figure 12-7 Scrambler/Descrambler

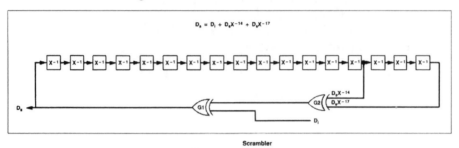

Encoder and Signal Point ROM

The encoder arranges successive data bits in groups of two (dibits), four (quadbits), or more, depending on the transmission speed and the modulation protocol. Then, depending on the value of each group of bits, the signal ROM assigns the appropriate amplitude and phase values to each signal. As discussed in Chapter 3, the assigned amplitude/phase values are based on constellation diagrams that are associated with the appropriate modulation protocols.

Modulator and Low-Pass Filter

The modulation rules depend on the protocol chosen. The now-obsolete Bell 103-type protocol uses frequency shift keying (FSK), with two frequencies for the originating modem and two frequencies for the answering modem. The 212A protocol uses differential phase shift keying (DPSK) modulation with carrier frequencies at 1200 Hz and 2400 Hz, while the V.22 protocols and most modern high-speed protocols use the combined phase/amplitude quadrature amplitude modulation (QAM).

Encoding of one or more bits per signal sample is made according to the appropriate protocol. Figure 12-8 shows the complex DPSK and QAM modulators. The corresponding demodulator in the receiving modem is the mirror image of the modulator. Both the modulation and demodulation are performed in the time domain on digital samples of the data signal.

Figure 12-8 DPSK/QAM modulator

$$z(t) = x \cos \omega_c t - y \sin \omega_c t$$

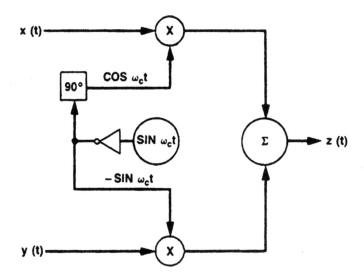

The signal sampling rate required for time-domain filtering is typically 7200 samples per second, which is equivalent to a low-pass filter with 3600 Hz cutoff frequency (slightly more than the bandwidth of a voice-band channel). The sampling process generates repeated harmonics (multiples of a frequency) of 3600 Hz. To avoid ringing (caused by these harmonics), which could result in inter-symbol interference, the signals must be filtered by an analog low-pass filter. A so-called raised cosine filter, with the shape similar to the shape of a cosine, is chosen as the optimum filter. Since the QAM signal consists of x and y components, there are actually two filters provided to shape each component. In modern designs, filtering can be performed in the time domain, rather than in the frequency domain, by convoluting the two signals.

Band-Pass Filter and D/A Converter

The digital signal is first filtered by convoluting its digital values with the appropriate functions corresponding to a band-pass filter. The signal is then transformed by a D/A converter into an analog signal, which—after band-limiting so it does not interfere with signals modulated on adjacent carrier frequencies—is sent over the telephone line to the remote modem.

Compromise Equalizer

The compromise equalizer pre-distorts the signal to be sent over the telephone line or some other transmission facility, with a mirror image of the expected attenuation and delay distortion. The fixed-form equalizer compensates for the average and not for the actual distortion of the transmission facility, as the actual transmission path is not known in advance.

Adaptive Equalizer

Most of the building blocks in the receiving section of a modem are equivalent to similar blocks in the transmitting section. A unique building block found only in the receiving section of the modem is the adaptive equalizer.

In the receiving section of the modem, the residual transmission channel distortion is compensated by means of an adaptive equalizer. The residual transmission channel distortion is equal to the actual distortion minus the compromise equalizer in the transmitting modem.

An adaptive equalizer, often referred to as a transversal filter, adjusts itself during the handshake sequence between the two modems. The equalizer behaves like a tunable filter, which corrects for the transmission channel attenuation and delay distortion by producing a complementary response. The flattening of the combined attenuation and delay characteristic reduces the inter-symbol interference during data transmission.

Figure 12-9 shows a block diagram of an adaptive equalizer with the individual delay and gain blocks. A signal passes through a series of delay blocks, each of which are T/2 seconds in duration, where T is the time between signal elements in seconds. Part of the signal after phase-reversal is fed back into each block to optimize the final shape. The variable-gain blocks, C_0 through C_{12}, that are associated with each delay block, are adjusted by the decision logic circuit, which tries to optimize the equalizer while receiving a known sequence of bits during the handshake before the actual data is transmitted. The gain adjustments made during the handshake period at the beginning of the transmission remain set for the remainder of the call.

Figure 12-9 Adaptive equalizer

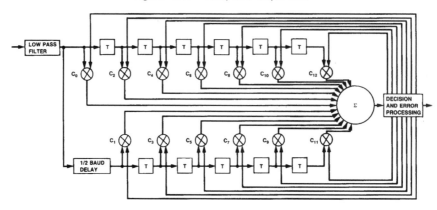

Tone Dialer

The tone dialer, which is shown in more detail in Figure 12-10, sends the standard dual-frequency tones as follows:

Digit	Frequency Pairs in Hz
0	941 1336
1	697 1209
2	697 1336
3	697 1477
4	770 1209
5	770 1336
6	770 1477
7	852 1209
8	852 1336
9	852 1477
*	941 1209
#	941 1477

The tones are generated first as pairs of 8-bit numbers, and then are changed into analog tones by a D/A converter.

Call-Progress Tone Detector

This circuit, shown in Figure 12-11, detects the dial tone, ringing, and busy signals. This information is then passed to the host computer by the communications software. This information is required when starting a new dialing attempt if the phone is busy. It is also displayed on the screen during the calling process. The main components of the tone-detector circuit are an automatic-gain control block, and a band-pass filter that detects signals between 345 and 635 Hz, where the call progress tones are located. The smoothing low-pass filter then averages the detected energy, while the output-control circuit decides what kind of call-progress tone was detected.

Figure 12-10 Touch-tone dialer

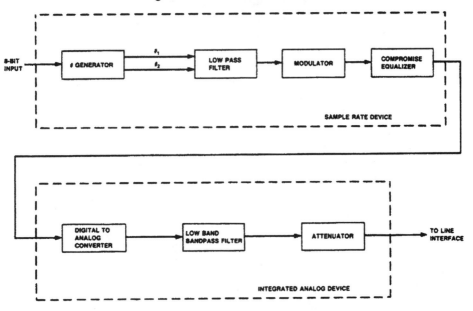

Figure 12-11 Call progress tone detector

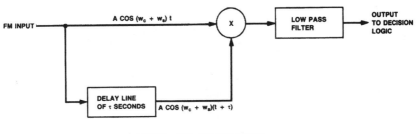

w_c = CARRIER = 1270, 1070, 2225, OR 2025

w_s = SHIFT FREQUENCY = ±100 Hz

$w_c \tau = \frac{\pi}{2} = 2n\pi; n = 0, 1, \ldots$

Data Access Arrangement (DAA)

A phone-line-protecting circuit is required by the Federal Communication Commission (FCC) to be included in each device that is electrically connected to the telephone network. This part of the modem contains high-voltage protection circuitry such that if the modem is inadvertently connected to a power line, it will not put 110 or 220 Volts on the telephone line and damage associated equipment or electrocute a person.

Complete Modem

The two or three chips that comprise the RC modem chip set, along with a few external components, are combined to make a high-quality modem. Thanks to digital design, the modems based on modem chip sets have low variability from sample to sample in their basic characteristics, such as the minimum signal-to-noise ratio required for errorless data transmission for a given type of modulation.

Figure 12-12 shows a general block diagram of a modem using the RC chip set. Figure 12-13 shows a nearly-complete schematic of such a modem. In addition to the modem chips, a complete modem also requires a power supply capable of providing +5 V, +12 V, and -12 V. In this particular implementation, -12 V is derived from +5 V with a voltage regulator (VR1). The modem also needs a quartz crystal (Y1) for the master clock, a dual operational amplifier (Z1), a transformer (T1), a relay (K1), and other components for the data access arrangement (DAA) device.

Figure 12-12 Block diagram of a complete modem

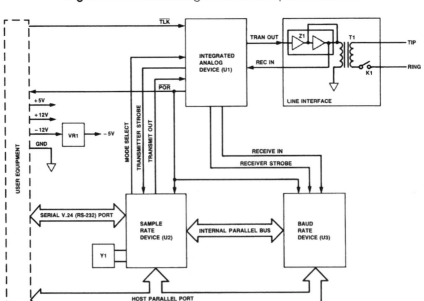

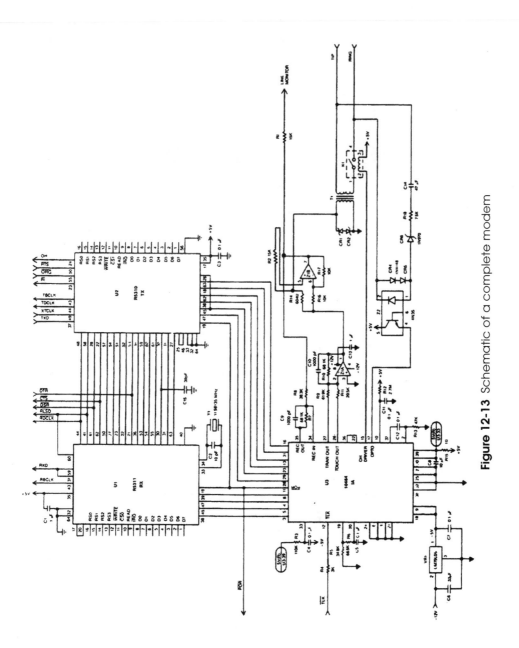

Figure 12-13 Schematic of a complete modem

The modem chip set supports both the CCITT/ITU-T V.24 and RS232-C serial interfaces. Functionally, the two interfaces are equivalent for most applications, but the V.24 does not specify the electrical interface and voltage levels. In a typical implementation of V.24, TTL voltage levels (0 V, +5 V) are used, while RS232-C uses bipolar levels, which are typically +12 V and -12 V.

Although data to and from the host computer is transferred via the serial RS232-C interface, the modem also provides a parallel bus interface to the host computer, which would be required in an internally-mounted modem. The parallel interface allows access to scratch memory in the signal processor, and gives it the capability to set the modem configuration and control assorted functions, independent of the serial data path.

AT Commands accepted by the Rockwell Chip Set

The Rockwell RC modem chip set supports data modem, fax class 1 and 2.0, MNP 10 voice and S-register commands, in accordance with selected options. Following is a list of these commands. You will find a subset of them implemented in the modem installed in your PC. Many of these commands are also described in more detail in Chapter 9.

Data Modem Commands

In response to the AT+FCLASS=0 command, the modem will accept the following basic AT commands. Default parameters support US/Canada operation. The upper-case character is the actual command; n is a numerical parameter variable.

A/	Re-execute command (does not require AT prefix)
A	Answer a call
Bn	Set CCITT or Bell mode
Cn	Carrier control
Dn	Dial (originate a call)
E	Command echo
Fn	Select line modulation
Hn	Disconnect (hang-up)
In	Identification
Ln	Speaker volume
Mn	Speaker control
Nn	Auto mode enable
On	Return to on-line data mode
P	Set pulse dial default

On	Quiet results codes control
Sn=xx	Write to the n-th S Register
Sn?	Read n-th S Register
T	Set tone dial default
Vn	Result code form
Wn	Error correction message control
Xn	Extended result codes
Yn	Long space disconnect
Zn	Soft reset and restore profile
&Cn	RLSD option
&Dn	DTR option
&Fn	Restore factory configuration (profile)
&Gn	Select guard tone
&Jn	Telephone jack control
&Kn	Flow control
&Ln	Leased line operation
&Mn	Asynchronous/synchronous mode selection
&Pn	Select pulse dial make/break ratio
&Qn	Asynchronous/synchronous mode selection
&Rn	RTS/CTS option
&Sn	DSR override
&Tn	Test and diagnostic mode
&V	Display current configuration and stored profiles
&Wn	Store current configuration
&Xn	Select synchronous clock source
&Yn	Designate a default reset profile
&Zn=x	Store phone number
+MS	Modulation select
%En	Enable/disable line quality monitor and auto retrain or fall back/fall forward
%L	Report line signal level
%Q	Report line signal quality
%TTn	PTT testing utilities
\Kn	Break control
\Nn	Operating mode
#CID	Caller ID detection and reporting

MNP 10 Commands

MNP 10 operation of the modem is supported by a set of MNP 10 commands.

)Mn	Enable/disable cellular power level adjustment
*Hn	Set link negotiation speed

-Kn	MNP extended services
-Qn	Enable fallback to V.22 bis/V.22
@Mn	Select initial transmit level
:E	Compromise equalizer enable

Fax Modem Commands

Fax functions operate in response to fax class 1 commands such as:

```
AT+FCLASS=1 or AT#CLS=1
```

or in response to fax class 2.0 commands such as:

```
AT+FCLASS=2 or AT#CLS=2
```

Fax Class 1 Commands

+FCLASS=n	Service class
+FAE	Data/fax auto answer
+FTS=n	Stop transmission and wait
+FRS=n	Receive silence
+FTM=n	Transmit data
+FRM=n	Receive data
+FTH=n	Transmit data with HDLC framing
+FRH=n	Receive data with HDLC framing

Fax Class 2.0 Commands

+FCLASS=n	Service class
+FCIG	Set the polled station identification
+FDT	Data transmission
+FET=n	Transmit page punctuation
+FDR	Begin or continue Phase C receive data
+FK	Terminate session
+FLPL	Document for polling
+FSPL	Enable polling

Fax Class 2 DCE Responses

+FCIG:	Report the polled station identification
+FCON	Facsimile connection response
+IDDCS:	Report current session
4FDIS:	Report remote capabilities
+FDTC:	Report the polled station capabilities
+FCFR	Indicate confirmation to receive
+FTSI:	Report the transmit station ID
+FCSI:	Report the called station ID

+FPTS:	Page transfer status
+FET:	Post page message response
+FHNG:	Call termination with status
+FPOLL	Indicates polling request

Fax Class 2.0 Session Parameters

+FMFR?	Identify manufacturer
+FMDL?	Identify model
+FREV?	Identify revision
+FDCC	DCE capabilities parameters
+FDIS	Current session parameters
+FDCS	Current session results
+FLID	Local ID string
+FPTS	Page transfer status
+FCR	Capability to receive
+FAA	Adaptive answer
+FBUF?	Buffer size (read only)
+FPHCTO	Phase C time out
+FAXERR?	Fax error value
+FBOR	Phase C data bit order

Audio and Voice Modem Commands

Voice-mode functions operate in response to voice/audio commands such as:

```
AT#CLS=8, AT#VBS=2 or AT#VBS=4
```

Audio modem functions operate in response to voice/audio commands such as:

```
AT#CLS=8 and AT#VBS=8 or AT#VBS=16
```

Sampling rate is determined by AT#VSR=11025 or AT#VSR=7200

Specific audio/voice commands

#BDR	Select bps rate
#CLS	Select data, fax or voice
#MDL?	Identify model
#MFR?	Identify manufacturer
#REV?	Identify revision level
#VBQ?	Query buffer size
#VBS	Bits per sample
#VBT	Beep tone timer
#VCI?	Identify compression method

#VLS	Voice line select
#VRA	Ring back go away timer (originate)
#VRN	Ring back never come timer (originate)
#VRX	Voice receive mode
#VSD	Enable silence deletion
#VSK	Buffer skid setting
#VSP	Silence detection period (voice receive)
#VSR	Sampling rate selection
#VSS	Silence detection tuner (voice receive}
#VTD	DTMF/tone reporting
#VTS	Generate tone signals
#VTX	Voice transmit mode

World Class (Non US/Canada) commands

World class functions are enabled by ATW commands such as:

*B	Display blacklisted numbers
*D	Display delayed numbers
*NCnn	Country select

Register Commands

These commands set or display the S registers stored in a modem's NVRAM, which control all modem parameters. The general format of these commands is:

ATSn=xx	Sets the register n to value xx, and
ATSn?	Displays the current value of register n.

Registers supported by the Rockwell chip set are as follows:

S0	Rings to auto-answer
S1	Ring counter
S2	Escape character
S3	Carriage return character
S4	Line-feed character
S5	Backspace character
S6	Maximum time to wait for dial tone
S7	Wait for carrier
S8	Pause time for dial-delay modifier
S9	Carrier-detect response time
S10	Carrier-loss disconnect time
S11	DTMF tone duration
S12	Escape-code guard time
S13	Reserved

S14	General bit-mapped options
S15	Reserved
S16	Test mode bit-mapped options (&T}
S17	Reserved
S18	Test timer
S19-S20	Reserved
S21	V24/general bit-mapped options
S22	Speaker/results bit-mapped options
S23	General bit-mapped options
S24	Sleep inactivity timer
S25	Delay to DTR off
S26	RTS-to-CTS delay
S27	General bit-mapped options
S28	General bit-mapped options
S29	Flash modifier time
S30	Inactivity timer
S31	General bit-mapped options
S32	XON character
S33	XOFF character
S34-S35	Reserved
S36	LAPM failure control
S37	Line connection speed
S38	Delay before forced hang-up
S39	Flow control
S40	General bit-mapped options
S41	General bit-mapped options
S42-S45	Reserved
S46	Data compression control
S48	V.42 negotiation control
S82	Break-handling control
S91	PSTN transmit attenuation level
S92	Fax transmit attenuation level
S95	Result-code messages control

Mathematical Modem Simulation

Modems can be designed and tested by connecting individual physical components and building circuits. An alternate way of designing and testing is by means of a mathematical modeling technique, which delivers similar results. Mathworks Inc. of Natick, Massachusetts, developed within their general-purpose mathematical simulation program called Matlab, a so-called Communications Toolbox. The series of programs included in the Communications Toolbox provide a user with a selection of simulated communication blocks, such as modulation converters that adhere to specific ITU-T V-series protocols, error control blocks with assorted error detection and correction algorithms, transmission channel blocks with noise and other impairments, input and output blocks, and measuring instruments. Even a simulated oscilloscope is included in the toolbox to observe the transmitted and received signals. Programs describing each block are written in the C language. Source code is included, which makes it relatively easy to modify and generate programs to satisfy specific transmission requirements.

Like every other programming tool, the Communications Toolbox can be of great value, but it can also be misused. Unless great care is taken in properly describing each block, incorrect results can appear following the garbage-in-garbage-out principle. Still, a well-simulated circuit can yield results in a fraction of the time it would take to build a hardware model. Minor changes can be accomplished by changing a few lines of code rather than by rebuilding a circuit. Figure 12-14 shows the library of modem blocks available for Matlab simulation. Each block corresponds to a program segment, which can be modified to fit the designer's exact specifications.

Figure 12-15 shows a display of a V.34 modem simulation with a simulated signal appearing on the simulated oscilloscope screen. The simulation can be started and stopped, results displayed, and parameters changed. It certainly beats working with a soldering iron!

Figure 12-14 Library of communications blocks relating to modem design

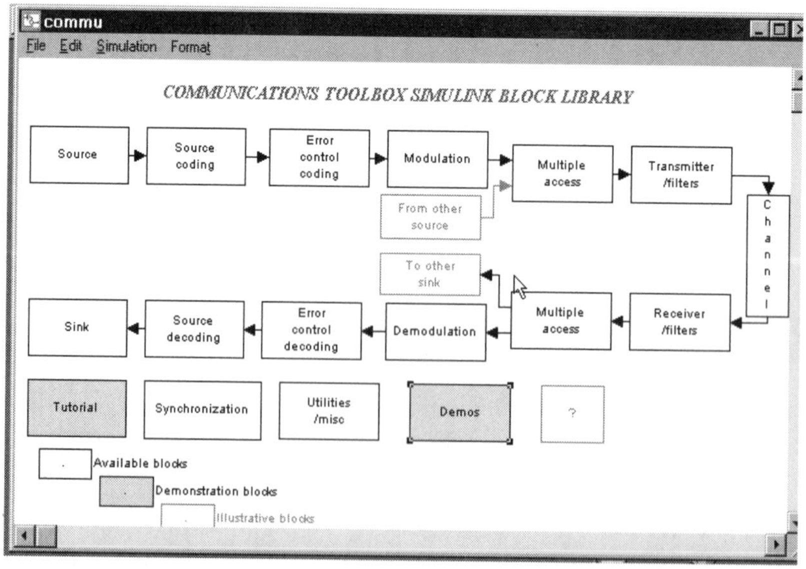

Figure 12-15 Mathematical simulation of a V.34 modem

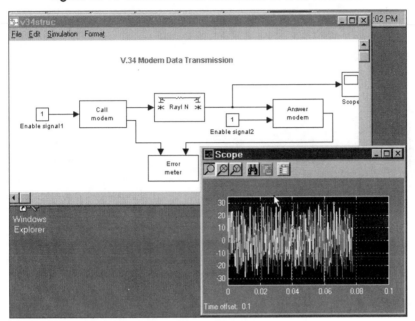

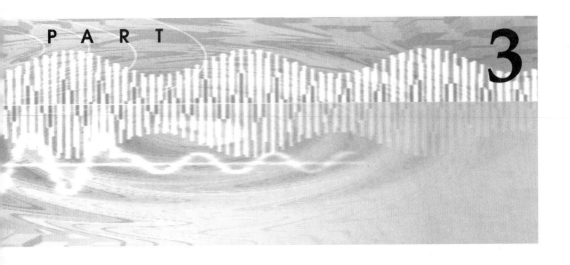

Specialized Modems

There are many types of modems other than those designed specifically for the personal computer market, which were discussed in previous chapters. Some of these modems are used only in the commercial and industrial market and fulfill the specific needs of large corporations. They often provide communication between micro, mini and mainframe computers. Many of these modems require special leased lines or broadband transmission facilities, which are expensive to build, lease and maintain, rather than the public switched telephone network (PSTN). Other special-use modems can be found in the commercial and PC markets where there is a need for higher transmission speeds or wireless communication.

Chapter 13 provides a historical perspective of commercial modem development and how voice-band transmission facilities for certain commercial applications are designed. It also discusses signaling and battery options used on transmission lines that are leased for commercial applications. The remainder of Chapter 13 describes voice-band commercial modems currently in use.

Chapter 14 covers wide-band and wireless modems for both commercial and personal computing applications, including ISDN, cable, wireless, and cellular modems. These four types of modems are finding increased applications in both commercial and personal computing markets.

Chapter 15 describes short-haul modems, often called limited distance modems, that are frequently used for data exchange between nearby locations. Short-haul modems include popular control devices that use house AC power lines to transmit their signals.

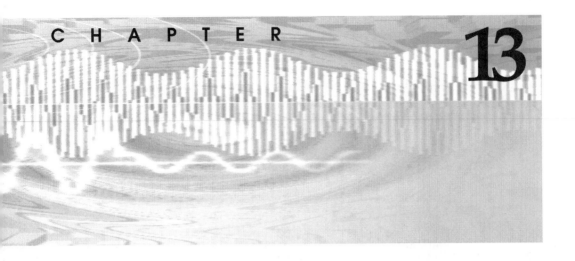

CHAPTER 13

Commercial Voice-Band Modems

The huge personal computer market, which started in the mid-1980s, drives the development of new modem features and pushes their prices down. This is not the case with the industrial and commercial markets. The large cost of conversion often keeps the existing network implementations without change for many years. As the saying goes, if it works, don't fix it. Therefore you can find many modems in the commercial market that are no longer in use in the PC market due to their obsolescence. High-quality modems required for the industrial and commercial markets are being produced in much smaller quantities than those made for the PC market. Small production runs and high distribution and maintenance costs result in prices that are several times higher than for equivalent PC modems.

PSTN and Private-Line Equalization

There are three types of communication services that can be obtained from a common carrier—your local telephone company such as General Telephone, Southwest Bell and Pacific Bell—or a long-distance provider such as AT&T or MCI. These three basic services are shown in Figure 13-1 and are part of the public switched telephone network (PSTN).

The first type of service is a switched—meaning dial-up—connection, where charges are computed by the telephone company depending on the time of day, distance and duration of the call. The second type of service uses leased non-switched circuits between two or more fixed locations. The non-switched circuit—meaning it is always connected to the modem—could also be privately owned, in particular if it covers only a short distance. For example, a span between two buildings or a university campus would be a candidate for leased or privately-owned transmission facilities. The third type of service is a hybrid service, which consists of a switched circuit leading to a node, which is then connected through leased lines to a second node. An alternate type of hybrid service would be a leased conditioned circuit that leads from a customer's premises to a local telephone office, the so-called *data access line*, from where the call would access the public telephone network.

Figure 13-1 Common carrier services

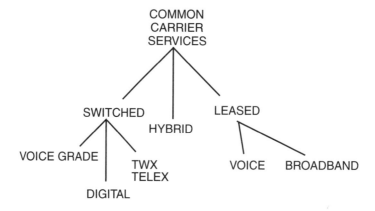

A dial-up connection made over the public switched telephone network (PSTN), the separately-billed call using Direct Distance Dialing (DDD), or the bulk billed calls using Wide Area Telecommunications Service (WATS), has the enormous advantage of flexibility and reliability. A call can be made to any other telephone subscriber in the world and, if a section of the telephone network breaks down, there is enough redundancy in the system to quickly reroute the call.

The disadvantage, especially on a data call, is the unpredictability of the transmission quality. The only guarantee that a common carrier or long-distance provider will make for a dialed voice connection is that the two parties can communicate freely by talking to each other, and that the perceived quality of the connection will be satisfactory. There is no specific quantitative guarantee of limits on attenuation or delay distortion, or freedom from noise bursts caused by fading of microwave links and switching transients or delay on a satellite circuit.

Attaching transmission measuring test equipment to a switched telephone line that is used for data communications will typically show a wide distribution of noise and distortion, even on calls that are dialed between the same two locations, but are made at different times. The variability is caused by the different routes that are made for each call, depending on traffic conditions, the time the call is made, and status of transmission facilities.

Other disadvantages of using a switched service for data communications is a relatively long time to connect, typically 2 to 5 seconds, and the presence of network signals during the connection. In particular, calls made in Europe have periodic beeps that indicate charging units, which may interfere with data transmission. Also, only two-wire connections are available on the switched public network, although this may change in the near future. Still, for personal computer users, a dialed connection is a satisfactory and inexpensive solution for data communications. A big advantage for North American nonbusiness telephone subscribers is that there is no charge on many local calls in U.S. and in Canada.

If a corporation has a number of fixed field locations between which frequent data calls are made, use of privately-owned or leased transmission facilities is often an economic advantage. Leased-line services are available from local telephone companies and long-distance carriers, including AT&T, MCI, and others. Leased lines consist of local loops that connect the subscriber to the nearest central office, and to inter-exchange channels, which connect central offices.

The advantage of a privately-owned or leased point-to-point or multiple-point transmission facility is its complete predictability—each call is always routed the same, since the call does not need to be switched over various routes. In addition, the user can select from a number of common carrier offerings that, under tariffed schedules, guarantee specific transmission parameter values. If users, after installing a leased transmission link, find that it does not meet the published specifications, then they can notify their common carrier and the facility will be brought up to the guaranteed standards.

The main disadvantage of a leased or private line is the high initial and monthly cost. The initial cost will, however, be quickly amortized for high-usage links. Other disadvantages include less flexibility and a significant time-to-repair in the event of problems. In applications where reliability is of utmost importance, the user may want to lease two circuits instead of one, and automatically switch to the alternate circuit in case of trouble.

Following is a list of the primary types of leased and privately-owned line facilities that are used by corporations and other commercial businesses for data communications:

- Leased voice-grade lines with or without line conditioning—used with two- and four-wire modems, for both asynchronous and synchronous transmission.
- Wide-band analog lines—used for telephone channel multiplexing and for high quality audio broadcasting. One of these offerings is called Telpac. This type of service has only a limited application for data transmission.
- Dataphone Digital Service (DDS)—typically used for 56 kbps transmission and for multiplexing several lower-speed data channels.
- Fiber optic links—used for computer-to-computer and other high-speed applications.

Voice-Band Transmission Facilities

Table 13-1 lists leased voice-band transmission facilities offered by common carriers under applicable tariffs. The lowest-priced offering is the so-called 3002 channel, which has characteristics similar to a typical dialed connection. The higher-priced offerings are called C1, C2, C4 and D1. The guaranteed characteristics are implemented by common carriers using line conditioning, which consists of inserting fixed attenuation and delay equalizers that are selected after measuring the non-equalized line.

Telephone company engineers have computer programs at their disposal that will select the best possible combination of fixed equalizers to compensate for the measured delay and attenuation distortion of a telephone circuit and to satisfy a given tariff. In cases where a transmission facility cannot be equalized to the tariffed requirements, the common carrier will substitute a different transmission facility that can be brought up to specification. It should be noted that leased circuits are, in general, a part of the public switched telephone network. The circuits are just patched through in various telephone offices, instead of going through switching equipment. Therefore leased lines are subject to the same amplitude and frequency restrictions as any switched lines. The purpose of these restrictions is to protect users of the PSTN from mutual interference.

It is interesting to compare the guaranteed parameter values shown in Table 13-1 with the typical expected values of transmitted and received levels, line losses, signal-to noise ratios, and threshold ranges of signal-to-noise levels required by various types of modems for error-free operation. Table 13-2 shows the typical parameter values that can be found on voice-grade telephone lines.

Table 13-1: Leased Voice-Band Transmission Facilities

	Non-Conditioned 3002 Channel		With C1 Conditioning		With C2 Conditioning		With C4 Conditioning		With D Conditioning
Frequency Range in Hertz (Hz)	300-3000		300-3000		300-3000		300-3200		
Attenuation Distortion (Net Loss at 1000 Hz)	Frequency Range	Decibel Variation	Frequency Response	Decibel Variation	Frequency Response	Decibel Variation	Frequency Response	Decibel Variation	
	300-3000	−3 to +12	300-2700	−2 to +6	300-3000	−2 to +6	300-3200	−2 to +6	
	500-2500	−2 to +8	1000-2400	−1 to +3	500-2800	−1 to +3	500-3000	−2 to +3	
			300-3000	−3 to +12					
Delay Distortion in Microseconds (µs)	Less than 1750 µs from 800 to 2600 Hz		Less than 1000 µs from 1000 to 2400 Hz Less than 1750 µs from 800 to 2600 Hz		Less than 500 µs from 1000 to 2600 Hz Less than 1500 µs from 600 to 2600 Hz Less than 3000 µs from 500 to 2800 Hz		Less than 300 µs from 1000 to 2600 Hz Less than 500 µs from 800 to 2800 Hz Less than 1500 µs from 600 to 3000 Hz Less than 3000 µs from 500 to 3000 Hz		
Signal to Noise (dB)	24		24		24		24		28
Non-Linear Distortion Signal to 2nd Harmonic (dB)	25		25		25		25		35
Signal to 3rd Harmonic (dB)	30		30		30		30		40
Maximum Impulse Noise	15 counts in 15 minutes at 90 dBrn								
Type of Service	Point-to-Point (two points) or Multipoint						Point-to-Point (two or three points)		
Channel Mode	Half of Full Duplex								
Local Loop Termination	Two or Four Wire								
Maximum Frequency Error	±5 Hz								
Maximum Bit Error	Approximately 1 bit error per 100,000 bits transmitted								

Table 13-2: Typical Transmission Parameters on a Voice-Grade Facility

	Leased	DDD Local	DDD Long Distance
Transmit Level	0 dBm	-10 dBm	-10 dBm
Receive Level	-15 dBm	-25 dBm	-27 dBm
Line Loss	15 dBm	15 dBm	20 dBm
S/N Ratio	40 dB	39 dB	35 dB

The leased circuits are connected either on a point-to-point basis or are part of a multi-point network. The two methods of operation are shown in Figures 13-2 and 13-3.

Figure 13-2 Point-to-point network

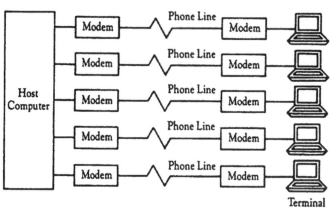

A series of point-to-point networks
where each terminal is linked to the host
computer by its own phone line can be
very expensive.

Figure 13-3 Multi-point network

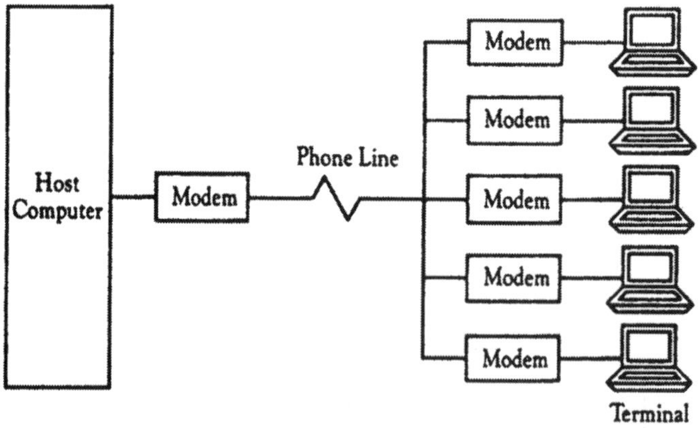

One solution for reducing line and
modem costs is a multipoint network
where several terminals share the same
phone line. In this case, a multipoint
configuration eliminates four phone
lines and four modems.

Signaling and Battery Options

At the customer premises, a telephone loop connected to the PSTN, in
addition to being able to carry voice or data information, also accepts
and carries signaling information, provides current from the central-
office, and sends ringing current, when a subscriber is called by another party. All these functions may or may not be provided on a leased

transmission facility and must be requested at the time the circuit is being set up.

The simplest leased-line circuit is the so-called *10 dB line*, which provides 10 dB of loss between both ends and does not provide battery or signaling options. The battery is required for telephones that still use the old-fashioned carbon-based microphone or its equivalent, and for a pulse dialer and some line-powered modems, but is not necessary for a modem-to-modem connection. Ringing current and signaling capabilities may or may not be needed, depending on the particular setup.

When setting up a private network using leased-line transmission facilities, one should always consider how the connection between modems will be established. The problem does not arise in the personal computer market where a modem behaves like a regular telephone set and users can dial the remote terminal from their computer and establish a connection. Many commercial modems used on leased circuits do not have dialing capabilities or cannot automatically answer a call from a remote modem. A separate dialing unit may then be required. Another option that may be needed for certain applications, typically for limited-distance modems or test circuits, is a metallic connection, which assures DC continuity. Circuits satisfying one or more of these requirements can be leased from a common carrier if both ends are within the same telephone exchange.

Listing of Commercial Voice-Band Modems

As with the modem market for personal computers, the commercial modem market offers many choices and options, albeit at a higher cost. What you will often find in the commercial market are 1980-vintage modems that satisfy specific needs and are not being replaced. Many of these modems are still being manufactured for that market and can be found in current equipment catalogs.

Commercial modems cover a whole gamut of applications. They range from the typical modem use, which is the exchange of data between terminals and computers, to such applications as reading utility meters, controlling traffic lights, connecting point-of-sale terminals, and controlling ATMs and vending machines. Many of these applications operate in half-duplex mode, use the entire bandwidth of the transmission channel, and can therefore take advantage of the faster transmission rates. Development of inexpensive modem chipsets led to unobtrusive built-in designs, where the user is hardly aware of the presence of the modem as an integral part of the equipment.

Features of Commercial Modems

Similar to personal computer modem choices discussed in Chapter 7, the commercial market also has many options, the most important being compatibility with existing equipment. Frequently the only solution in the commercial market for expanding an existing data network is to purchase equipment from the same manufacturer who supplied the previously-installed devices.

Most commercial modems, except for a few that use proprietary circuitry, follow one of the established standards. These standards are either the former Bell System de-facto standards, or the recommendations of the world's standards organization, the ITU-T.

The few proprietary commercial modems that do not follow established standards can be found mostly in high speed and wireless applications, and in the limited-distance and specialized-applications modem markets, where they are used on private or leased transmission facilities. Proprietary modems fill certain small market niches of high-speed high-performance (and high price) devices using techniques such as multiplexing, data compression and specialized error correction. As these devices are mostly used for point-to-point and multi-point connections over leased transmission facilities, compatibility with other modems is of limited importance. Such specialized applications are discussed in more detail in Chapters 14 and 15.

Bell and ITU-T Commercial Modem Standards

Similar to the standards listed for PC modems in Chapter 4, Table 13-3 lists standards that are mostly found in the commercial modem market.

Table 13-3: Voice-Band Commercial Modem Standards

Standard	Speed in bps	Remarks
Bell 103	300	US Standard
Bell 202/212	1200	US Standard
Bell 201	2400	US Standard
Bell 208	4800	US Standard
Bell 209	9600	US Standard
ITU-T V.26	2400	Old European standard
ITU-T V.27	2400/4800	Fall-back mode of V.29
ITU-T V.29	9600	Four-wire standard for leased and private lines
ITU-T V.33	14400	Leased or private lines only

Table 13-4 lists the approximate equivalents between the Bell and ITU-T commercial modem standards. Although Bell and ITU-T modems may use the same modulation schemes and bps rates, they may differ in other parameters such as timing of handshake sequence, fallback transmission rates, and so on.

Table 13-4: Equivalent Bell and ITU-T Voice-Band Modems

Speed (bps)	Bell Standard	ITU-T	Mode
1200	202	V.23	Half-duplex
2400	201	V.26	Half-duplex
4800	208	V.27	Half-duplex
9600	209	V.29	Half-duplex

Some commercial modems that use these standards should be capable of communicating with personal modems, provided that the software discrepancies, if any, are resolved and the proper subset of requirements are selected. For example, a commercial modem may not be able to interpret the AT command set, or its fall-back transmission rate may be different from that of a corresponding personal computer modem. Also, commercial modems often provide a wider array of options and higher reliability standards than their personal computer counterparts. These features are usually reflected by a wide price differential between personal computer and commercial modems.

Following is a short description of the Bell and ITU-T commercial standards, which were not described in Chapter 4. The remainder of this chapter describes applications of modems using those standards.

Low-Speed Voice-Band Commercial Modems

The low-speed modems are based on the Bell 103/108-type designs, and feature FSK modulation, full-duplex operation over the PSTN, and 0 to 300 bps asynchronous transmission rates. All of these modems provide an RS232-C serial interface. Following is a list of features that differentiate several closely-related products:

Bell 103 and 212 Standards

These standards, which are used in both commercial and PC markets, were partially described in Chapter 4.

Specific Bell 103/212 Implementations

103JLP Modem—This 300 bps asynchronous modem derives its power from the telephone line. It is equipped with a manual Talk/Data switch. The modem has a manual originate mode and a manual/automatic answer mode.

103J Modem—This modem has the same basic features as ITU-T 103JLP but requires an external power supply. The modem is equipped with an analog and digital loop-back test capability, and also has a manual Talk/Data/Test switch.

212A LP Modem—This modem has all the standard features of a Bell 212A-type modem, DPSK modulation, and the capability to operate in full-duplex mode at 1200 bps over a two-wire PSTN. In addition, it derives its power from the telephone line, so no external power connection is required. Similar to many other modems, it requires only a subset of the RS-232-C signals, namely pin #2 (TD), #3 (RD), #5 (CTS), #6 (DSR), #8 (CD), #20 (DTR), #22 (RI) and #7 (GND).

EC212A/D Modem—This is an auto-dial modem that has special error control. Data is framed in groups of up to 128 bytes per frame. Each frame contains parity bytes that are computed using the CRC polynomial algorithm from the frame data (see description of error protocols in Chapters 4 and 10). A detected error causes a re-transmission of the defective block of data. The auto-dial feature of the modem allows storing of five numbers having up to 30 digits each. The numbers are stored in non-volatile memory (NVRAM) and are therefore not lost if power fails or when you turn off the modem.

Bell 202 Standard

The success of the Bell 103/108/113-type 300 bps full-duplex modems in the 1960s led to the development of a faster, asynchronous 1,200 bps modem that uses a similar technology—frequency shift keying (FSK) modulation—but is capable of only simplex (one-way) operation on two-wire transmission facilities. However, on four-wire leased circuits, the same modem could operate in full-duplex mode. When operating on two-wire circuits, the Bell 202-type modem can send data at 1200 bps in one direction in half-duplex mode on the primary channel and, as an option, can send data in the opposite direction on the reverse channel at a slow speed of only 5 bps.

Although the primary channel uses the simple and inexpensive circuitry required for the FSK modulation method, as in the Bell 103/108/113 type modems, the carrier frequencies used in the two modem types are different. In the 202-type modems, the frequency of 1200 Hz is assigned to binary 1, and 2200 Hz is assigned to binary 0. Because of the simple modulation scheme, the modem encodes only one bit of information for each signal element. Its modulation rate in Baud is therefore equal to its transmission rate in bps.

The optional reverse channel uses an even simpler modulation method, called On/Off Keying (OOK), which consists of using a tone of 387 Hz to indicate a binary 1 and no tone (silence) to indicate a binary 0. The reverse channel, when provided, can be used to confirm transmission of a data block on a primary channel, or it can request re-transmission of a block of data in the event of a detected error. The reverse channel is also used to request that a terminal or computer change the transmission direction for the primary and secondary channels. The command would normally be sent from the computer to the modem via the RTS lead of the serial interface and would be confirmed by the CTS lead. Figure 13-4 shows the frequency assignments for the primary and secondary channels of the Bell 202-type modems.

Figure 13-4 Frequency assignments for Bell 202 modem

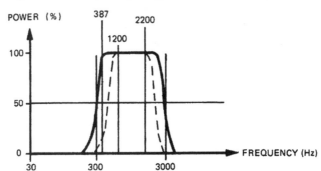

Although the Bell 202-type modem is obsolete by current standards, it is still being manufactured and can be found in many commercial sites—in particular, in four-wire leased-line applications. Its cost is usually comparatively low, since it was typically purchased many years ago and its performance may still be adequate for the particular application.

Specific Bell 202 Implementations

202S/T Modem—This 202S/T modem has all the basic features of a 202-type modem, namely 1200 bps asynchronous half-duplex operation on a two-wire facility on the PSTN, or full-duplex operation over four-wire private lines. Self- and remote-test capability, and a Data/Talk switch to alternate between voice and data transmission are also provided.

202S/D Modem—This modem is similar to the 202S/T, except it has the provision for an automatic dialer and a call-progress tone detector.

This modem is also compatible with certain parts of the ITU-T V.23 protocol, so it can be used on a European connection.

202T Modem—This modem is specifically designed for private line operation. It has an anti-streaming capability to prevent a network hang-up. It has a full complement of internal tests.

Bell 201 Standard

This 2400 bps synchronous modem is still used on many four-wire leased lines, and it finds occasional use on the PSTN. It operates in two-wire half-duplex or four-wire full-duplex mode. Although frequently equipped with auto-dialing features, it does not follow the Hayes AT standard, which is popular in the personal computer market. This modem uses DPSK modulation similar to the 212A modem, and it assigns two bits to each relative phase change. The 201-type modem can operate on a 3002 unconditioned leased line without adaptive equalizers. Therefore, it can respond to a polling carrier within only 15 ms, which makes it an ideal candidate for multi-point private networks. Its modulation rate is 1,200 Baud. The phase assignments of the Bell 201-type modems are shown in Table 13-5. They differ from phase assignments of the 212A-type modem. The 201-type modem uses a single carrier frequency of 1800 Hz.

Table 13-5: Bell 201 Relative Phase Assignments

Dibit	Relative Phase Change (Degrees)
00	45
10	135
11	225
01	315

Specific Bell 201 Implementations

201B/C Modem—This modem features 2400 bps synchronous operation over four-wire private lines. It provides an RS232-C and an ITU-T V.24 serial interface and has a full complement of test features. An 8-position switch is used for feature selection. The modem uses DPSK modulation and operates in half-duplex mode over a two-wire PSTN or leased lines.

201C/D Modem—The 201C/D modem has all the features of the 201B/C modem. In addition, it can detect call-progress tones such as busy, ringing and dial tones, and can auto-dial. Automatic detection of IBM's EBCDIC and ASCII characters is also provided.

201C/LS Modem—The 201C/LS modem has all the features of ITU-T

201B/C modem. In addition, it supports IBM's 3270 terminal communication protocols for mainframe computers.

Bell 208 Standard

The 4800 bps synchronous modem can be used on the PSTN and on two-wire and four-wire private-line circuits. The modulation method is DPSK, with three bits assigned to each signal change. The modulation rate of the modem is 1600 Baud. The carrier frequency is 1800 Hz. The modem operates in half-duplex mode on the PSTN or on a two-wire private line, and in full-duplex mode on a four-wire private line. The relative phase assignments of Bell 208-type modems are shown in Table 13-6. They differ from those of the ITU-T V.27 modems. Although the two modem types are similar, they are not capable of communicating with each other.

Table 13-6: Bell 208 Relative Phase Assignments

Tribit	Phase Change (Degrees)
001	22.5
000	67.5
010	112.5
011	157.5
111	202.5
110	247.5
100	292.5
101	337.5

The 208 demodulator multiplies the carrier frequency by eight and derives the phase information by comparing the interval between zero crossings of the demodulated signal with the stored phase-change values from Table 13-6. A feedback signal is then created, which is related to the difference between the actually measured values and the values from the table. This difference signal affects the decision points, where decision is made whether the signal is e.g., a 000 or a 010. The basic handshake time for a 208-type modem is 50 ms, but can be set to 150 ms for longer circuits by changing some internal strappings.

Specific Bell 208 Implementations

208A/B Modem—The 4800 bps synchronous modem is equipped

with an adaptive equalizer on the receive side and a compromise equalizer with selectable characteristics on the transmit side. The modem includes an anti-streaming feature, which prevents one modem from disabling all other modems on a multi-point network. To do it, the CTS delay is selectable between 8.5, 50 and 150 ms, depending on the polling characteristics of the network.

The modem can operate in half-duplex mode on the PSTN or in full-duplex mode on a four-wire private circuit. Modulation is 8-phase DPSK with a modulation rate of 2,400 Baud. The carrier frequency is 1800 Hz, which is in the middle of the voice-band. The modem provides an Answer-Back tone of three seconds duration at 2025 Hz. The transmitter output level is adjustable.

There is an internal version of this modem available under the designation Sync-Up 208A/B. The internal modem fits into a standard full-size slot of a personal computer. There is a software package called Sync-Up BSC, which allows a personal computer with an internal 208A/B-type modem to emulate various IBM terminals, in particular the IBM 2780/3780 and the IBM 3270.

Bell 209 Standard

This 9600 bps standard has been largely supplanted by the ITU-T V.29 standard. It provides for operation on four-wire leased and private circuits.

ITU-T V.26 Standard

This 2400 bps modem standard is designed to operate in full-duplex mode over four-wire leased telephone lines. Modems using this standard are similar to Bell 201 modems. They use differential four-phase modulation (DPSK) and encode two bits for each phase change. The modulation rate is 1200 Baud. There are two alternatives, A and B, which assign different coding schemes to a specific dibit-to-phase translation. Before setting up a connection using a V.26 modem at each end, it should be determined whether all modems on the data network use the same alternative.

ITU-T V.27 bis and V.27 ter Standard

The V.27 standard is a European equivalent of Bell 208. It is frequently implemented as a fall-back mode of the 9600 bps V.29 standard, although some modems, such as the Penril Datalink 4800, use V.27 4800 bps standards exclusively. The V.27 bis standard is optimized for leased lines, while the V.27 ter standard can be used on the PSTN. The carrier frequency is 1,800 Hz, and three bits are assigned to each signal element. The modulation rate of the modem is therefore 4800/3 = 1600 Baud. The standard also provides for a fall-back mode of 2400 bps.

ITU-T V.29 Standard

This 9600 bps modem is still a popular standard on four-wire leased- and private-line circuits. The main characteristics of the V.29 modems are fall-back rates of 7200 bps and 4800 bps, the capability to operate in full- or half-duplex mode (depending on the transmission rate), combined amplitude and phase modulation, synchronous operation, provision of an automatic adaptive equalizer, and optional inclusion of a multiplexer to combine 2400 bps, 4800 bps and 7200 bps data streams. The carrier frequency is 1700 Hz. Each signal element consists of four bits. Therefore the modem operates at $9600/4 = 2400$ Baud.

Some manufacturers have adopted this standard to operate over the PSTN by using various proprietary measures, including error correction schemes. Modems that use this standard may also operate over two two-wire lines on the PSTN.

Specific V.29 Implementations

9600FP Modem—The 9600FP modem has an extremely short response time—the RTS-CTS delay—of only 8 ms. The short response time assumes that no training time for adaptive equalizers is required. The short delay is of importance on multi-point networks, where several modems are polled in succession.

9600 Trellis A/B Modem—The special feature of this modem is error detection and correction by means of trellis coding. Use of trellis coding and of adaptive equalizers makes this modem a better choice than the previous modem for data transmission in half-duplex mode over the PSTN .

ITU-T V.33 Standard

The V.33 recommendation is not an official standard, but it has met with a preliminary approval of the international committee. Several manufacturers make modems that follow this recommendation. The modem can operate at multiple transmission rates—specifically at 14.4 kbps, 12.0 kbps, and 9.6 kbps—which are user selectable. At 9.6 kbps the modem complies with the V.29 standard. The modem uses QAM for its modulation method, with six bits assigned to each signal sample and a seventh bit used for trellis encoding. Therefore at its highest speed, the modem signaling rate is 2400 Baud. Although the V.33 standard is only specified for private four-wire transmission facilities, some half-duplex versions of V.33 modems exist, which can operate over a two-wire PSTN.

Setting Modem Options

Setting up a commercial modem requires setting many options, far exceeding the choices required for setting up a personal computer modem. The number of options increases when the modem is used on a leased point-to-point or multi-point private network. It is particularly important that all modems involved in a data connection use the same options.

Following are some of the options, which can be configured with jumper or switch settings, or with software commands. These options must be set in addition to standard options such as transmission speed, parity, and synchronous or asynchronous operation. For details of specific options applicable to your modem, consult the appropriate manual.

Wire/Carrier Options

There are usually three options for which a modem can be set.

- The four-wire constant-carrier option is used when the modem is a master station on a four-wire multi-point network, or when it is used on a point-to-point four-wire connection.
- The switched four-wire option is used when the modem is in a remote station or is a slave station on a multi-point connection.
- The two-wire switched option is selected when the modem is used on a two-wire dialed connection.

Fallback Option

This option provides a choice of standards at which the modem will communicate when operating at a lower speed than originally specified. A typical choice for a 9600 bps modem would be 4800 bps as the fallback speed. The choice is between the 4800 bps Bell 208 or the ITU-T V.27 bis/ter standards. It is important that all modems on the network use the same fallback option.

Transmit Level

The transmit level should be set—for dial-up lines on the PSTN—to -10 dBm +/- 1 dB. A higher transmission level is in general not allowed, as it could interfere with other users. On leased or private lines, the transmit level will typically be set between -15 dBm and -1 dBm. The level should be set to the lowest value at which error-free transmission can be achieved.

Carrier Detect Level

The carrier detect level is normally set between -20 dBm and -45 dBm. Choosing a lower carrier-detect threshold will make the modem more sensitive, but will cause occasional false readings due to noise. Again, the detect level should be chosen as low as possible while ensuring error-free transmission.

Anti-Stream Timer

This option prevents a modem on a multi-point private network from shutting down the whole network by permanently setting the Request to Send (RTS) signal to the On state. This is similar to a stuck microphone on a fire or other public safety radio network. The timer will turn the RTS signal off when its duration exceeds a preset time.

Echo-Suppressor Disabler

A modem can usually generate tones, which disable echo suppressors, on a dialed connection that exceeds 2400 km (1500 miles) in length. This option should be selected if full-duplex mode of operation over long dialed-up connections is used. A working echo suppressor would otherwise block transmission in one direction, effectively turning a full-duplex modem into a half-duplex modem.

Signal-Quality Option

Many high-speed modems produce a signal on pin 21 of the serial interface, which indicates when a certain error rate has been exceeded. The signal-quality option sets the threshold for this indication. Typical choices are one error in 1000 bits, or one error in 100,000 bits.

Dial Backup Option

Some modems, such as the IBM 5866 and AJ 2441-1, sense line-quality and, if the connection is not satisfactory, will redial. The user can also select an alternate number to call if the primary connection is interrupted. The modem can also redial if a leased line breaks down. The phone numbers to dial, and time-out periods, must be preset before a connection is established.

Modem Address

When a modem is used in a private line multi-point application, each modem needs a specific address. The master unit usually has the letter M in its address. Other units use numerical codes.

Software for Commercial Modems

The large variety of commercial modem types, and adherence to many different Bell, ITU-T and proprietary standards, makes it difficult to obtain support from the major communications software houses. Many of the software programs mentioned in Chapter 10 will work with commercial modems, although some modifications may be required to account for additional or missing features. Because of the need to communicate with mainframe computers, multi-platform communications programs such as Kermit are frequently used in the commercial environment. In general, a commercial modem manufacturer will also provide a software package to support a specific modem.

An example of such a hardware/software combination is the SyncUp SNA 3770 modem card. The modem card plugs into any IBM PC or compatible computer, and works in the Bell 201, 212 and 208 modes. Depending on the mode selected, the modem will operate over a two-wire PSTN or over two- or four-wire leased circuits. The modem card emulates (as far as the main-frame computer is concerned) a 3770, a 3777-3, or a 3776 terminal. The modem will operate in accordance with the SDLC synchronous data-link control protocol at 2400, 4800, 7200, or 9600 bps. It can also operate in asynchronous mode at 300 bps. The modem provides automatic dialing and automatic answer capabilities. This feature allows a personal computer that is equipped with this modem to operate as an attended or unattended remote job entry (RJE) terminal to a mainframe computer.

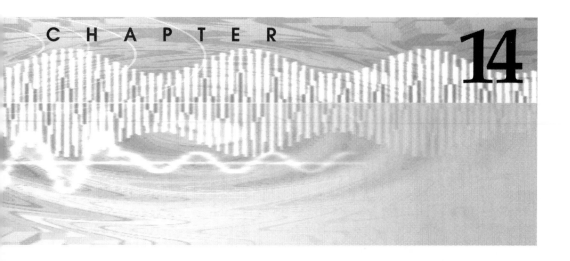

ISDN, Cable, LAN, Wireless, Cellular, and 56K Modems

This chapter discusses modems that satisfy certain specific needs, such as high transmission rates, the capability to connect to local area networks (LANs), and the capability for wireless communication. The ISDN modem is currently coming into widespread use because of its reasonably low cost and reasonably-high transmission rates. The cable modem, with its much-higher transmission rate, is slowly being introduced. If successful, the cable modem may take over a large part of the computer communications market. Wireless modems will fill a certain niche for point-to-point communications, while cellular modems will appeal to the more mobile population. The last section in this chapter describes the operation of 56K modems, which are taking over the personal computer market.

ISDN Modems—Setting Up an ISDN Connection

Integrated Services Digital Network, or ISDN for short, is one of the current technologies used by the telephone companies to provide a telephone subscriber with digital access to the telephone network at a reasonable cost. ISDN replaces the local analog loop—between the telephone subscriber and the local telephone office—with a digital connection without physically replacing the copper wires.

Starting in the 1960s, the telephone network evolved from a purely analog system to a digital transmission and switching system. Only the local loop between the subscriber and the local telephone office remained analog, because of the large cost to replace the older analog equipment with newer digital equipment.

In the mid 1980s, the CCITT (ITU-T) recommended a standard for establishing ISDN connections. Because of incompatible implementations of this standard that were made by major switching-equipment manufacturers, local telephone companies were initially reluctant to adopt ISDN. In the early 1990s, an agreement was finally reached between the local carriers and equipment manufacturers on a set of ISDN features called N1-1. Only then did the telephone companies start offering ISDN service to the general public.

Currently, the typical pricing of ISDN service is about $25 per month for basic service, plus usage fees of between 1 and 2 cents per minute. Some telephone companies charge a slightly higher per-minute rate when both B channels are used at the same time. The terminology of ISDN (B + D Channels) is described later in this section.

ISPs typically charge $25 to $30 per month for unlimited access using ISDN. The cost of an ISDN "modem"—officially called an ISDN terminal adapter (TA)—is around $350. The reason for the quotes around modem is that the original definition of a modem—converter of digital signals into analog signals (D/A) and vice versa (A/D), no longer applies, at least to the data traffic. Data from the computer is fed into a digital ISDN *pipe* without being converted into analog signals. Still, an ISDN connection can also carry a voice channel, which then requires A/D and D/A conversion. Considering the reasonable cost and ability to double or quadruple data transmission rates, many organizations as well as individual PC users are now subscribing to the ISDN service.

ISDN Terminology

ISDN, as mentioned previously, stands for Integrated Services Digital Network. The word *integrated* refers to an end-to-end digital connection. Functionally, an ISDN line is split into several so-called B channels (or bearer channels), and one D channel (or data-signaling channel). Each B channel can carry voice or data at rates of up to 64,000 bps, while the D channel carries signaling for the B channels at 16,000 or 64,000 bps depending on implementation. In general, the B and D channels cannot be used independently.

The D channel carries dialing and call-setup information necessary to establish a connection, to request network services, to route data over B channels, and to terminate the call when completed. The

D channel signaling travels over a separate, dedicated network. It is similar to the out-of-band signaling provided internally for the PSTN. The result of using separate data and signaling paths is that the time necessary to establish a call is always less than five seconds, and often less than one second, even on long-distance calls.

There are two types of ISDN service, as defined by the ITU-T:

1. Basic Rate Interface (BRI) service, which consists of two B channels and one D channel (2B+D). The D channel operates at 16,000 bps. This is the service of interest to the residential PC user.
2. Primary Rate Interface (PRI) service, which consists of 23 B channels and one D channel (23B+D) in North America and Japan, and of 30 B channels and one D channel (30B+D) in Europe. The D channel operates at 64,000 bps. The PRI 23B+D service is the ISDN equivalent of a 1.544 Mbps T1 digital-carrier line. The PRI service is of interest to ISPs and larger commercial users.

The ISDN BRI service comes to a subscriber's premises as two pairs of wires. Each ISDN BRI line has two phone numbers assigned to it (one for each B channel). An ISDN BRI line can be used as two separate telephone voice/analog lines, or one voice/analog line and one data line, or two data lines. The voice/analog line can be used for a regular voice call or for fax transmission. The two B channels can be combined through a process called *bonding* into a single 128,000 bps data line.

The ISDN service must be ordered from your local telephone company. With constant changes in the field—where one company constantly invades the historical turf of another company—your ISDN supplier in the future may even be your power, cable or long-distance phone company. You must also purchase an ISDN terminal adapter (TA), often called an ISDN modem.

Installation of the TA is similar to that of a regular modem. An internal unit will plug into an available slot in your PC, while an external unit hooks up to a serial port. A noticeable difference from a regular modem is that the jack and cable going to the ISDN outlet on the wall are thicker than the regular cable. ISDN uses RJ45 modular 8-contact jacks and plugs, rather than the 4-contact RJ11 jacks and plugs used for regular telephone connections. Also, on the back of the TA you should find two, rather than one, RJ11 jacks, since each ISDN line is associated with two phone numbers.

A typical setup for a TA with associated equipment is shown in Figure 14-1.

Figure 14-1 ISDN terminal adapter and associated equipment setup

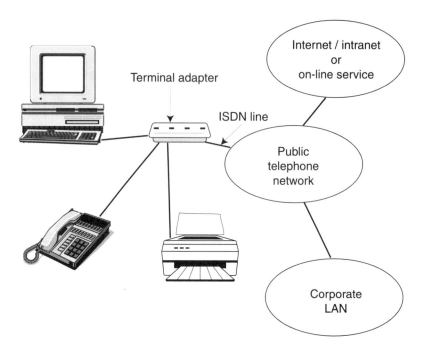

Terminal Adapter

The Terminal Adapter (TA), the ISDN equivalent of a modem, adapts the variable transmission rates from the computer to a fixed transmission rate of a B channel—64,000 bps. This function is normally transparent to the user. The TA also provides regular modem service for voice or fax transmission by converting it into a *digital* stream.

Notice that a regular modem at the transmitting end converts digital data into *analog* signals for transmission, while a TA, when working as a modem, converts *analog* voice or fax into *digital* pulses for transmission over the digital ISDN line. A TA does not have to modulate data for transmission—it just sends it over the digital 64,000 bps B channel. The TA can also provide for *dynamic bandwidth allocation* by combining B channels (two for BRI, more for PRI) to suit an ISDN line's bandwidth capability as its input dynamically changes.

A terminal adapter also has all basic functions of a regular data/fax modem. It can be used to send and receive faxes, and it will accept basic AT commands.

ISDN Software

ISDN service should be transparent to PC applications to allow common data communications software, such as described in Chapter 10, to be used. This includes terminal emulation and browser software.

You will also get proprietary configuration software when purchasing a Terminal Adapter. This software configures the TA to operate with the service and equipment at the telephone company end, and the equipment at your end. It may provide a graphical point-and-click interface, which will display configuration options, diagnose errors, and suggest solutions. A typical screen in a setup program for a 3COM TA is shown in Figure 14-2. It lets you enter all parameters such as the remote switch type and assigned phone numbers.

Figure 14-2 3Com software for ISDN setup

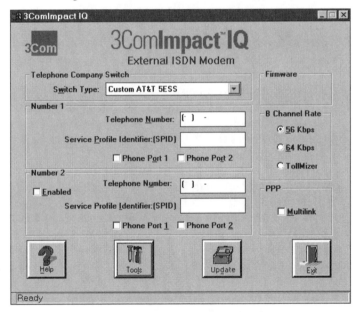

All information to be put in the setup screen should be provided by the telephone company when ISDN service is being ordered.

Cable Modems—Promises and Reality

Cable television (CATV) has gained access to a large percentage of homes in North America and the rest of the world. Home access via coaxial cable, rather than a pair of copper wires, provides a much wider bandwidth. A telephone voice connection requires less than 4 kHz of bandwidth, while a television channel occupies 6 MHz, or 1500 times as much. With 50 or more channels piped into a typical CATV system, the bandwidth required is over 300 MHz. With proper repeaters, coaxial cable systems can easily support several GHz of bandwidth. Why then can we not hook up high-speed modems and other equipment to this high-bandwidth pipe?

The answer is that it is not that simple. CATV systems were originally designed for broadcasting as a one-way (simplex) operation, where information travels in only one direction—towards the subscriber. Only recently-designed CATV systems provide for two-way communication. It should also be noted that repeaters in a CATV system are optimized to amplify the frequencies currently allocated to television channels. Furthermore, the current TV transmission is purely analog, while data transmissions for optimum performance should be carried on a digital system. And finally, there is very little switching involved in a CATV system, because most subscribers receive the same television channels. While a telephone connection allows you to reach to any ISP, the CATV cable modem connection only allows you to reach one specific ISP—the one who will sign a deal with the cable company.

However, once these problems are resolved, a two-way cable system will be able to support a variety of services such as the *plain old telephone service* (POTS), ISDN, video phone, and high-speed data access to the Internet.

In one cable modem implementation, the cable company supplies cable television, regular phone, and Internet services. To operate a cable modem, the subscriber will first need a cable access unit (CAU) installed at the cable entrance to the house. The CAU bridges the cable and separates the low-frequency phone access from video signals on the downstream path (to you), and injects telephony signals into the upstream path (from you). The CAU features standard telephone interface connectors (RJ11) to accommodate a conventional telephone, and a coaxial connection for the cable interface.

A cable modem that is located near a computer will connect the subscriber's computer to the CAU in order to communicate with online services. The modem typically offers throughput speeds of up to 10 Mbps per user in the downstream direction, and 700 kbps or more in the upstream direction.

In another cable modem implementation, currently offered by a TV cable service provider, a special network card has to be installed in the user's PC. The card is connected to the cable modem, which is located next to the computer. The incoming cable is split in two, one going to TV sets as before, the other feeding the cable modem. In this implementation, no regular phone service is provided over the cable.

When users turn their computers on, they are immediately connected to the Internet, with no dialing required. The current charges for this service, as offered in Monmouth county, New Jersey, with unlimited Internet access, are $40/month for regular (TV) cable subscribers, $60/month otherwise. In addition, there is a $175 installation fee.

When providing phone service, the cable operator will need some additional equipment at the head-end of the CATV system—a *concentrator* to combine voice-band telephone channels and pass them to a switch—and the cable control unit (CCU) to communicate with the CAUs and act as a protocol converter from the cable plant to a local telephone office. It allows the cable operator to integrate telephony and enhanced data services with the existing CATV signals. Also at the head-end is the cable router, which provides an interface between a hybrid fiber/coax transmission system and local and remote TCP/IP networks such as the Internet. Video (TV) will still be delivered as usual to a subscriber.

Use of a CATV system for bi-directional data transmission will require a major investment in the CATV infrastructure to be successful. Still, some small companies, such as Terayon Corporation, are generating many press releases while they develop equipment and conduct field trials of cable modems. The same applies to large companies, such as Motorola, who do not want to be left behind in the event that cable-modem technology takes off.

It should be noted that all development in cable modems is strictly proprietary, so equipment from one manufacturer will not operate with equipment from another manufacturer unless they collaborate. Basic specifications of the CyberSURFR™ cable modem, as released by Motorola Corporation, are shown below. The modem uses Ethernet connectors similar to the LAN Ethernet modem described in the following section. However, it is specifically designed to work with CATV systems.

CyberSURFR™ Cable Modem, General Specifications

Features

- High-speed upstream and downstream Ethernet network communications for subscriber's personal computer, workstation, MacIntosh computer or other network device running TCP/IP protocol
- Modem configuration file automatically provisioned within cable-router database
- Supports communications for multiple network devices
- Standard 10BaseT Ethernet connectivity
- TCP/IP protocol transport system

- Upstream noise robustness in hostile noise environments
- DES-based encryption in both upstream and downstream communications
- Auto-provisioned in head-end cable-router database
- Easy to install and operate
- Low cost per subscriber

Description

The CyberSURFR Cable Modem is a component of the Motorola Cable Data System, which connects a subscriber's personal computer or other TCP/IP-addressable device to a hybrid fiber/coaxial (HFC) system. The Cable Data System is specifically designed for high-speed communications for online services, Internet access, telecommuting, and other emerging services for home and business PC users.

The CyberSURFR connects to the subscriber PC using standard 10BaseT Ethernet. A single CyberSURFR supports communications for multiple personal computers with IP addresses. An Ethernet wiring hub interface provides the connections for more than one PC.

The CyberSURFR Cable Modem supports IP communications from the subscriber's PC to host computers and servers. The CyberSURFR Cable Modem is not assigned an IP address and does not require an IP subnet, conserving operator IP addresses. The modem performs the function of filtering and forwarding packets both to and from the PC attached to it, giving the appearance that the PC is directly connected to a LAN in the cable head-end.

RF transmitters and receivers in the CyberSURFR Cable Modem provide the physical layer communications over the HFC system. The data from a subscriber PC is transmitted upstream on a 768 kbps shared packet data channel, which uses a 600 kHz carrier. Downstream, the subscriber shares a 30 Mbps channel, which uses a 6 MHz carrier and provides a maximum of 10 Mbps throughput to each subscriber. Throughput varies depending on Internet access, channel load, PC processor and configuration, and head-end equipment load. The CyberSURFR Cable Modem operates in the upstream spectrum of 5 MHz to 42 MHz, and downstream between 65 MHz and 750 MHz.

High performance communications across HFC systems is ensured by the CyberSURFR Cable Modem's ability to identify and correct errors caused by transient ingress noise through forward error correction (FEC) techniques. In addition, the frequency agility feature embedded in the cable router enables the CyberSURFR to switch automatically to an alternate frequency if a sustained high noise level is experienced on the existing frequency.

Secure communications is supported through the use of Data Encryption Standard (DES) in both upstream and downstream transmissions. User data is encrypted, providing complete privacy to subscriber transmissions over the shared HFC system. When the CyberSURFR is powered up, a registration authorization is performed to ensure that a subscriber is using a valid system CyberSURFR Cable Modem.

The CyberSURFR Cable Modem is easily installed by connecting it to the subscriber PC using a standard 10BaseT Ethernet connection. The installation is completed by simply plugging the CyberSURFR Cable Modem into the cable in the home. The CyberSURFR Cable Modem will then automatically go through the registration and authentication process. Four LEDs indicate connectivity status and also act as a diagnostic aid. The TCP/IP address for the subscriber's network device is either configured statically in the IP communications software or the address can be configured dynamically by using a DHCP server.

Management of the CyberSURFR Cable Modem is provided through an SNMP proxy agent in the Cable Router and the Enterprise MIB variables. MIB II variables are also supported. Software in the CyberSURFR Cable Modem can be updated to the latest version through a downline load process, executed by the cable router, to allow an operator to maintain consistent software throughout the subscriber base. Interface specifications of the CyberSURFR modem are shown below.

CyberSURFR™ Cable Modem Interface Specifications

- 10BaseT Ethernet Connector
- HFC Drop Connector: Female "F" Type

RF Specifications:

Transmitter
- Bandwidth: 600 kHz
- Data Signaling Rate: 768 kbps
- Symbol Rate: 384k symbols/sec
- Modulation: 9/4-DQPSK
- Transmit Frequency Range: 5 MHz - 42 MHz w/dynamic frequency agility
- Oscillator Stability: <= 1ppm
- Input Impedance: 75 ohms (nominally)
- Dynamic Range: 24 - 55 dBmV

Receiver
- Bandwidth: 6 MHz
- Data Signaling Rate: 30 Mbps
- Symbol Rate: 5M symbols/sec
- Modulation: 64 QAM
- Receive Frequency Range: 65–750 MHz w/frequency selectable
- Channel Plans: Standard, IRC, HRC
- Input Impedance: 75 Ohms (nominally)
- Minimum CNR (at receiver): 30 dB
- Sensitivity: +5 to -15 dBmV
- Group Delay Tolerance: 130 ns

LAN Modems

In business environments, the communication connection from a computer is often not only to a telephone network but also to a local area network (LAN). To accommodate these needs, there are special modems that can be used as regular telephone modems with an RJ11 jack, but can also transfer high-speed LAN traffic between a computer and a LAN through a coaxial connector. Examples of such modems are two units from Xircom Corporation, one for Ethernet LAN and another for Token Ring LAN.

Ethernet Modem

The following describes a 33.6 kbps modem with Ethernet LAN capabilities in a single PC Card from Xircom Corporation. The modem provides quick access to information on the office network, Internet, or other online services.

Modem Features

Global Access

Modems with the GlobalACCESS logo have features designed for the worldwide traveler. Xircom's Country Select software, *Guide to International Connectivity*, ensures connections to foreign networks. Xircom's International Modem Travel Kit allows the modem to connect to over 160 telephone systems worldwide.

LAN Access

Xircom's suite of network drivers provides access to the most popular network operating systems, including Novell NetWare, Microsoft Windows 95, and Windows NT. Credit Card Ethernet+Modem 33.6 features include *advanced look-ahead pipelining*, and an SNMP agent that allows a portable PC to be managed using SNMP-based network management software. Xircom's installation software provides for easy setup.

Cellular Communications (U.S. and Canada only)

Communicate from any location using Xircom's Credit Card Modem 33.6 or Ethernet+Modem 33.6 with a Motorola or AT&T cellular phone. Connection requires a Xircom Cellular Connection Kit.

Minidock Connector System (U.S. and Canada only)

Xircom's MiniDock connector system combines both modem and Ethernet LAN connectors in a single durable, reliable device. Modem and LAN LEDs provide status information at a glance. General modem specifications are shown below.

Xircom Ethernet Modem—General Specifications

Power Requirements:

Credit Card Modem: +5 VDC, 260 mA typical
CreditCard Ethernet+Modem:

- (10BASE-T): +5 VDC, 320 mA typical
- (10BASE-T and 10BASE-2 combo): +5 VDC 430 mA typical

Physical Characteristics:

- Type II PC Card.

Memory Size:

- 256K Flash memory, 64K RAM (CreditCard Modem)
- 32K network buffer.

Memory Allocation (Ethernet+Modem):

- 4K shared memory

Certification:

- FCC Part 15, Class B and FCC Part 68
- Canada DOC
- CE Mark
- Additional approvals

Ethernet Standards:

- IEEE 802.3

Data Modem Transmission Speeds:

- Full duplex at 33,600, 31,200, 28,800, 26,400, 24,000, 21,600, 19,200,16,800, 14,400, 12,000, 9600, 7200, 4800, 2400, and 1200 bps.

Fax Speeds:
- 14,400, 12,000, 9600, 7200, 4800, and 2400 bps.

Fax Standards:
- V.17, V.29, V.27ter, Group 3, EIA/TIA Class 1 and Class 2.

Error Control:
- V.42/MNP Levels 2-4.

Data Compression:
- V.42bis (4:1) or MNP Level 5 (2:1)

Command Set:
- Hayes, Microcom compatible AT commands

Non-volatile RAM:
- Two user-definable profiles; up to four 36-digit telephone numbers; LAN adapter ID and serial number.

Hardware Compatibility:
- Supports all popular PC Card compliant PCs including AST, Compaq, DEC, Dell, Hewlett-Packard, Gateway 2000, IBM, Micron, NEC, Sharp, Texas Instruments, and Toshiba.

Software Compatibility:
- Network Operating Systems (Ethernet+Modem)
- Supports all popular network operating systems including Novell NetWare, Microsoft LAN Manager, Artisoft LANtastic, Banyan VINES and DEC PATHWORKS. Packet driver support for TCP/IP.

Operating Systems:
- Supports all popular network operating systems including Microsoft Windows 95, Microsoft Windows 3.x, Microsoft Windows NT, Microsoft Windows for Workgroups, DOS, OS/2.

Card and Socket Services:
- Supports Card and Socket Services including Award/VMI, Cardlite/Cardview/AMI, Cardware/VMI, CPQDOS - Cardsoft (Compaq), Databook, IBM, Phoenix 3.x, SystemSoft.

CreditCard Token Ring+Modem 33.6

The following describes a 33.6 kbps modem with Token-Ring LAN capabilities in a single PC Card from Xircom Corporation. The modem provides fast access to information on the office network, the Internet, and other online services. The basic features and specifications for model CTM-33CTP are shown below.

CreditCard Token Ring+Modem 33.6 Features and Specifications

- Combined Token-Ring 16/4 adapter and fax modem
- Simultaneous LAN and data/fax modem capabilities on a single PC Card.
- Requires only one PC Card slot
- 33.6 kbps data and 14.4 kbps send and receive fax modem
- Automatic cable-sensing
- Can be configured for modem-only operation
- Flash-ROM makes software upgrades quick and easy
- Automatically configures PC, modem, and network operating system setup
- Send and receive faxes directly from Windows. Includes terminal-emulation software
- Comprehensive suite of drivers including Windows 95 and Windows NT

Cable: Unshielded twisted pair (UTP)
-         Shielded twisted pair (STP)

Connector: RJ-45 for UTP
-         UTP/STP converter (balun)
-         RJ-11

Standards:
- Type II PC Card
- Physical Characteristics:
- Size: $54 \times 86.5 \times 5$ mm
- Power Requirements:
- +5 VDC @ 400 mA   LAN + modem operation
- +5 VDC @ 300 mA   LAN operation + modem in sleep mode
- +5 VDC @ 200 mA   Modem-only operation
- +5 VDC @ 80 mA   Modem-only in sleep mode

Memory Size:
- 128K RAM for LAN buffers
- 128K Flash-PROM for modem
- 32K RAM for modem

Certification:
- FCC Part 15, Class A and FCC Part 68
- Canada DOC, CE Mark

Data Modem Transmission Speeds:
- Full-duplex at 33,600; 31,200; 28,800; 26,400; 24,000; 21,600; 19,200; 16,800; 14,400; 12,000; 9600; 7200; 4800; 2400; and 1200 bps

Fax Speeds:
- 14,400; 12,000; 9600; 7200; 4800; and 2400 bps

Fax Standards:
Control: V.17, V.29, V.27terbo; Group 3, Class 1 and 2

Error Control:
- V.42/MNP Levels 2-4
- Data Compression:
- V.42bis (4:1) or MNP Level 5 (2:1)

Command Set:
- Industry standard AT command set

Hardware Compatibility:
- Supports all popular PC Card-compliant PCs including AST, Compaq, DEC, Dell, Gateway 2000, Hewlett Packard, IBM, NEC, Sharp, Texas Instruments, Toshiba and Zenith Data Systems Compatible with both PC Card and CardBus-equipped notebooks

Software Compatibility:
Network Operating Systems:
- Supports all popular network operating systems including Novell NetWare, Microsoft LAN Manager, Artisoft LANtastic, Banyan VINES, and DEC PATHWORKS

Operating Systems:
- Supports all popular operating systems including Microsoft Windows 95, Microsoft Windows, Microsoft Windows NT, Microsoft Windows for Workgroups, DOS, and OS/2

Card and Socket Services:
- Supports Card and Socket Services including Award/VMI, Cardlite/Cardview, Cardware/VMI, CPQDOS-Cardsoft (Compaq), Databook, IBM, Phoenix 3.x, and SystemSoft

Wireless Modems

There are several methods to send and receive data without a direct wire connection to the telephone network. The first and the most obvious method is to use a modem, which connects to a cellular phone, which in turn communicates with the rest of the world via the cellular and the wired telephone network. This method is discussed in the later section of this chapter.

The second method of wireless communication is to install a small radio frequency transceiver (transmitter/receiver) inside the modem itself, equip it with a pop-up antenna and communicate via proprietary wireless networks. Such modems are manufactured by Motorola and U.S. Robotics. They operate at frequencies of 896-902 MHz for transmitting and 935-941 MHz for receiving. A typical transmission rate among wireless modems is 9600 bps.

The two major wireless data networks are operated by RAM Mobile Data under the name of RAM Mobile Data Network and by Motorola under the name of Ardis Network. The two networks are not compatible with each other. They each cover most of the U.S. metropolitan areas. The wireless networks operate with proprietary software and hardware so that a specific modem type will only work with specific software, connect to a specific wireless network, which in turn uses a specific e-mail and Internet service provider. Wynd Communication Corporation supports e-mail for the RAM Mobile Data Network while Ikon Office Systems supports it for the Ardis network. The wireless modems with the associated software work with any computer supporting Windows or Windows CE, as well as other operating systems. The computer can be a laptop or even a small palmtop like those manufactured by Hewlett Packard, Casio, Philips and Compaq.

The third method of wireless communication is a point-to-point or a multi-point private network, where each location is equipped with a wireless modem. A recent example of such an application is the Pathfinder mission to Mars, which landed on the red planet on July 4, 1997. The small vehicle named Sojourner Rover, which travels independently on the Mars surface, communicates with the main space probe using a 9600 bps wireless modem.

Another example of point-to-point or multi-point data transmission is a wireless transmitter/receiver, model GINA 5000N/NV from GRE Corporation, which is described next. Such modems operate in the 902 to 928 MHz band assigned in North America for industrial, scientific and medical (ISM) use. Outside of North America, this band cannot be used for private data communications; the 2.4 GHz band is used instead. Wireless point-to-point modems operate under Section 15 of the FCC, which limits the maximum transmitted power as a function of frequency.

The GINA 5000N receives and transmits data in the ISM band of 902 to 928 MHz in half-duplex mode. The GINA 5000N uses a standard RS232-C serial-data interface, which can be driven asynchronously at rates from 1200 through 19.2 kbps (optional 38.4 kbps). The GINA 5000N does not require any synchronization between it and the DTE device (computer). It automatically synchronizes on any speed up to 19.2 kbps without any setting. It is a plug-and-play transparent link to any device with an RS232-C serial interface. There is no special setup required. The GINA 5000N is a highly-secure spread-spectrum system with no specific built-in protocols. It does not packetize or perform any error correction, which allows it to be completely transparent. The GINA 5000N, being completely transparent, allows any custom communication protocol to be used. Synchronous units are available as an option.

Introduction to Wireless Data Communications

Following is an introduction to principles of wireless data communications, courtesy of GRN Corporation and Chuck Hartley:

> To communicate from "here" to "there" by wireless (i.e., without an interconnecting cable), it is necessary to have a carrier convey information. In the beginning this consisted of simply "keying" the carrier on and off and it was an extension of the telegraph. Later, as we learned to transmit more complex information, voice, data, etc., the modulation process became more complicated. Common modulation techniques are AM and FM. Amplitude modulation, or AM, means to make the carrier stronger or weaker in unison with the modulating information and frequency modulation, or FM, means to change the frequency of the carrier higher or lower with the modulating information.

> Sidebands

> When a carrier is modulated, the frequencies of the modulating signal will add to and subtract from the carrier frequency, setting up sidebands, or multiples of the carrier frequency, on either side of the carrier. These sidebands occupy valuable space in the spectrum and many modulation techniques are being used to attempt to conserve space. This modulation method is called narrow-band transmission.

> Spread Spectrum (SS)

> Recently, a modulation scheme called spread spectrum (SS) was de-classified by the military for civilian use. SS can be a very wide-band type of transmission, but it uses the band in a way that, theoretically, lessens the total interference. There are various kinds of spread-spectrum technologies, with *direct sequence modulation* being a more common mode of transmission. Direct sequence is a phase-shift scheme, resulting in a wide-band FM-like signal. The

spreading is accomplished by a pseudo-random noise (PN) code that is modulated onto a carrier. Since the PN code is not totally random—it has a definite pattern—the idea is that several patterns can be interleaved in a wireless band of frequencies, resulting in little interference. The information to be conveyed is embedded within the PN code. The GINA wireless modem uses this modulating technique.

Demodulation

After the carrier has been modulated, it must be transmitted over a distance to a receiving system that reverses the modulating process. First, the modulated carrier is detected and un-spread, the PN code is deciphered, and the embedded information is validated. The GINA 6000 employs an additional step—at the transmitting end, the embedded data is assembled into packets, plus some additional information to ensure accuracy, before it is transmitted. At the receiving end, after the embedded information is recovered, the packets are disassembled. This packetized communication is termed *modified* X.25 and its purpose is to ensure accurate data transfer. The packet assemble and disassemble process is referred to as PAD.

Antenna Theory

To get the carrier signal from here to there involves an antenna system. The height, type and efficiency of this system will determine the distance between transmitting and receiving stations. UHF wireless and light are alike in many ways, especially in their propagation characteristics. They each tend to travel in a straight path, called line-of-sight, but because the wireless signal is at a lower frequency, it bends some and will go a bit further than light. This increase, at 900 MHz, is approximately 1.18 times the line-of-sight distance (an 18% increase).

Because we live on a rounded world, the line-of-sight distance will vary with height above ground level (AGL). At 6 feet AGL, line-of-sight is about 3 miles. At 10 feet AGL, it is about 3.9 miles. At 25 feet AGL, the line-of-sight is about 6 miles, but the UHF wireless distance is around 7 miles. The approximate formula for calculating antenna height H versus line-of-sight distance D in meters or other linear units is:

$$D = 3569 \times \sqrt{H}$$

Space Attenuation

Wireless signals get weaker as distance increases. In the 900 MHz region, the attenuation is 96 dB for the first mile and decreases by 6 dB each time the distance doubles. Two miles would equal 102 dB, four miles would equal 108 dB, eight miles would equal

114 dB, and 16 miles would equal 120 dB. These numbers are important in determining how strong the received signal will be and if a proposed link is practical.

Antenna Types

The next step is to design an antenna system. There are many options in choosing a proper antenna, and a very common external array is a 50-ohm Yagi. The Yagi is a directional antenna that has a relatively-wide transmitting and receiving angle. It can be mounted to provide either vertical or horizontal polarity, and is offered with several gain figures. The higher the gain, the narrower the angle. Another popular antenna is a panel type used mostly with cellular systems.

There are times when it is necessary to install a non-directional—or omni-directional—antenna. These also come in various gains, with 6 dB being a good choice. A gain limit is specified by the FCC to reduce range and interference. This limit is termed *effective isotropic radiated power*, or EIRP, and the value will be +36 dBm, or about 4 watts. It refers to the total output power level, in dBm, of the radio (+30 dBm) plus antenna gain (+6 dB), must not exceed a total of +36 dBm. Therefore the antenna-gain figure is important in meeting the EIRP technical requirement.

Coaxial Cable

How the connection from the antenna to the GINA wireless modem is accomplished can be as important as the antenna. If much of the wireless signal is lost before it reaches the antenna or the radio, the system won't work very well. The coaxial cable transmission line must be chosen to match the 50-ohm antenna impedance, and must limit losses due to cable length. A reasonable choice for up to 100 feet of cable is the Belden 9913, which has a loss figure of about 4.5 dB per 100 ft. As an example, the loss, expressed in dB, is linear with distance, so the power loss for a 50-foot cable would be 2.25 dB.

Data Quality

The GINA wireless modem needs -95 dBm of clean received signal to provide a minimum bit error rate (BER) of 1 error out of 1,000,000 bits sent. It is wise to provide for some degree of insurance, or *fade margin*, so the designed signal strength at the radio should be -80 dBm, allowing for a 15 dB margin.

System Calculations

By putting all of these calculations together, it is possible to determine the viability of a wireless link before installing the equipment. The process is simply to add the gains and subtract the losses, with the goal being a received signal better than -80 dBm.

For example, suppose that it is necessary to send data five miles over flat terrain. The height of the receive and transmit antennas should be between 45 and 50 feet AGL for best results.

Note

A smaller (15 ft AGL) antenna can work, but signal losses and ground reflections do occur, which will reduce the overall performance.

GINA's transmitter RF power is +30 dBm. The receive and transmit antennas each have a gain of +6 dB. By adding those numbers, the total system power is +42 dBm. Five miles of space loss is approximately -110 dB, with 2.7 dB of coaxial cable losses (60 feet) at each of the transit and received ends, so the total loss is about -116 dB (rounded up). Then -116 dB and +42 dBm combined will equal -74 dBm, which is the calculated received signal strength. This -74 dBm is a stronger signal by 6 dB than the minimum design -80 dBm signal—meaning it is four times better—and the path will work very well. Remember, -74 dBm is "less negative" than -80 dBm, so it is "more."

If the antenna's height above the ground is lowered to 15 feet at each end, then an obstruction loss of at least 10 dB will occur. Now the received signal is - 84 dBm, or about 60% weaker than the design minimum. The system will still work, but it is more susceptible to interference and fade.

Site

There are some important details that must be considered before the antenna is installed. Check the proposed antenna site for other radio transmitting antennas—the GINA antenna must be as far away from them as possible. If the antenna is a directional antenna, such as a Yagi or panel type, it must be pointed away from any other transmitting antennas. If an omni-directional antenna is used, it should be mounted above or below the field of other transmitting antennas. Stay away from locations that are sites for TV stations, radar stations, paging systems, all high-powered pulsing modes of RF energy, and other spread-spectrum systems. Taking these precautions can save a lot of time and effort. The external antenna parts are vulnerable to many problems, and they must be installed for easy repair. If the antenna system fails, the wireless system fails!

Installation Tips

If it is necessary to provide a tower to elevate the antenna, then the structure must safely support the weight of any service personnel. When a mast is used, it is desirable that the top be reachable with a self-standing ladder to orient the antenna. Lightning protection is very important in some parts of the country, and the National

Electrical Code (NEC) and/or the local building codes should be consulted for proper grounding procedures, especially on tall buildings. Consider a coaxial cable lightning arrestor as well as an antenna that has a grounded matching network for added protection. PolyPhaser Corp. offers an excellent application guide regarding lightning protection.

The antenna is usually mounted by U-bolts, with either vertical or horizontal polarization. Vertical is typical for wireless, although sometimes horizontal is used to reduce interference. Either will work, but the polarity must be the same at each end of the link for successful operation. Point the antenna as near as possible toward the far end of the path.

When routing a coaxial cable, leave a service loop (slack) at the antenna so there will be sufficient length of cable to replace a faulty connector, if necessary. Secure the coax so that there is no mechanical stress at the antenna connection. Follow the super structure, with the cable to its base, to the building. If the cable requires suspension from the base to the building, use a stranded wire to support the cable weight. The support will prevent a migration of the cable's inner conductor to the shield.

Upon entering a building, leave a drip loop so that water will not follow the cable inside. If the cable lays on a roof or the ground, protect it with a conduit to guard against crushing. Inside, at the radio, use a short flexible patch cable, with the appropriate fittings, from the radio to the main coaxial run for stress-elimination at the radio's SMA connector. An electrical test should be performed at this time. After the test, water-proof all outdoor fittings, ground connections, and cable entry points into the building.

Testing

When the mechanical installation is completed, a voltage standing wave ratio (VSWR) test should be done to determine that all electrical connections are correct and that the antenna is properly matched. A test instrument called an inline wattmeter is connected between the radio and the coaxial cable going to the antenna. The transmitter is turned on and its output power is measured on the meter. Next, the reflected power, or VSWR, is read. The VSWR should be less than 5% of the transmitter power to be acceptable.

Each radio site will be checked in this way. If the VSWR is acceptable and the antennas are pointed correctly, the link is ready to be tested using real data. The data can be two computers using G-TALK, two terminals, or any device that will send back a known signal when polled using a DGH module, which is a bit error rate (BER) test set—also called a BERT. Or just put the system online and see if it works!

RS232 Data

The GINA 5000 RS232 port is wired as a DCE to accept DTE, with the transmit and receive data lines and CTS lines all active. The data flow and transmit functions are controlled internally by the radio in a half-duplex mode.

The GINA 6000 (X.25 packet) is also wired as a DCE, operating in a half-duplex mode, but the RTS and CTS lines are active along with the transmit and receive data lines. Before a connect can be initiated, it is necessary to make RTS high. One way to do this, with limited hardware data-flow control, is to disconnect the DTE RTS wire and strap the CTS and RTS pins together at the DB-9 interface. Pin 7 to 8 provides additional flow control using the XON/XOFF commands, which are supported in the firmware.

The VOICE option, if it is installed, is a push-to-talk and release-to-listen operation, the same as in two-way radio communication (half-duplex). A bonus with this option is the convenience of easy transmitter keying for RF testing.

Test Equipment

To achieve satisfactory performance with an external-gain antenna, it is necessary that all components—the accessory parts, transmission line, connectors, filters, and so on, and the antenna—function properly. A common RF test instrument is the inline VSWR watt meter. A suitable low-cost meter is the Comet 900N.

RF adapters will also be needed because most of the 900 MHz equipment has N-type RF couplers, and GINA uses a reverse-type SMA connector. An 8' to10' RG58 cable with a reverse SMA male connector on one end and an N-type male connector on the other end, plus an N-type female (a barrel) adapter would be handy. The preference should be connectors that allow the easiest access to the antenna system without compromising performance.

Miscellaneous gender-menders and data cables are also necessary. The GINA has a DB-9F RS232-C serial interface, and most asynchronous data equipment use DB-25 connectors. In case of trouble, a DATA TRAKKER will simplify diagnosis, but it requires the DB-25 connectors, as does the BERT and most breakout boxes.

To configure the GINA 6000, a PC with a terminal program such as G-TALK or ProComm is required. A convenient alternative is a Termiflex ST./2000 palm-sized dumb terminal. To test the quality of a link from the far end, a DGH smart module will respond with a message when addressed, and affords a quick-and-easy way to send and receive data over the radio path—it's a poor-man's BERT. However, a BERT is needed when a high-quality data circuit must be guaranteed. A volt-ohm meter (VOM) is also an important item on every test-equipment list. Specifications of the GINA 5000N wireless modem from GRE Corporation are listed below.

Wireless Modem Specifications (GINA 5000N)

- Mode of operation: 1.2 to 19.2 kbps Half-Duplex RS232 (DB-9F) (Optional 38.4 Kbps)
- Channels: 21, selectable by dip switch
- Transmitting Power: 725 mW (+28.6 dBm)
- Receiver Sensitivity: -100 dBm
- Control: CTS
- Data Format: Transparent (any data format)
- Dimensions: 1.52"H × 4.17"W × 5.0"D
- Dynamic Range: -100 dBm
- Frequency Range: 902 to 928 MHz
- Modulation: Bi-Phase Shift Keying (BPSK)
- Power Requirements: 10.5 to 13.8 VDC
- Radio Technique: Spread Spectrum (direct sequence)
- Nominal Range: 800 feet indoors 500 to 1500 feet outdoors 12+ miles - direct line-of-site
- Voice Option Interface: RJ11 w/handset
- Weight: 16 oz.

GINA and GRE are registered trademarks of GRE America, Inc.

Cellular Modems

Another method of wireless communication is by means of a cellular modem. Such modems will often be used in conjunction with a laptop computer and a PC-CARD (formerly known as a PCMCIA card). Installation of a modem with a PC-CARD interface was discussed in Chapter 8.

The modem connects to the computer—via the PC-CARD or serial socket, depending on the design, and to the cellular phone—with a special connector. Such connectors are included inconnectivity kits made for many cellular phones. For example, Xircom Corporation makes a Cellular Connection Kit for analog cellular telephone systems in the U.S. and Canada for their card modems. As shown below, the direct-connect cellular cable allows you to connect a PC-CARD modem to one of the following Motorola or AT&T cellular phones:

Kit Model	Phones Supported
CCK-33MOT	Motorola MicroTAC Lite
	MicroTAC Lite II
	MicroTAC Lite XL
	MicroTAC UltraLite
	MicroTAC Alpha
	DPC550
	DPC650/Piper
	TeleTAC 200
CCK-33ATT	AT&T 3740

A typical connection between the laptop computer, the PC-CARD modem, and a cellular phone is shown in Figure 14-3.

Figure 14-3 Cellular phone connection to a laptop computer

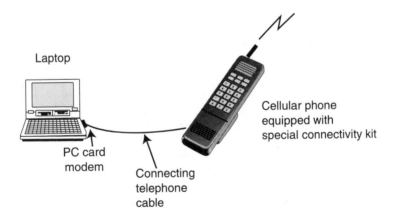

The transmission and reception is done by the cellular phone and is therefore subject to limitations of cellular traffic. You will not be able to communicate when traveling in a tunnel or when too far away from a cellular repeater tower. Occasional signal fading can also affect data transmission. To minimize disruptions due to fading, cellular-capable modems provide additional error protection with either the Microcom MNP10 or with other proprietary protocols such as the AT&T Enhanced Throughput Cellular (ETC) protocol featured in the BitSurfer line of cellular modems from Motorola.

Another cellular modem manufacturer, Megahertz Corporation, a subsidiary of U.S. Robotics, features a special connector called XJACK™, in their PC-CARD modems. The XJACK connector is an easy way to connect to a standard phone line with no custom cables to carry or lose. It simply pops out to hook to a standard phone cable, then locks back in for travel.

Megahertz Corporation offers a full line of PC CARD modems for notebook computers with PCMCIA Type II and Type III slots. The XJACK connector is used for standard phone transmission, and a separate cable port connects directly to a cellular phone. The modems also provide Caller ID and TAD (Telephone Answering Device) capability. Models with TAD and Caller ID include data/fax/voice software, which operates through a computer's sound card.

All Megahertz cellular modems feature Digital Line Guard to protect users from connecting to high-voltage digital/PBX phone lines.

What Makes Those 56K Modems Tick?

As discussed in Chapter 4, starting in 1996 both the U.S. Robotics and Rockwell International in collaboration with Lucent, have been publicizing their two incompatible modem technologies, called respectively x2 and K56Flex. To take advantage of these high speed technologies the Internet Service Providers (ISP) have to be equipped with the 56K modems and be digitally connected to the telephone network. Technical information about specific aspects of these technologies is still scarce, in particular an explanation of how those modems seemingly violate the Shannon theorem discussed in Chapter 2, which should limit the maximum transmission speed over a telephone line to approximately 35,000 bps. Evidently the modem developers are waiting for the patents to be granted before divulging too much information. The following is based on the few published papers released in 1997.

I will refer to the 56K technology, rather than to x2 or K56Flex, which under optimum conditions allows download transmission rates of up to 56,000 bps. This rate is currently limited to a maximum of 53,000 bps due to an obscure FCC regulation. The x2 and K56Flex technologies use the same basic method of data transmission with slightly different rules and implementations.

The 56K technology is based on the current nearly universal use of digital transmission on the Public Switched Telephone Network (PSTN). The only analog part of the network is usually the subscriber loop, a pair of copper wires between the subscriber and the local (Class 5) telephone office. As shown in Figure 14-4 at the Class 5 office the subscriber loop is terminated with a hybrid circuit, which converts the transmission path from 2-wire, where both directions of transmission occur over the same pair of wires, to 4-wire, where the two directions of transmission are physically separated. The Class 5 office will also feature analog-to-digital (A/D) and digital-to-analog (D/A) converters, which translate the analog signals from and to the subscriber into digital signals for further transmission. The A/D and D/A converters sample the analog signal 8,000 times per second and convert each sample to 8 bits. This sampling rate should result, based on the Nyquist criterion, in a maximum transmitted frequencies of 4 kHz (half the sampling rate). The 8 bit sampling will properly decode and encode 256 levels (0-255) of an analog signal. To improve the quality of A/D and D/A conversion of speech, the major concern of the telephone company, these 256 levels are not uniformly distributed but there are more levels at low amplitudes and less at high amplitudes. The non-linear conversion used in North America follows the so called μ Law. It differs from the non-linear A Law conversion used in Europe. The 8 bits of information sent or received 8000 times per second result in a data stream of 64 kbps. Twenty-four such channels are usually combined in a T1 carrier resulting in a combined transmission rate of 1.536 Mbps.

The A/D and D/A non-linear conversion at the Class 5 office introduces quantization noise, the equivalent of the difference between the original smooth analog wave shape and, after passing through the A/D and D/A converters, its step-wise approximation. The quantization noise results in a signal-to-noise (S/N) ratio of approximately 35 dB, which should limit the maximum transmission rate on a voice band channel, based on the Shannon theorem, to approximately 35,000 bps. This theorem calculates the maximum transmission rate for an information channel of a given bandwidth in the presence of random noise.

In an analog non-56K modem the digital signals from your computer are translated into analog signals. At the Class 5 office they are then converted into digital signals, travel as digital signals over the PSTN to the Class 5 office next to the ISP, and are converted there back to analog signals. The ISP receives the analog signals, converts them in the local modem to digital signals which after passing through a router enter the Internet network as packets of data. This process is then repeated in the opposite direction from the ISP to you.

If you and your ISP have compatible 56K modems the process is somewhat different. As shown in Figure 14-5 The ISP's 56K modem is now connected via a digital line (T1 or ISDN) to the Class 5 office without an intermediate analog loop. This avoids at least at one end an A/D and a D/A conversion and the accompanying quantizing noise. In the direction from the ISP to you the ISP sends an 8-bit

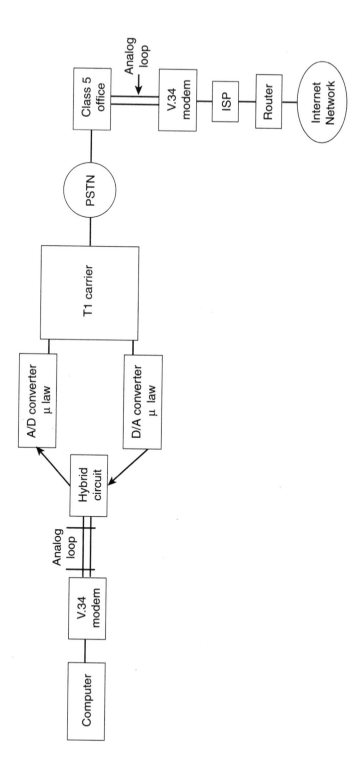

Figure 14-4 PC-to-ISP connnection with a non-56K modem

digital signal, which is sent without A/D or D/A conversion over PSTN to your local Class 5 office. This 8 bit signal, which arrives at a rate of 8,000 times per second undergoes a single D/A conversion and is sent to you over the analog loop as one of 256 analog signal levels. The receiving 56K modem interprets the received signal levels and translates them into 8 bit data packets for your computer.

Figure 14-5 PC-to-ISP connection with a 56K modem

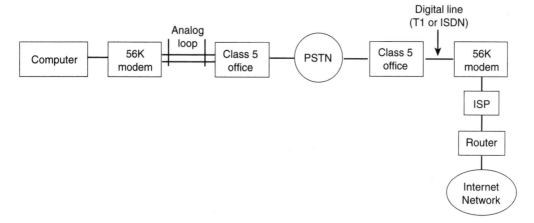

To correctly interpret the 256 levels received at the local computer requires dynamic equalization in the modem to account for the local loop, the hybrid circuit and the D/A converter in the Class 5 office. With the current technology it can be done only in the downstream direction (Class 5 to you). Therefore the upstream transmission is still at the maximum rate of only 33,600 bps. However, speeding up the downstream direction is more important from the point of view of the user. Connection to the Internet normally involves uploading short queries and downloading large files.

The reason that the Shannon theorem discussed in Chapter 2 is no longer limiting the transmission rate to 35,000 bps is that the noise received by your modem is primarily quantizing noise, which is predictable at each of the 256 possible signal levels. The amount of random noise subject to the Shannon theorem is much lower than the quantizing noise and will for an average loop raise the Shannon limit to 64,000 bps.

Finally, the reason that the 56K modem does not operate at 64,000 bps (8 bits times the sampling rate of 8000 per second) is that the T1 carrier "borrows" some bits for frame synchronization and for call progress and signaling information thus lowering the effective transmission rate to 56,000 bps. Also due to the non-linear conversion in the Class 5 office some signal levels are so close together that they can not be distinguished by the 56K modem.

You can see that the 56K technology in its present form can not be used between United States and Europe because of different non-linear A/D and D/A non-linear encoding rules. It also can not be used behind a digital PBX which introduces additional A/D and D/A conversions and in some other instances, where additional A/D or D/A conversion is performed. If the 56K modems decide during the call setup that the 56K technology can not be applied they fall back automatically to V.34 rates of 33,600 bps or less.

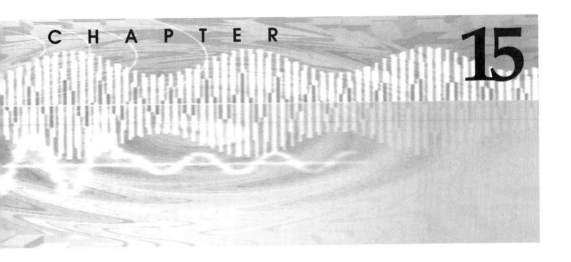

15

Short-Haul Modems

This chapter discusses a special class of modems that permit the exchange of data between computers and other devices over short distances. Short-haul modems provide a quick and inexpensive solution in such cases

Short connections—from a few feet to a few miles—do not include the public switched telephone network (PSTN) or local area networks (LANs) discussed in previous chapters. Using short-haul modems is in many ways less expensive than setting up a LAN, but is less flexible and often has a lower transmission rate than a LAN.

Short-haul modems use proprietary technology, meaning that a modem made by one manufacturer will probably not be able to talk to a modem made by another. We must distinguish between one class of short-haul modems, the limited-distance modems (LDMs), which use dedicated transmission facilities such as copper wires or optical fiber, and AC modems, which use existing AC house wiring as the transmission medium.

Limited-Distance Modems

A limited-distance modem (LDM) is a special type of short-haul modem that provides an inexpensive alternative for data transmission between permanent nearby locations at distances generally not exceeding 25 km (16 miles). Typical applications for LDMs include linking various manufacturing plants within a company, or linking facilities located on a university campus that may be distributed over a limited geographical area.

To take advantage of limited-distance modems, the local facilities should be connected using privately-owned twisted-pair copper wires, coaxial cable, or optical fiber cable, neither of which are a part of the PSTN. To be able to use limited-distance modems, the cables (except for glass fiber) should provide DC continuity between the modems. Cables with DC continuity are often referred to as metallic lines. A leased telephone circuit or a circuit on the PSTN will, in general, have no DC continuity.

A limited-distance modem extends the permissible distance between two RS-232-C devices, which is otherwise (according to specifications) only guaranteed for distances of up to 16 m (50 ft). The connecting copper wires should not have any embedded loading coils of the type commonly found on long local loops. As discussed in Chapter 1, such coils (usually 44 or 88 mH each) are installed by the telephone company every 1.7 km (6000 ft) to improve the voice frequency response of long loops.

The inductance of these coils combines with the internal capacitance of the copper cable to form the equivalent of a low-pass filter. The result is a relatively-flat frequency response in the voice band, up to about 2500 Hz, but high attenuation at frequencies above the voice band. High attenuation above 2500 Hz would be disastrous for higher data transmission rates. Similarly, if an out-of-service telephone cable is being used for a limited-distance modem, bridged taps that remain from previous installations of former telephone subscribers should be removed. A bridged tap appears when telephone service is discontinued and the telephone loop that leads to the subscriber's premises is left bridged across the telephone line. The added capacitance of this bridge causes phase distortion, which also diminishes the quality of data transmission.

Things to remember when planning to lease a line from a telephone company is to ensure that the line is not loaded, is without attenuator pads, and has end-to-end DC continuity. The process of leasing wire pairs from one local telephone company has little in common with the same process from another local telephone company since the Bell-system divestiture, because each telephone company has developed its own terminology and procedures. Terms such as *metallic circuit, physical circuit, unloaded pairs,* or *data-grade* lines are typical. The customer should remember that the telephone company services its customers from the local telephone office. This means that the short-haul circuit length will be the distance from each site to its local telephone office, in addition to the distance between the two offices—so it's not the direct straight-line distance between the sites. This is critical when using short-haul modems at their maximum operating range.

A limited-distance modem typically consists of a box equipped with a 25-pin D-shell connector for the RS-232-C interface or a 34-pin connector for the V.35 interface at the terminal/computer side, and a screw-down terminal block to connect a pair of wires leading to the remote location. A limited-distance modem that uses fiber-optic cable as the transmission medium would need a special fiber connector instead of the screw-down terminals. There are a variety of limited-distance modems that have a wide range of transmission rates that will accommodate interfaces other than RS-232-C.

Limited-distance modems do not use intermediate repeaters. The maximum cable length, maximum transmission rate, and type of cable required (gauge and capacity) are closely interrelated. A limited-distance modem can usually be configured by setting switches that make the modem appear as though it were a DTE (Data Terminal Equipment) or a DCE (Data Communications Equipment), so no null modems are required to complete the connection.

Installing an RS-232-C limited-distance modem consists of attaching a 25-pin DB-25 connector to the terminal/computer serial port, either directly or through a short cable, at one end, and attaching a copper cable to screw terminals at the other end, as shown in Figure 15-1. If you do not know whether your modem is DTE or DCE, one easy way to find out is to observe the LED indicators that are provided on most limited-distance modems. When in the stand-by mode, both transmit and receive LEDs should be off if the modems are connected properly. If the indicators are on, change the position of the DTE/DCE switch on one of the modems.

Figure 15-1 Limited distance modem hook-up

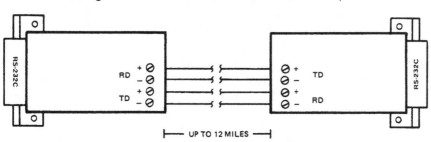

Surge Protection

An important feature to look for when selecting a limited-distance modem is to determine—for safety reasons—whether the modem provides electrical insulation between the computer/terminal and the transmission line. There should either be line-insulation transformers or opto-couplers ahead of the copper wires that lead to the remote modem. Lack of such safety features may lead to destruction of both the modem and computer in the event of a power surge. High-voltage transients caused by lightning and other electrical disturbances can easily destroy the delicate circuits. The induced voltages are primarily caused by the sharp rise in voltage associated with lightning. The leading-edge rise-time of a typical lightning bolt is on the order of kilovolts-per-microsecond.

Additional external over-voltage protection devices can also be connected between the limited-distance modem and transmission line. These devices provide protection beyond the insulation normally found in limited-distance modems. The protection scheme generally includes two to three separate stages, which limit high-voltage transients. These devices can be high-speed gas tubes, semiconductor devices, and metal-oxide varistors (MOVs). Each successive stage is slower than the previous stage, but can handle a higher energy. The protecting devices would normally be installed at both ends of a transmission line. Specialized voltage-protection devices can also be installed on dial-up telephone circuits, where the devices protect the tip and ring line (local-loop wires that have positive and negative DC voltages). The cost of a voltage-protection device is usually less than $100.

Lack of Standards

A minor disadvantage when using limited-distance modems is that there is currently no standard in modem design, and these modems use proprietary circuitry. Therefore modems at both ends of a communication link must be of the same model and be made by the same manufacturer.

Most limited-distance modems are current drivers. For example, the modem sends a current of typically 20 mA over a two-wire metallic line, similar to an old-time telegraph circuit. The reason for using current drivers is that a current driver can transmit a signal over longer distances on twisted-pair cable than an RS-232-C driver with bi-directional voltage signals. The twisted-pair cable exhibits a fairly low impedance at baseband frequencies, which is more suitable to propagating current than voltage. The current driver simply translates the serial interface voltages into current spurts that have reversing directions.

Considering that the DC resistance of a typical limited-distance link is 1500 Ohms, one can see that the voltage imposed by the modem on the metallic line can be up to 0.020 x 1500 = 30 V (or 20 mA across 1500 ohms). This is much higher than the signal voltage on a regular telephone line, which is on the order of 100 mV. It is one of the reasons why a limited-distance modem cannot operate on the PSTN—the high voltage would interfere with adjacent circuits, causing crosstalk and distortion. The FCC limits for signal power for data transmitted over the PSTN, specified in dBm (0 dBm = 1 mW) at various transmission rates, are listed in AT&T Publication 43401, and are shown here in Table 15-1 for reference. These limits apply to circuits that are leased from common carriers, but do not apply to private metallic circuits, which are outside of the FCC jurisdiction.

Table 15-1: Maximum Power Levels Permitted by AT&T 43401

Transmission Speed (bps)	Power Level (dBm)
2400	-4
4800	-11
9600	-18
19200	-25

Point-to-Point and Multipoint

Point-to-point communication is the term used to describe the link from one host or hub location to one remote location. In multidrop/multipoint operation, a computer using a single port communicates with several remote terminals. Multidrop modems enable remote sites to utilize a common transmission facility to the computer or central site. One should be aware that the number of multipoint locations will affect the distance at which LDMs can function.

Power to an LDM is either provided by an external power source, typically an AC transformer plugged into a wall outlet, or in more modern designs it is derived directly from the RS-232-C interface.

Prices of limited-distance modems range from $100 to $200 for current drivers, to $500 to $1000 for units using internal modulators. Prices of fiber-optic-based modems are in the $1000 to $2000 range. The prices depend on the number of features, such as internal diagnostics and flexibility of use. Because LDMs are nearly always used on privately-owned circuits, there are usually no other recurring charges in the operation of such data links. Due to the low initial and operating cost for a limited-distance modem, it can pay for itself in a few months.

A description and specifications for a typical current driver limited-distance modem, model LRM-1 from Hull Speed Data Products Corporation, are listed below.

Features and Specifications of LRM-1 Limited-Distance Modem

Features:

- 2-wire Full-duplex
- Asynchronous LDM
- >1 mile at 9.6 kbps
- Line powered

The LRM-1 limited-distance modem is designed for asynchronous local terminal-to- computer connection applications. The LRM-1 is a 2-wire, full-duplex, 9.6 kbps private-line current driver with a range of over a mile at 9.6 kbps. The LRM-1 utilizes existing unused wire pairs, reducing installation costs. The use of CMOS technology allows the LRM-1 to derive power from the connecting device, which eliminates the need for an external power source. The data and control leads of the DB-25 RS232-C interface provide the power source.

Specifications:
- Data Rates: Up to 9.6 kbps
- Connectors: (1) DB-25S, female
- Interface: RS232-C; 2-position terminal strip
- Transmission Range: Up to 5.9 miles
- Transmission Line: 2-wire unconditioned, DC continuity.
- Application: Full-duplex, point-to-point
- Data Format: Asynchronous
- Control: DSR & DCD turn-on with DTR; CTS follows RTS.
- Power: Derived from data and control leads

Dimensions:
- 0.78 inches high (1.98 cm)
- 2.00 inches wide (5.08 cm)
- 3.50 inches deep (8.89 cm)
- Weight: 6 oz. (170 gm)

In addition to limited-distance modems that use simple current drivers to duplicate RS232 signals, there are also limited-distance modems that use modulation methods similar to those found in regular telephone-type modems. The advantage of such units is that they can operate at lower voltages, so they will not interfere with regular voice circuits carried in adjacent cables on a leased telephone circuit. The disadvantage is higher cost. An example of one such top-of-the-line modem is the Motorola DA 56 DSU/CSU, whose specifications are shown below. The limited-distance modem includes automatic line equalization, 56 kbps operation with maximum distances of 3.4 miles (using 26-gauge wire) to 11.6 miles (using 19-gauge wire), and a standard V.35 interface. The unit can be used as an LDM or a DSU/CSU on a Dataphone Digital System (DDS) circuit.

Motorola DA 56 DSU/CSU Specifications

Operation:
Full-duplex, 1-channel, synchronous, point-to-point or multi-point DSU/CSU or limited distance modem (LDM)

Data Rates:
56 kbps

Customer Interface:
Physical: 34-pin (F) M-block
Electrical: V.35

DDS Line Requirements:
Dataphone Digital Services (DDS I or DDS II)

LDM Line Requirements:
19-, 22-, 24-, or 26-gauge 4-wire unloaded twisted pair

Line Interface:
RJ48S

Fiber Optics Applications for LDM

Fiber optics technology is also reaching into the domain of limited-distance modems. Transmission over optical fibers has a number of advantages, and a few disadvantages, over using metallic conductors. The advantages include a tremendous available bandwidth on the order of hundreds of MHz, low loss, freedom from environmental electromagnetic interference, invulnerability to surreptitious tapping, and last but not least, savings on scarce resources such as copper.

The disadvantages are the relatively high cost, specialized installation equipment, and need for experienced personnel to install and maintain it. A lot of work has been done by telephone companies to facilitate glass-fiber installation. Special splicing connectors have been developed, which precisely align two ends of a glass fiber. The glass-fiber ends are then secured in a plug using epoxy, while an elastic boot limits the bending radius of the fiber at the connector's entrance. A modern glass-fiber connector can be installed in less than 15 minutes by an experienced technician.

Figure 15-2 compares the effective loss, as a function of frequency, of two kinds of glass fibers with a pair of 22-gauge copper wires and with coaxial cable.

Figure 15-2 Optical fiber loss versus frequency

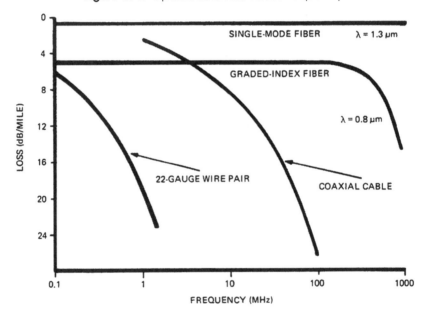

Examples of limited-distance modems using fiber optics include models 2280 and 2290, manufactured by Canoga Perkins Corporation of Canoga Park, California. The 2290 modem operates at transmission speeds of up to 6.312 Mbps; the 2280 modem operates at rates of up to 12 Mbps. Both modems have V.35, RS-422, RS-423, T1, TTL, and DMR11/DMC11 interfaces. The transmitter uses a light-emitting diode (LED) the receiver uses a PIN diode. Both modems use proprietary coding schemes. The maximum distance between modems is 2.5 km for standard configuration and 6 km with a long-distance option.

Typical applications, which can benefit from the high transmission rate of optical-fiber modems, include satellite downlinks, T1 multiplex extensions, high-speed graphics devices, and CPU-to-CPU data transfers.

AC Modems for Control Applications

There is a family of AC modems/controllers that operate over standard AC house wiring, which are sold by Radio Shack and other companies. The AC modems consist of controllers, timers, switches and actuators. A sending device, activated either manually or by a timer, sends a modulated signal over the house wiring to a receiving device, typically an AC on/off switch or a dimmer. The system allows for up to eight device addresses. The sending device can also be connected to a serial port on a computer and can be programmed through the computer.

The controllers are often used to turn lights on and off for home security while away. I use the controllers to tape radio programs. While a VCR allows for taping TV programs at specific times, there is no corresponding commercially-available device for taping radio programs. Although it is possible to remotely tape a radio program by hooking up the radio to a HiFi VCR, it is much easier to do it with AC modems/controllers.

For example, in the New York City metropolitan area, there is a radio station (WNYE-FM) that re-transmits British Broadcasting Corporation (BBC) World Service programs during the early morning hours between 1:00 am and 6:30 am. I am usually not up at this time. However, when I want to tape a specific BBC program during this period, I use the AC controllers as shown in Figure 15-3.

Figure 15-3 Taping a radio program with an AC modem/controller

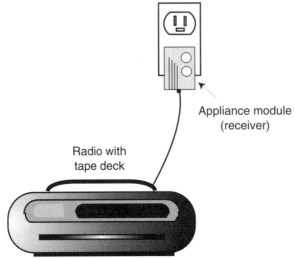

Appliance module
(receiver)

Radio with
tape deck

AC Modem / controller
with timer
(sender)

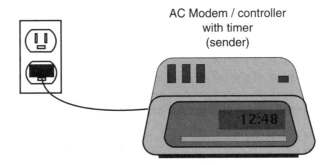

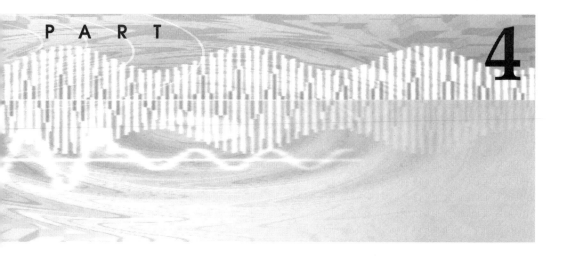

Testing and Troubleshooting

Part 4 describes how to test and troubleshoot modems and the transmission facilities that connect them. It also shows what to do if data gets lost or garbled somewhere between the local computer or terminal and its destination.

Chapter 16 discusses modem-testing, testing the interfaces between modems and computers, and testing modems using a simulated transmission path. It also covers common modem problems and their solutions.

Chapter 17 describes testing of transmission facilities—in particular, those that are leased from common carriers.

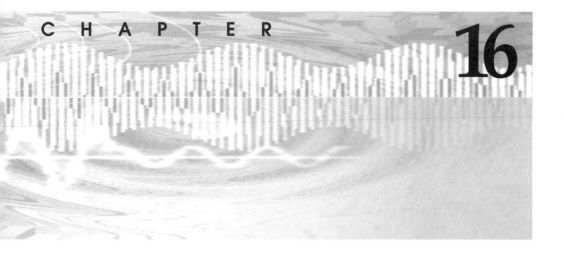

16

Testing and Troubleshooting Modems and Interfaces

Prior to deregulation of the United States telecommunications industry, the best course of action for a failed data connection was to call the local telephone company for repair. Today, things are different—most data communications equipment is no longer leased, but is owned, maintained and operated by the customer, not the phone company.

Furthermore, the jurisdiction over various data links is often blurred. Therefore the responsibility for at least the initial diagnosis of a problem and its resolution shifted from the service provider to the customer. Otherwise, the local telephone company, the long distance carrier, the ISP, and the equipment supplier will likely all announce that their equipment or service is not at fault, leaving the customer holding the bag. In fact, if the problem can be traced to the customer's premises, calling the local telephone company may not only *not* solve the problem, but by adding insult to injury, can often result in a substantial service charge.

When data transmission fails, the failure is caused by one or more of the following data communications system components:

1. Local user
2. Local computer

3. Local communications software
4. Local cabling
5. Local interface
6. Local modem
7. Telephone loop between user and the local telephone office
8. Toll-connecting trunk to the toll office
9. Toll trunks between toll offices
10. Toll-connecting trunk to the remote-end office
11. Telephone loop between remote-end office and service provider
12. Remote modem
13. Remote interface
14. Remote cabling
15. Remote communications software
16. Remote computer
17. Remote user

Methods for diagnosing the problem, localizing it to one of the seventeen components listed here, and fixing it, are the subjects of this and the following chapter. These methods range from sending the famous telegraph phrase exercising every upper-case letter and digit, "THE QUICK BROWN FOX JUMPS OVER THE LAZY DOG 0123456789," to use of the most sophisticated test equipment. Much of the testing and diagnosis can be done by means of test circuits, which are built into many of the better modems, by use of diagnostics software, and even by proper interpretation of indicator lights provided on external modems.

In the event of a problem, the first things to check should be obvious, but are often overlooked (because they are so obvious). Do these first:

- Check that power and signal cables are plugged in and connected.
- Verify that power indicator light(s) are on.
- Ensure that transmission parameters on transmitting and receiving terminals—in particular, bit rate and parity—are the same.

If these items check OK, the next step is to try to interpret the problem and try to find possible explanations. In fact, no fully-automatic instrument or procedure could ever replace the power of logical deduction based on previous experience, and observation aided by instrument readings. For example, transmission errors on only a few calls—even between the same two locations—are probably caused by the switched section of the telephone network (numbers 8 through 10 in the previous list) over which the communications user has very little control. A lower transmission rate, use of error detecting and correcting protocols, or selecting a higher quality modem may be the best solution. On the other hand, if the trouble only occurs on calls to a certain location, then the remote end of the communication path (numbers 11 to 17) may have to be investigated further.

What follows is a review of assorted instruments, indicators and tests that should help to localize and resolve most data-communications problems. The chapter concludes with a number of modem-troubleshooting hints.

The chapter will first describe the various indicator lights that can be found on most external modems, followed by a description of self-test procedures and interface signal tests.

Modem Indicator Lights

All external modems, except for those that are at the bottom of the manufacturer's product line, will have a row of light-emitting diode (LED) indicator lights on the front panel, as shown in Figure 16-1. For internal modems, software can simulate the same indicator lights on the computer screen. Proper interpretation of these lights can help in localizing many problems encountered during a communication session. Because the function of these indicators can vary from modem to modem, your specific modem will probably not have all the indicator lights described here. The names of indicator lights can also be slightly different among different modems, but will likely have names similar to those used in this chapter. The more expensive commercial modems may have, in addition to or as a replacement of LEDs, liquid crystal display (LCD) screens, which display messages and often prompt the user for input.

Figure 16-1 Front and back panels of an external modem

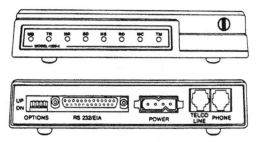

Modem Ready (MR) Indicator

MR indicates whether the initial power-up test failed, or if one of the self-tests—started with the &T commands, discussed later in this chapter—is currently active. On power-up, if the RAM read/write test fails, MR goes off and the modem stops further activity. If the RAM test passes but the computed PROM checksum is incorrect, MR will flash at a rate of 853 ms on, 853 ms off (50% duty cycle) while preventing additional activity. If both tests pass, MR remains on. In general, if the MR indicator flashes when a self-test is not active, or stays off, first check all connections. If they are OK, it is time to get another modem.

When a self-test (&T1, &T3, &T6–&T8) is running, MR also flashes at the same rate as when the PROM checksum is incorrect. It flashes while the remote digital loopback (RDL) process, which is part of the self test, is active and until the test is terminated.

Data Terminal Ready (TR) Indicator

Interpretation of this indicator depends on certain configuration settings determined by modem switches or jumpers. In general, the TR indicator shows the condition of the Data Terminal Ready (DTR) lead (pin 20) on the RS232-C interface. The data terminal designation applies in this case to the computer or to the terminal, whichever is connected to the modem, and which has announced—by setting the DTR lead on—that it is ready to start sending data. The next step in the handshake sequence between the modem and the computer is setting the Data Set Ready DSR lead on by the modem, indicating that the modem is ready to receive data from the computer. If the TR indicator is not on, then most of the modem features are disabled and the modem will not operate. If that happens, try a different modem, or consult the user manual for further detail.

If the Ignore DTR option (&D0) is in effect, TR remains on regardless of the state of DTR. Otherwise, TR tracks the state of DTR.

Send Data (SD) or Transmit Data (TD) Indicator

The TD indicator monitors the TD lead of the serial interface and flashes as data is being transmitted from the local computer/terminal to the modem. SD tracks the signal on pin 2 of the DTE interface within 1 ms. A mark signal on pin 2 corresponds to the LED being off. It is also off when the DTE is not sending characters to the modem. By observing the flashing rate, one can deduce the approximate transmission speed, at least at transmission rates below 4,800 bps. The SD/TD and the RD indicators described next are important diagnostic tools, because they indicate whether data is being exchanged between the computer and modem—if both directions of data transmission are operating properly—and whether the modem operates at the correct transmission rate.

Receive Data (RD) Indicator

This indicator monitors the RD lead (pin 3) of the serial interface and flashes as data is being received from the remote device. RD is on when a signal representing a space is sent to the DTE from the modem over the RD line (pin 3). RD is off when the signal represents a mark. By observing the flashing rate, one can deduce the approximate transmission rate of the received data, which in general should be the same as that of the transmitted data.

High-Speed (HS) Indicator

On multiple-speed modems, this indicator is on when the highest transmission rate is in operation. The HS indicator may be controlled by a switch setting, or it may derive its status from one of the serial-interface leads, which is set according to the transmission rate. On some modems, HS remains off at the lowest speed (such as 300 bps), turns green at 4800 bps, and turns red at 9600 bps or higher. In some Hayes modems, HS is on at 4800 bps or higher. The HS indicator is often activated by pins 12 and 23 on the DTE serial interface.

Modem Check (MC) Indicator

The MC indicator lights up when the modem is offline. The light is off when a connection has been established through a successful handshake, and the modem is ready to send or receive data. On most modems, this indicator will flash to show an error during an automatic test.

Make Busy (MB) Indicator

This indicator is usually associated with self-tests. For example, MB will be on during the analog loopback self-test. During that test, the modem sends a test pattern through the modem's transmitter section, which is then immediately returned to the modem receiver section without involving the phone line. The Make Busy circuit in the modem prevents the modem from transmitting the pattern on the phone line while the test is being performed.

Carrier Detect (CD) Indicator

The CD indicator lights up to indicate the reception of a valid data carrier from the remote terminal. If nothing appears to be happening on the screen during a transmission, then this indicator will show whether the connection is still alive. When dialing an ISP, a BBS, or another computer, the first indication that a connection has been established is when the CD turns on.

CD indicates the state of the modem carrier detector. CD is off during the idle state. During the handshaking state, CD turns on after the CONNECT result code is sent to the DTE. Once the online mode is entered, CD tracks the detected carrier within 20 ms (LED on means carrier power is detected). CD is not affected by the value of S9 (Carrier Recovery Time), or the value of the &Cn (CD Always On option) command.

Auto Answer (AA) Indicator

The AA indicator turns on when the modem is configured to automatically answer incoming calls. This is one of the indicators that should always be checked when leaving the house while the computer is on. If AA is on and you use the same phone line for voice and data calls, then all incoming calls will be greeted by a happy beep, which the caller may not know how to interpret. On the other hand, if a data call is expected, then AA should be on before you leave. When the modem is configured for the Answer and Originate mode, and it detects an incoming call, it will keep the AA light on during the ring. If it is configured for Answer and Call mode, then the AA light will be off during the ring.

AA indicates the status of S-register S0 and the ring detector. If S0 = 0, AA will be off in the absence of a ring signal, and will be on while a valid ring signal is being detected.

Off-Hook (OH) Indicator

This indicator turns on when the modem is directly connected to the phone line, is receiving a dial tone, is dialing, or is transferring data. The OH indicator shows the state of the line relay. It is off when the modem is on-hook (disconnected) and vice-versa.

This indicator is quite useful because most data communications software packages turn off the modem speaker when a data call is in progress. Watching the OH and CD indicators helps in assuring the user that the modem is still on-line. The OH indicator is off when the modem hangs up. Only then should you pick up the phone to establish a voice connection.

Error Control/FAX (ARQ/FAX) Indicator

This indicator has a different meaning, depending on whether it is in the data or fax mode. When in data mode, it turns on when the calling and the called modem successfully negotiated an error-control protocol. If the indicator flashes, it means that a block of data is being re-transmitted. In the fax mode, the indicator flashes to show that a fax message is being sent or received.

Use of &T Commands for Self-Testing

Many modems, in particular those in the middle and upper price range, have various built-in self-test functions that should aid in diagnosing many data transmission problems. Some of these functions are similar or are the same as those found in test instruments costing thousands of dollars. The most common built-in test functions found in many modems include a power-up self-test, local analog loopback test, local digital loopback test, and a remote digital loopback test.

Power-Up Self-Test

This test varies from modem to modem. It usually checks whether power-supply voltages are within an acceptable range and whether there is continuity between the major components. Passing this test means that the basic modem circuitry is operating satisfactorily. The test is performed each time the modem is powered up, and may also be executed by the user in some modems.

Loopback Tests

These tests are based on CCITT/ITU-T recommendation V.54 and consist of selectively connecting, in a back-to-back fashion, various sections of the data transmission path. The tests check the local and remote modems, the local terminal, and the connecting telephone lines. They help in determining which element is at fault. High error rates or a total inability to communicate may be the fault of either the local or remote terminal/computer, the local or remote modem, or the telephone line.

The modem diagnostic and test facilities should enable you to determine the source of the problem. The loopback tests are activated in Hayes-compatible modems by the AT&T*n* commands, where *n* can be 0 through 8. Each test can be manually terminated by means of the AT&T0 command, or it can be timed for 1 to 255 seconds by the timer resident in the internal modem register S18. Following is a list of modem self-test AT commands:

&T0   Terminate any test in progress

&T1   Initiate local analog loopback test

&T2   Initiate end-to-end self-test with internal pattern generators
       at both ends

&T3   Initiate local digital loopback test

&T4   Enable response to remote digital loopback test request

&T5   Disable response to remote digital loopback test request

&T6   Initiate remote digital loopback test

&T7   Initiate remote digital loopback self-test with internal pattern
       generator

&T8   Initiate local analog loopback self-test with internal pattern
       generator

For example, sending the command ATS18=15&T8 <CR>, where
<CR> stands for carriage return, to the modem will start the local ana-
log loopback self-test with the internal pattern generator (typically an
alternating string of 1s and 0s) and will finish the test after 15 seconds.
This and other tests return result codes 000, followed by OK, if the test
passed successfully, or a three-digit number showing the number of
errors encountered during the test, followed by OK.

Following is a description of these self-tests, shown in the order
in which they should be performed.

Local Analog Loopback Test (&T1)

This is the first test to try. The test configuration is shown in Figure 16-
2. The test checks the local terminal/computer, the connecting cable,
and the local modem. The modem "loops back" the data to the screen
of the local terminal. To start the test, disconnect the phone line and
type AT&T1 <CR> while in local command mode. The panel-test indicator MR, if provided, should now be on and the modem should
change to online mode. Type THE QUICK BROWN FOX JUMPS
OVER THE LAZY DOG and see whether the message appears correctly on the screen. To end the test, type +++ to put the modem back into
local command mode, upon which the modem should respond with
OK. Then type AT&T0 <CR> to end the test.

If the typed text does not appear correctly on the screen, or not
at all, first check all connecting cables and the DIP switch/jumper settings. In particular, the Echo on/off switch, if provided, should be in
the position that echoes to the terminal in the command mode. In a
software-controlled modem, send the command ATE1 instead of setting switches.

Figure 16-2 Local analog loopback test

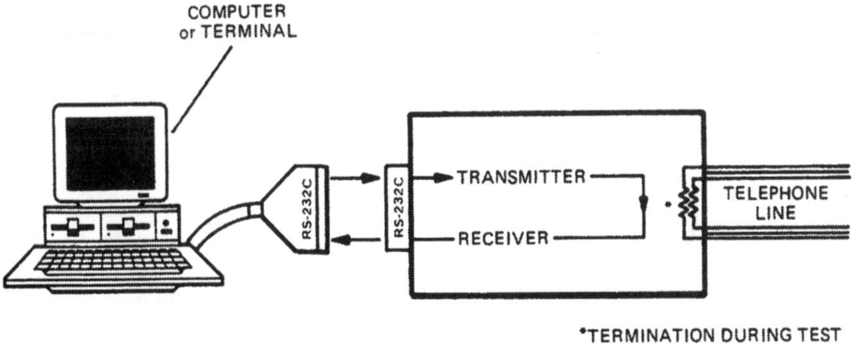

Next, check the cable between the serial-interface jacks on the computer/terminal and the modem. The pin configuration could be incorrect, the terminal may be configured as DCE instead of DTE, or pins 2 and 3 in the cable could be reversed (see description of a null modem in Chapter 6). Check that the Data Terminal Ready (DTR) lead in the serial interface is on. If the terminal/computer does not provide the DTR signal to the modem, which may be the case, then there should be a switch, a jumper, or an AT command to force the DTR lead to the on status. As a last resort, if the problem of not being able to see your own typing echoed on the terminal screen still persists, substitute another modem.

Local Analog Loopback Self-Test (&T8)

This test checks the ability of the local modem to send and receive data and should be performed if the previous (&T1) test passed. Figure 16-3 shows the test configuration. The internal test generator sends a test pattern, and the internal counter counts errors and sends the count (if any) to the terminal. The test can be initiated in the local command mode, for example, with the command ATS18=15&T8 <CR>. After 15 seconds the message:

```
000
OK
```

should appear on the screen to indicate no errors, or:

```
nnn
OK
```

to indicate nnn errors. Failure of this test usually points to a defective modem.

Figure 16-3 Local analog loopback self-test with pattern generator

TEST PATTERN
GENERATOR → TRANSMITTER

ERROR
DETECTOR ← RECEIVER

LAMP

TELEPHONE
LINE

*TERMINATION DURING TEST

Remote Digital Loopback Test (&T6)

All digital loopback tests must be performed while the modem is configured for asynchronous operation. Before beginning, place the modem in the asynchronous mode with the &Q0 command (asynchronous/synchronous mode selection).

This test is an extension of the local analog loopback test, but it also tests the telephone transmission facility and remote modem. The test configuration is shown in Figure 16-4. The test is initiated by first establishing a connection to the remote modem, then going into the offline mode by typing the escape sequence +++, then typing the AT&T6 <CR> command. Typing THE QUICK BROWN FOX... should result in the same sentence appearing on the terminal/computer screen. Problems showing up during this test, which have not showed up in a previous test, are most likely caused by failure of either the telephone line or the remote modem.

Figure 16-4 Remote digital loopback test

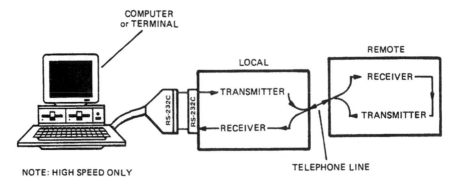

COMPUTER
or TERMINAL

LOCAL

REMOTE

RECEIVER

TRANSMITTER

RECEIVER

TRANSMITTER

RS-232C   RS-232C

NOTE: HIGH SPEED ONLY

TELEPHONE LINE

Remote Digital Loopback Self-Test (&T7)

This test is similar to the remote digital loopback test except that it uses the internal test-pattern generator instead of the terminal/computer. The test configuration is shown in Figure 16-5. By excluding the local terminal from the overall test, it focuses more on problems related to the telephone line and remote modem.

Figure 16-5 Remote digital loopback self-test with pattern generator

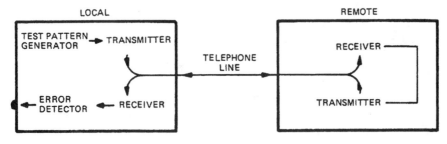

NOTE: HIGH SPEED ONLY

Local Digital Loopback Test (&T3)

This test checks the communication link, local modem, and remote modem (it permits a non-ITU-T-compatible remote modem to engage in a digital loopback test with the local modem).

The test is initiated by the remote terminal or remote modem. The test may require an additional telephone voice line to coordinate the test set-up. The test configuration is shown in Figure 16-6. The local modem loops back all data received from the remote modem. The remote modem can then compare the data sent with the data received.

This test is used primarily to allow a remote modem that is not equipped with digital loopback capability to perform such a test. The test is started from the local modem by first establishing a connection, going into the offline mode by typing +++, then typing the command AT&T3 <CR>. The remote terminal can now start sending THE QUICK BROWN FOX... and look for the correct sequence appearing on the remote terminal. The test is concluded by typing AT&T0 <CR> at the local terminal.

Figure 16-6 Local digital loopback test

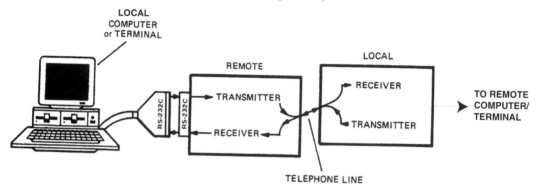

End-to-End Self-Test (&T2)

This test requires coordination between the local and remote terminals, because the remote modem must be instructed—by means of the same AT&T2 command—to send a test pattern and check for errors in the test pattern received from the local modem. As shown in Figure 16-7, both modems are sending test patterns, which are checked by their counterparts. Of course this test assumes that both the local and remote modems have the capability for sending such a pattern. This test is passed if both modems report no errors.

Figure 16-7 End-to-end loopback self-test with pattern generator

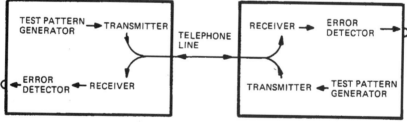

Testing Modems With Specialized Equipment

If the initial tests point to the modem as the probable cause of data errors, it may be advisable to test the modem by itself without any additional data channel components such as the computer or telephone line. Such tests measure the susceptibility of the modem to transmission facility impairments, as indicated by the signal-to-noise (S/N) ratio, and to other parameters such as delay and attenuation distortion. Such tests are beyond the scope of an individual modem and PC user, but are critical for manufacturers of communication equipment and many large organizations that use such equipment. Such tests are also performed in modem surveys conducted periodically by independent testing laboratories, or by publications such as *Byte* or *PC Magazine*. The surveys compare the relative performance of modem types made by various manufacturers.

When testing a specific condition, it is not always easy to properly interpret a test that is performed on a single misbehaving sample of a modem. The test results should be compared to similar tests performed earlier on the same modem or performed on the same kind of modem known to be in good working condition. The main reason for the difficulty in deciding whether the modem is or is not working properly from a single measurement is that modem manufacturers very seldom provide figures such as S/N ratio or susceptibility to distortion in their specification sheets.

The conceptual layout of a modem performance test is shown in Figure 16-8. The layout simulates a realistic data connection, but the variable components, such as line noise and distortion, are replaced by reproducible line simulators and transmission impairment generators.

A line simulator introduces attenuation and delay distortion that corresponds to a switched telephone line; an impairment generator introduces white (broad-band) and impulse (spike) noise. The modem test sets shown at both ends of the connection exercise all modem functions, starting with handshake protocols, and continue with the transmission of pseudo-random sequences of data. The final result, as recorded by the modem test set, is the bit error rate for a range of S/N ratios, and other distortions that are selected with the impairment generator. Signal and noise levels are monitored with the level meters shown in Figure 16-8.

Figure 16-8 Modem noise-sensitivity test setup

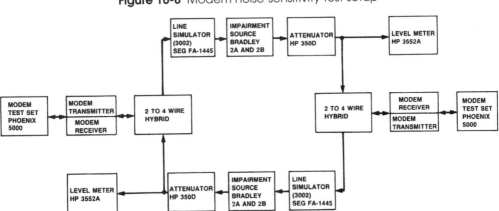

The conceptual layout of general-purpose instruments, as shown in Figure 16-8, has been considerably simplified by using specialized test instruments. Telecom Analysis Systems Inc. (TAS) in Eatontown, New Jersey developed a complete line of such instruments. One of their products is the TAS Series II GT Automatic Modem Test System. The product includes all hardware and software required for thorough modem testing. It is used for testing standard modems, PC internal modems and fax modems.

The TAS Series II system emulates analog transmission impairments, and also emulates digital transmission facilities provided by the telephone company, such as T1, and digital loops, such as the Asymmetric Digital Subscriber Line (ADSL) discussed in Chapter 18. The TAS Series II Universal Central Office system emulates various telephone exchange formats. TAS Series II GT meets modem testing specifications from EIA/TIA, ITU (CCITT), ETSI, and other standards organizations.

At the heart of a Series II GT Modem Test System is the TAS 1200 Telephone Network Emulator. This instrument emulates the local telephone exchange at each end of a PSTN connection, as well as the analog and digital transmission facilities between local exchanges. The emulator allows you to vary central office conditions, analog facility impairments, and digital facility impairments.

The local loop is simulated by the TAS 240A Loop Emulator. It provides explicit subscriber loop emulation to match the unique testing requirements of high-speed echo canceling modems. The dual analyzer performs the tests that are critical to evaluating high-speed modem performance, including bit/character/block error rates, message error rates, file transfer throughput, and call connect reliability.

The instruments are controlled by a Windows-based program called TASKIT. It includes control programs for individual instruments and an automatic test executive that provides coordinated control of all modem test system components. The instrument control programs allow you to create, edit, and save instrument configuration files. The TASKIT Automatic Test Executive (TASKIT/Auto) combines instrument configuration files into complete modem test procedures.

The following figures show some display screens of TASKIT programs. Figure 16-9 shows the loop simulation test, with choices corresponding to a real-life telephone loop. Options include wire-gauge, length of each section in kft, and bridged taps that remain from previous subscriber installations.

Figure 16-10 shows a simulated telephone connection between two telephone offices. Transmission parameters such as noise, distortion and echo characteristics can all be adjusted.

Figure 16-9 TASKIT loop simulation

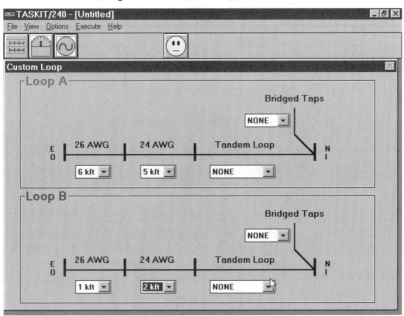

Figure 16-10 TASKIT modem setup simulation

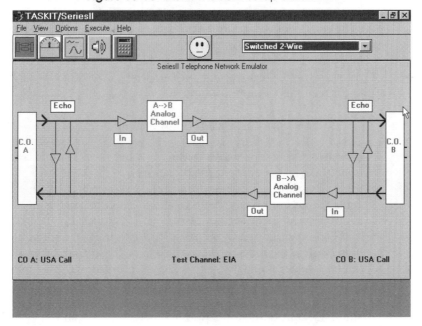

The Ultimate Modem Handbook

Testing RS-232-C Signals with a Break-Out Box

Modem testing with specialized instruments is not for the average PC user, due to the cost and complexity of such instruments. However, there is an inexpensive solution to testing modems beyond merely observing modem LEDs. Modem indicator lights, although helpful, are not provided on internal modems, and those provided on external modems only monitor a few selected signals out of the 25 leads that constitute the complete RS-232-C interface.

In order to be able to diagnose modem-related problems in the serial interface, it is recommended that you use a so-called break-out box. A break-out box is one of the basic tools for testing an RS-232-C interface between a computer and modem, and to help in localizing interface-related problems.

A break-out box belongs in any organization that depends on transferring data, in order to perform at least preliminary checks or diagnose a data-transmission problem. Using a break-out box requires a certain amount of understanding and ingenuity for interpretation of results, but considering its low cost, is an excellent investment.

A break-out box has built-in male and female DB-25 jacks into which serial interface cables are inserted. Each of the 25 leads can be patched through, opened up, or bridged for observation. Many break-out boxes feature a row of LEDs to allow the status of any lead to be continuously monitored. On data leads #2 and #3, there will frequently be a green and a red LED, or one LED that can turn either red or green to indicate lead polarity. When the polarity changes, as it does during the transmission of data, the LED alternates between red and green, and therefore appears orange during a fast exchange of data.

A break-out box is also equipped with a row of small jacks associated with each of the 25 leads. By means of wire jumpers plugged into those jacks, one can reconfigure a serial cable or, in a few minutes, make an experimental cable with any non-standard pin arrangement. To calm one's nerves, one should remember that the RS-232-C standard guarantees that no damage to the connected equipment will result from an incorrect connection of leads in a serial cable.

The complexity of break-out boxes ranges from simple home-brewed devices, to a standard, commercially-available passive break-out box, to a data analyzer that stores received and transmitted data and filters the stored data in search of certain tell-tale patterns. There are even software substitutes for break-out boxes—programs that simulate some of the functions of a break-out box and display the status of certain leads on the computer screen. What follows is a short description of the various levels of sophistication in break-out box design.

Simple, Home-Built Break-Out Boxes

Figure 16-11 shows a home-brewed break-out box, which helped me solve many serial-interface problems. The circuit uses a red LED and a green LED to distinguish between the positive and the negative signal polarities. Touching the alligator clip to the control leads, one at a time, will indicate their status. The On status, which is a positive voltage, should turn the green LED on; the Off status, which is a negative voltage, should turn the red LED on. Similarly, attaching the alligator clip to lead #2 (Transmitted Data) or to lead #3 (Received Data) should alternately turn both LEDs on when data is being received or transmitted. By rearranging the jumper wires, the break-out box can also be made into a null modem or a cheater cord, as discussed in Chapter 6.

Figure 16-11 Break-out box, home-brewed version

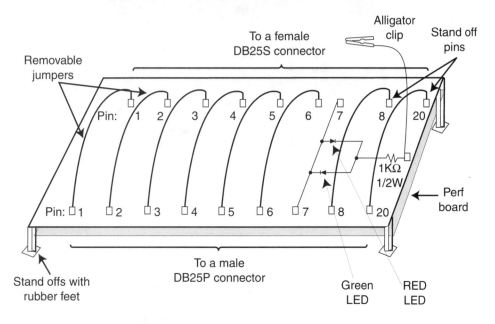

Commercial Break-Out Boxes

Figure 16-12 shows a commercial version of a break-out box. A big advantage of this type of device, compared to previous implementations, is that the status of several leads can be monitored at the same time. Also, each of the 25 leads can be patched through or can be opened up by means of miniature DIP switches. Short jumpers can be plugged into the panel to connect an oscilloscope or voltmeter to any of the leads, or jumpers can be used for cross-connecting any of the leads. A battery is often included to allow forcing individual leads to the on or off state. In addition, LEDs continuously monitor the status of data and control leads #2 (TD), #3 (RD), #4 (RTS), #5 (CTS), #6 (DSR), #8 (CD) and #20 (DTR). A high-impedance transistorized circuit powered by a built-in battery minimizes the load on the monitored leads while operating the LEDs at full brightness.

Figure 16-12 Break-out box, commercial version
with full implementation

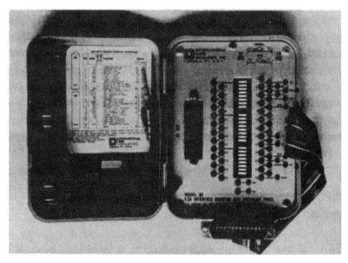

Observing the important indicators associated with leads #2 through #6, and #8 and #20, should provide the following results:

Condition A: Computer not powered, or computer powered, but communications software not loaded. All indicators are off.

Condition B: Communications software program is started, but connection not yet established. The RTS, DSR and DTR lights should be on. At this point, you can press keys on the computer or terminal keyboard and the TD light should blink.

Condition C: Connection to the remote terminal is established. The RTS, CTS, DSR, CD and DTR lights should be on permanently, the RD light should blink when receiving incoming messages, and the TD light should blink when sending messages.

When the observed indicators do not agree with conditions A, B or C, consult Table 6-1—which shows the sequence of events and signals occurring on the serial port between the computer and the modem—to make an initial diagnosis of the serial interface problem. The first event that occurs is the DTR (Data Terminal Ready) lead going high. If the DTR indicator is off, the problem is on the computer side.

When DTR is on and everything else is working properly, the next signal to go high is DSR (Data Set Ready). The missing DSR light may indicate a defective modem or a broken lead.

The next two leads to go high are RTS (Ready to Send lead from the computer) and CTS (Clear to Send lead from the modem). At this point, the connection should be established and the remote terminal should start sending a carrier signal. The successful reception of the carrier should turn the CD light on. If it is not on, it may indicate that the connection was not established or that the local or remote modems are defective.

After successfully establishing a connection with the remote terminal, the RTS, CTS, DSR, CD and DTR lights should all be on continuously, and the RD and TD lights should blink when data is being received or transmitted, respectively. The blink rate gives some indication of the transmission speed. If the LEDs are blinking but nothing appears on the computer screen, the fault probably lies in the computer and not in the modem or communication path.

There are two other types of commercially-available break-out boxes. One type is more and one is less expensive than the basic model described previously. The more expensive version, which sells for around $200, contains a simple asynchronous data generator and a built-in error counter. It allows you to send a pseudo-random data stream, loop it back, and check for its accuracy as it comes back. This type of instrument is not very popular today because it duplicates some of the modem's built-in &T tests described earlier in this chapter, and is therefore a redundant expense.

Another type of popular break-out box, which sells for less than $10, is a small connector that is equipped with male and female DB-25 jacks at both ends, and a row of red and green LEDs that monitor the status of the principal data and control leads. The unit I recently purchased at a computer show, for around $6, monitors the following leads: TD (2), RD (3), RTS (4), CTS (5), DSR (6), RLSD (8), QM (11), SRS (19) and DTR (20). It is shown in Figure 16-13.

Figure 16-13 Break-out box, inexpensive commercial version

Protocol Analyzer with a Break-Out Box

A considerably more flexible, but also more expensive ($10,000 to $15,000) test option is a *protocol analyzer*, which includes a break-out box. In fact, it could be called the ultimate break-out box. A protocol analyzer not only supervises all serial-interface leads, but it also captures their status over a period of time and analyzes the data according to any specified criteria. Some of this analysis occurs online by means of data filters, which select only data that meets certain requirements—for example, by looking only at the first 1024 bits that have a certain flag pattern. The data is captured into the analyzer's memory and saved to a diskette or hard disk, where it can be analyzed offline. Use of data filters is the only way to diagnose intermittent problems, because continuous recording of data over an extended period of time would likely overflow any storage medium and make the problem difficult to find if it was captured at all.

A good example of such an instrument is the Hewlett Packard J2301B Protocol Analyzer, also called the LAN/WAN/Internet Advisor. This portable instrument looks like a modified laptop computer. When the instrument is attached to a serial interface, it will automatically determine all transmission parameters, such as transmission speed, number of bits, parity, and so on. and analyze the stored data.

Figure 16-14 shows the decision tree that the instrument uses to decide how to interpret data captured at the serial interface. Operational options are displayed in a sequence of menus, with menu items also assigned as keyboard soft keys for quick keyboard access. The J2301B will perform bit error rate testing, search for and display any given data pattern, show a side by side display of signals on various leads, and permanently store the captured data in its built-in RAM, the built-in diskette drive or hard disk drive.

The HPJ2301B analyzer supports transmission speeds of between 50 bps and 45 Mbps. The data can be transmitted in ASCII, EBCDIC, Baudot, IPARS (the airline reservation code), and in several other formats as well. Whatever the format, it will be properly decoded by the instrument and displayed on the screen. The instrument supports several serial interfaces, with different optional plug-in break-out boxes applicable to each interface. The interfaces supported by the manufacturer are RS-232/V.24, V.35, V.36/RS-449/V.10/V.11, DDS 4-wire and T1, ISDN, DS3/E3, OC-3/STM-1, ISDN BRI/PRI, and X.21.

Figure 16-14 Decision tree of a protocol analyzer

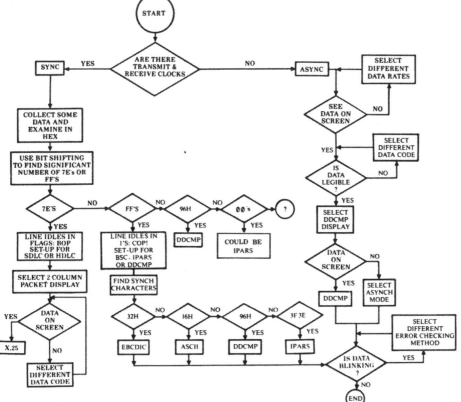

The J2301B protocol analyzer can be used in either the *immediate* mode, where all selections are made while the interface is being tested, or it can be used in the *programmed* mode. A program that controls all instrument settings, filters, and performs preliminary analysis can be written in advance and stored on a diskette. The program can then be executed by the instrument at a later date. This approach provides self-documenting results that are easily repeated.

The following example shows a short program, written in a procedural language supplied with the analyzer, to measure the time-delay between the RTS lead turning on and the CTS lead turning on, and display all control signals during that part of the handshake sequence.

```
BLOCK 1:
     START DISPLAY
     WHEN LEAD RTS GOES ON THEN GOTO BLOCK 2

BLOCK 2:
     BEEP
          AND THEN
     HIGHLIGHT
          AND THEN
     START TIMER 1
     WHEN LEAD CTS GOES ON THEN GOTO BLOCK 3
BLOCK 3:
     BEEP
          AND THEN
     HIGHLIGHT
          AND THEN
     STOP TIMER 1
          AND THEN
     STOP TEST
```

A protocol analyzer, as seen from this short description, is an instrument that belongs in any large organization involved in data communication. Use of the instrument assumes a certain amount of training, but it will pinpoint difficult-to-find and often intermittent problems that could not be found by any other method.

As with other expensive instruments, there is usually an inexpensive alternative for the PC user. Many terminal-emulation programs, such as Smartcom for Windows from Hayes Corporation, include a *data analyzer* mode, an option to show modem commands. When this option is enabled, the computer will display all data exchanges between the computer/terminal and modem. The computer will also display all AT commands sent to the modem, all modem responses, all handshaking results, and so on. Although not as flexible and comprehensive as the previously-described specialized instruments, it is basically free and therefore accessible to every modem user.

An example of a session with an ISP using this option is shown in Figure 16-15. Computer-to-modem commands are shown as a raised line, while modem responses appear as a lowered line. Special non-ASCII characters are shortcuts for control characters such as carriage return (small c/r) and line feed.

Figure 16-15 Smartcom in data analyzer mode

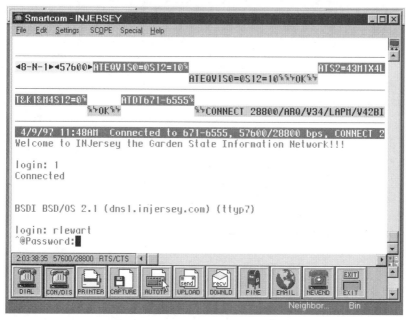

Modem Test Software

Hardware and software can often be made to perform similar tasks. Much general and diagnostic information about the modem you are using can be obtained not only from expensive test instruments but also from assorted diagnostic software. Some of this software comes free with DOS and Windows operating systems, while some of it is available as shareware or commercial software at a reasonable cost. The use of MSD program that comes bundled with DOS was discussed in Chapters 6 and 8. The program WhatCom from Data Depot Inc., which you will find on the enclosed CD-ROM, is described next in more detail.

All modem diagnostic programs such as MSD or WhatCom should be run in pure DOS mode (not in a DOS window under Windows). As mentioned in Chapter 8, there are two ways to get into the pure DOS mode from Windows 95:

Method 1:   Click the Start button and select "Restart the computer in MS-DOS mode?" from the pop-up menu.

Method 2:   Hold the F8 key down when the computer first starts or reboots. A menu appears with the title "Microsoft Windows 95 Startup Menu." Choose option 6, "Command prompt only."

After getting a C:\> prompt, copy all files in the WHATCOM directory from the enclosed CD-ROM to a newly created directory with the same name on your hard disk. One of the files in the directory should be WHATCOM.EXE. Execute the program by typing:

```
whatcom
```

A logo screen appears for a few seconds when the program starts. When you execute the program for the first time, it will prompt for your name and company name, which will then appear in successive runs. WhatCom then scans your computer system for any installed serial ports and displays the status, UART chip type, and whether a mouse, modem, or unknown device is attached, and if any IRQ conflicts were detected.

The program will:

- Prevent and resolve IRQ I/O conflicts
- Find out if a serial port is present and what UART chip is being used
- Find out whether a mouse or modem is attached to a COM port
- Identify the modem's characteristics and command responses
- Find out if the port and modem properly sends and receives data
- Find out if the modem is attached to an active phone line

The opening diagnostic screen of the WhatCom program, with a list of options, was shown in Figure 8-2 in Chapter 8. These options are described next:

- External/Internal Modem Installation Advice.
Tells you how you should install an internal or external

modem, and which I/O address and IRQ line to use. Figure 16-16 presents a set of recommendations for an internal modem. Recommendations are based on the availability of COM ports and IRQs.

Figure 16-16 WhatCom internal modem recommendations

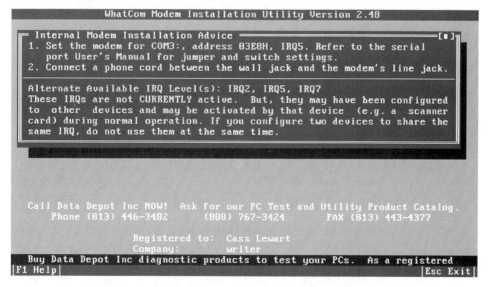

```
            WhatCom Modem Installation Utility Version 2.48
 ┌─ Internal Modem Installation Advice ══════════════════════════[■]┐
 │ 1. Set the modem for COM3:, address 03E8H, IRQ5. Refer to the serial
 │    port User's Manual for jumper and switch settings.
 │ 2. Connect a phone cord between the wall jack and the modem's line jack.
 │
 │ Alternate Available IRQ Level(s): IRQ2, IRQ5, IRQ7
 │ These IRQs are not CURRENTLY active.  But, they may have been configured
 │ to other  devices and may be activated by that device (e.g. a  scanner
 │ card) during normal operation. If you configure two devices to share the
 │ same IRQ, do not use them at the same time.
 └──────────────────────────────────────────────────────────────────┘



    Call Data Depot Inc NOW!  Ask for our PC Test and Utility Product Catalog.
       Phone (813) 446-3402      (800) 767-3424       FAX (813) 443-4377

                    Registered to:  Cass Lewart
                    Company:        writer
   Buy Data Depot Inc diagnostic products to test your PCs.  As a registered
 │F1 Help│                                                        │Esc Exit│
```

- Modem Test.
  Performs a data-transmission loopback test on the serial port chip and modem, an interrupt test on the serial chip, a self-identify test on the modem, and a dial-tone test on the phone line. Provides troubleshooting advice if a test fails.
  Figure 16-17 shows the results of tests performed on my modem. Notice that, in this example, the dial-tone test failed. The simple reason was that my phone was disconnected.

- Modem Information.
  The program will also analyze your modem and provide its specifications based on responses to the ATI commands discussed in Chapter 9. Figure 16-18 shows how WhatCom sees my modem. As with MSD, you cannot always trust the results 100%. If the modem manufacturer did not enter correct information in the modem's ROM, neither WhatCom nor any other program will get the right information.

Figure 16-17 WhatCom modem test results

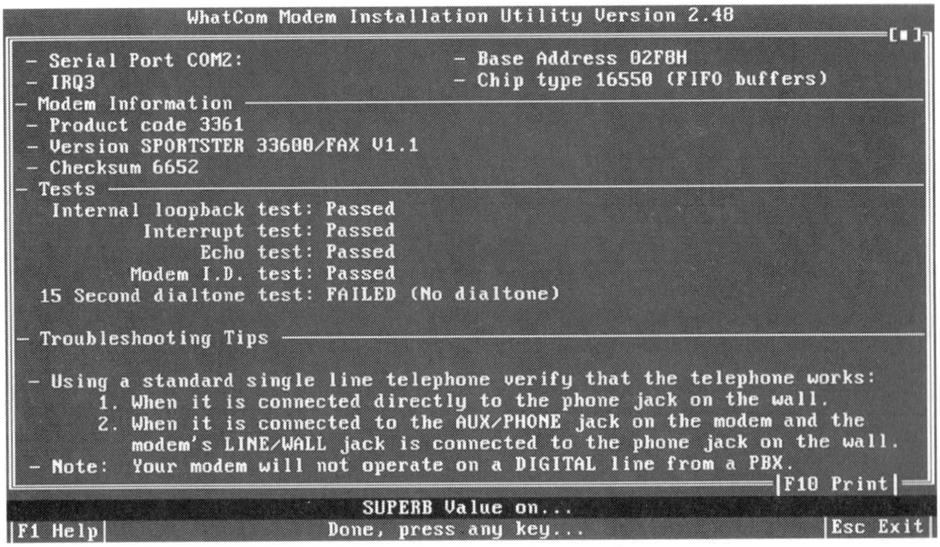

```
        WhatCom Modem Installation Utility Version 2.48
                                                                [■]
  - Serial Port COM2:              - Base Address 02F8H
  - IRQ3                           - Chip type 16550 (FIFO buffers)
 - Modem Information
  - Product code 3361
  - Version SPORTSTER 33600/FAX V1.1
  - Checksum 6652
 - Tests
    Internal loopback test: Passed
           Interrupt test: Passed
                Echo test: Passed
         Modem I.D. test: Passed
    15 Second dialtone test: FAILED (No dialtone)

 - Troubleshooting Tips

  - Using a standard single line telephone verify that the telephone works:
       1. When it is connected directly to the phone jack on the wall.
       2. When it is connected to the AUX/PHONE jack on the modem and the
          modem's LINE/WALL jack is connected to the phone jack on the wall.
  - Note:  Your modem will not operate on a DIGITAL line from a PBX.
                                                          |F10 Print|
                    SUPERB Value on...
 |F1 Help|          Done, press any key...              |Esc Exit|
```

Figure 16-18 WhatCom modem information

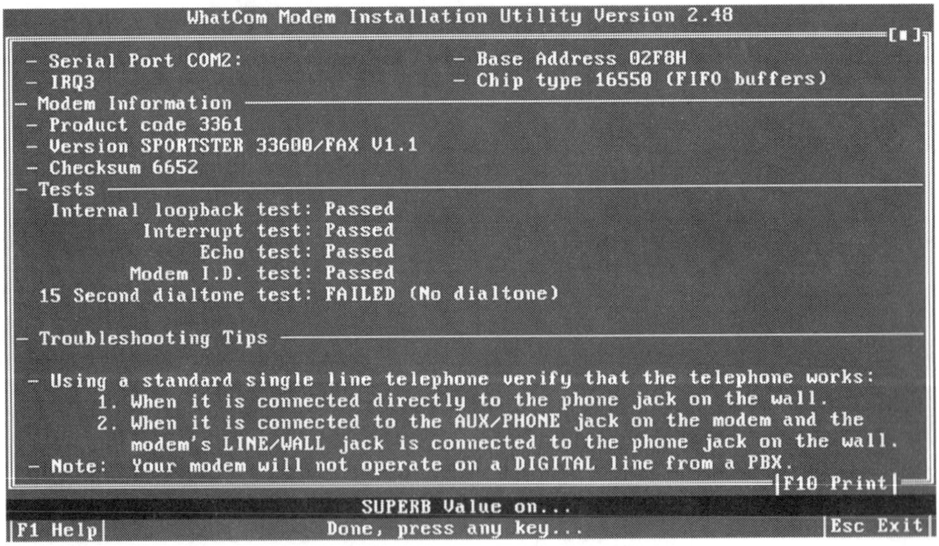

```
        WhatCom Modem Installation Utility Version 2.48
                                                                [■]
  - Serial Port COM2:              - Base Address 02F8H
  - IRQ3                           - Chip type 16550 (FIFO buffers)
 - Modem Information
  - Product code 3361
  - Version SPORTSTER 33600/FAX V1.1
  - Checksum 6652
 - Tests
    Internal loopback test: Passed
           Interrupt test: Passed
                Echo test: Passed
         Modem I.D. test: Passed
    15 Second dialtone test: FAILED (No dialtone)

 - Troubleshooting Tips

  - Using a standard single line telephone verify that the telephone works:
       1. When it is connected directly to the phone jack on the wall.
       2. When it is connected to the AUX/PHONE jack on the modem and the
          modem's LINE/WALL jack is connected to the phone jack on the wall.
  - Note:  Your modem will not operate on a DIGITAL line from a PBX.
                                                          |F10 Print|
                    SUPERB Value on...
 |F1 Help|          Done, press any key...              |Esc Exit|
```

- Terminal Tests.
  Allows you to send any command to the serial device (modem, mouse, and so on) and displays responses in text, ASCII, decimal, or hexadecimal format. The terminal emulator shows the status of the serial port registers.

- View Database.
  A self-learning feature that lets you store and edit modem characteristics in a database, allowing WhatCom to recognize your type of modem even if it is non-standard.
- A Second Check of Serial Ports.
  Reruns the first test to check serial ports in the event you forget to turn power on or attach a modem.
- Return to DOS.

Troubleshooting Specific Problems

As an introduction to this section, I will describe a recent experience with my modem. It shows how difficult it is to diagnose and solve a communications problem. A solution requires perseverance, patience and luck.

My ISP moved all their equipment a few months ago from the vicinity of Holmdel, New Jersey, where I live, to Newark, New Jersey, about 35 miles north of my location. I can still dial through using an FX (Foreign Exchange) line in Holmdel and be connected to Newark, but the telephone connection leaves much to be desired. While my old generic, internal 28.8kbps modem, from an unknown company named Pragmatic, worked fine with my ISP prior to the move, and worked fine with other ISPs and many local BBSs, it refused to work reliably with the FX line. It would either fail to connect at all, or it would connect and later drop the connection after a short period of time.

After a friend loaned me his U.S. Robotics 28.8 Sportster modem, I was able to dial through reliably on the FX line. I returned the modem to my friend, bought an identical modem, installed it, and for a couple of days had no problems. However, strange things soon started to happen. The modem would dial and connect, but then it would display the message No Dial Tone and quickly disconnect. One significant clue was that I could still dial using the Winsock dialer, or manually using the Hayes ATDT command, but could not dial when using the Smartcom for Windows communications program—which is my favorite—when it was configured for a Hayes-compatible 28.8K modem. I did not configure Smartcom for the U.S. Robotics modem because none of the options included a 28.8 Sportster modem.

My obvious conclusion was that my Smartcom program was sick, either infected by a virus or otherwise corrupted. But before reinstalling it, I tried my old Pragmatic modem with the sick (I thought) software. Surprise—it worked as well as it ever did before. I could do everything but dial the FX line to my ISP.

I then installed my U.S. Robotics 28.8 Sportster again and ran the WhatCom diagnostic program. It advised me to switch from Port 2 IRQ 3 to Port 2 IRQ 5. I changed jumpers on the Sportster modem card and I could now at least dial manually with the ATDT command. I then put Smartcom in the Data Analyzer mode described earlier in this chapter, where I could view the complete computer-to-modem handshaking sequence displayed on the screen. I noticed that the initialization string included an ATW1 command, which is supposed to enable the error-correction protocol for Hayes modems. However, this command is not implemented in the U.S. Robotics Sportster modem and results in an error condition. Not only does it return an error code, but it also corrupts the dialing string by making the modem expect a second dial tone—explaining the mysterious No Dial Tone message I was getting.

Next I configured Smartcom for a 14.4 Sportster modem. One of the choices, the W1 string, disappeared, and the modem worked fine again. Still, a few mysteries remained:

- Why did I have to switch the IRQ from 3 to 5?
- Why did the ATW1 command not affect the borrowed modem or the purchased modem in the first few days? The two modems appeared to have the same ROM date.

The likely answers are that I changed some other hardware setting in my computer configuration without realizing it, and that the modem manufacturer had different versions of the same program's ROM with the same date. Well, at least everything works now.

The following describes some common modem problems and appropriate solutions. The problems and solutions were collected from my own experiences, from various friend's experiences, and from browsing the Internet.

Before starting any diagnostics, be sure that the speaker in your modem is working, or that you have the audio monitoring circuit (described in Chapter 8) connected to your phone line.

Symptom: Dialing problems

Possible Causes:

Computer, modem or phone line are possible culprits.

Suggested Cures:

First learn how to dial manually before starting a data connection. Then perform further testing.

Put the communications software in local command mode, as described in Chapters 9 and 10. Then lift the handset on your telephone set and dial the number you wish to call (BBS or ISP). Type ATH1 to connect the modem, then hang up the handset. Then type ATO to put the modem in the online mode. You should now be prompted by the ISP/BBS for your User Name and password and should be able to exchange data with the remote computer.

Symptom: Modem does not dial

Possible Causes:

No power in the external modem.
Configured for the wrong COM port.
Wrong initialization string.
Phone line and phone jacks are incorrectly connected.

Suggested Cures:

Check power.
Try a regular phone on the modem line and listen for a dial tone.
Test COM port configuration with MSD or WhatCom.
Put the modem in local command mode and type ATDT followed by your own phone number. If you hear dialing followed by a busy signal, then the initialization string is incorrect. If you do not hear any dialing, the modem might be defective or incorrectly seated (for an internal modem), or the connecting cable might be defective, or there might be a conflict with the selected COM port.
Make sure that you have connected the LINE jack (on your modem) to the phone wall outlet, and the PHONE jack is connected to your telephone set. If not, reverse the leads.

Symptom: Modem does not wait for dial tone, and disconnects after making a connection.

Possible Causes:

The Wait register is set too low for your slow dial tone.

Suggested Cures:

Increase the value in the S6 register from the default of 2 seconds (the wait time for dial tone) by adding a command such as S6=5 (for 5 seconds) to the initialization string.

Symptom: You hear the dial tone and the modem dials, but does not break the dial tone.

Possible Causes:

Your phone system might not accept touch-tone dialing.
The line relay in the modem is defective.

Suggested Cures:

Change from touch-tone to pulse dialing in the software config-uration of your communications software (ATDP instead of ATDT), or request touch-tone service from the phone company.
Try another modem.

Symptom: Failure to connect after dialing.

Possible Causes:

Improper dialing code for your location.

Suggested Cures:

Make sure that you have included an "8," or "9," (be sure to include the comma, but not the quotes), if necessary, for reaching an outside line from an office phone. If you are calling from a hotel, you may need to check the instructions on the room telephone for the pre-fix to use (if any). If you call outside the U.S., you must use the inter-national access code.

Symptom: Call-waiting disconnects data calls.

Possible Causes:

Interference from an incoming call while you are online.

Suggested Cures:

Disable call-waiting by entering *70 or *71, whichever is appli-cable in your area, in the initialization string. For example enter, ATDT*70 in place of ATDT. If the *call-waiting disable* feature is not available from your phone company, increase the value in the S10 register from the default of 14 (in tenths of seconds) by adding the command S10=30 (3 seconds), for example, to the initialization string. The S10 register controls the wait time to disconnect after the carrier is dropped.

Symptom: Connection to ISP is established, but login fails.

Possible Causes:

Incompatible handshaking sequence due to an incorrect initialization string.

You or your ISP messed up your User Name or password.

Suggested Cures:

Set the initialization string to ATDT.

Check your User Name and password.

Symptom: Frequent busy signals.

Possible Causes:

Your ISP or BBS is overloaded with customers.

Suggested Cures:

Find another ISP or call during a less-busy time of day.

Symptom: Frequent disconnects.

Possible Causes:

Poor phone-line quality.

Modem too susceptible to line impairments.

Modem might have a bug in its ROM program.

Suggested Cures:

Try to convince the telephone company to improve your phone line.

Get a higher-quality modem.

Increase the value in the S10 register (the wait time to disconnect after the carrier drops) from its default of 14 (in tenths of seconds) by adding the S10=30 (3 seconds) command, for example, to the initialization string.

Try \A0 for a 64-character maximum MNP block instead of the default 256 character block.

Remove all commands that refer to data compression and error correction from your initialization string and disable MNP 4, 5, V.42, and V.42bis protocols.

Try using a lower data connection rate than the maximum rate of your modem.

Disconnect all other phones and answering machines that are connected on the same line to see if any of them interfere with your calls.

Check with your ISP/BBS if they have different phone lines and calling numbers for different types or speeds of calling modems.

Check with the modem support line, provide the version of your ROM (find it by using ATI commands) and find out if there are any problems unique to your modem.

If you have Internet access, post a question in one of the Usenet groups. One example is the very active comp.dcom.modems news group. You will get plenty of answers, some useful, some not so useful, from people all over the world who will try to help you.

Symptom: File transfers don't work.

Possible Causes:

Incorrect setting for flow control.

File transfer protocols in your communications software and on the server have different non-compatible versions.

Suggested Cures:

Change flow control from software to hardware control, preferably by selecting RTS/CTS.

Try a different file transfer protocol such as Zmodem, Ymodem or Kermit. Kermit is usually the most robust.

Change parity and block-size parameters in the file transfer protocol options.

Symptom: Fax connection is established but no data is received.

Possible Causes:

Incorrect Fax Class setting.

Suggested Cures:

You probably have a Class 2-capable modem, which your fax software is trying to use as Class 1. If this is so, change Fax Modem Class to 2 in your fax program.

Symptom: Fax modem fails to answer incoming faxes, although it can still send faxes.

Possible Causes:

The receive fax option was not set.

Suggested Cures:

Check to see that the Automatic Receive option of your fax program is enabled. If it is, your fax program might be using the wrong

COM port or might have some other configuration problem that will need to be corrected.

A good diagnostic tool is the call-back feature in many telephone offices. If you dial 550-xxxx, where xxxx represents the last four digits of your phone number, then hang up for a second, then hang up again (or press the Flash button on your phone, if it has one), you should get a call-back from the dialer in your local telephone office. The code (550) may differ from exchange to exchange. The 550 code works in my area of New Jersey.

You may also try to manually set the fax modem to the Answer mode by typing ATA. When an incoming call is received, the modem should answer and automatically change from the offline command mode to the online mode.

The list of symptoms and their cures could easily be expanded to fill an entire book. It is likely that you will add many more symptoms and cures applicable to your specific situation. You can also find more troubleshooting help by browsing the Internet sites listed in Appendix A.

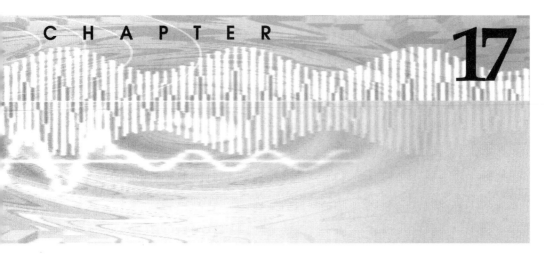

Transmission Facility Testing

Often, after running the tests described in Chapter 16, you will conclude that there is a problem somewhere in the data path that involves the terminal, modem or transmission facility. If the problem cannot be resolved locally, and all tests point to the transmission facility—meaning the telephone line or leased or privately-owned circuit—then the transmission parameters of that facility should be tested.

However, there is always the question of what you can do with the results—in particular, if the blame for poor performance is put on the switched telephone line. The telephone company does not guarantee any specific transmission parameters on such circuits. The only guarantee is that both parties in a telephone conversation can understand each other, and the only recourse is to ask for dropping of charges for a specific non-satisfactory call. The only time that conducting your own facility-transmission test will help in convincing the telephone company that they are responsible is when dealing with a leased-line circuit. For leased line, the telephone company explicitly guarantees specific minimum requirements—meaning characteristics such as attenuation, delay distortion, harmonic distortion, and noise—as discussed in Chapter 13.

The transmission parameters that affect data transmission can be measured individually with test instruments called *transmission measuring sets*, which are made by several major instrument manufacturers—most prominent being Hewlett Packard Company. They come in a wide price range—from a few hundred to tens of thousands dollars—and with a large array of features. Obviously such tests are beyond the scope of the average PC user, but are often performed by technicians and consultants who work for large corporations. However, there is now a simple method, currently being offered by a modem manufacturer, that allows for an easy telephone line and modem evaluation. The method is described in the last section of this chapter.

Tests using specialized instruments are performed either manually or automatically, and with or without a graphical output. Some instruments perform a single test, while others perform a number of tests. Digital testers measure the modem-channel-modem performance by determining the bit error rate and the block error rate. Analog testers measure various transmission parameters that are associated with the distortion of data. The inter-relationship of these measurements is sometimes difficult to grasp, such as the relation between amplitude distortion, delay distortion, and bit error rate.

In general, although such relations exist, they cannot be expressed quantitatively or explicitly. An experienced individual can usually determine, based on reviewing the modem specifications, that a certain combination of these parameters will provide satisfactory transmission, or that an improvement in any parameter by itself will result in additional improvement of data transmission performance, and vice versa. In the event of a problem, a combination of protocol, digital and analog testing should quickly pinpoint the problem.

First we will cover the basic transmission measurements of single-impairment parameters, then follow this with a description of specialized measuring instruments that combine several parameters to arrive at single transmission-performance figures.

Attenuation Distortion Measurements

An attenuation tester is the most basic type of transmission measuring instrument. It consists of a variable frequency generator in the transmitting section and of either a frequency selective or flat voltmeter in the receiving section. The generator should cover the frequencies of interest—typically 100 Hz to 10,000 Hz—and the output impedance should be selectable to suit typical transmission lines—namely 50, 75, 150, 600 and 900 Ohms. The input impedance should be selectable in the same steps, or High (above 10 kOhm) for bridging measurements. The output voltage should be stable within 0.1 dB over the entire frequency range, and should be calibrated in dBm (decibels above 1 mW). Using Ohm's law, one can easily translate dBm, the units of electrical power at a given impedance level, into Volts.

Formula 17-1 Units of electrical power

Review of units of power:

$$dB = 10 \log \frac{P2}{P1}$$

dB	Power Ratio
0	1.00
1	1.26
3	2.00
6	3.98
10	10.00
20	100.00

Units of power:

dB   Unit of relative measurement

dBm   Unit of absolute measurement; 0 dBm = 1 milliwatt

dBrn   Unit of absolute measurement; 0 dBrn = 1 picowatt = –90 dBm

Ohm's Law:

$$V = I \times R$$
$$P = V \times I = \frac{V^2}{R}$$
$$V = \sqrt{P \times R}$$

e.g., Power = 0 dBm = 0.001 W

R = 600

$$V = \sqrt{0.001 \times 600} = 0.775 \text{ V}$$

V   Voltage in Volts

I   Current in Amperes

R   Resistance in Ohms

P   Power in Watts

Entering values in the above formulas shows that, for example, a 0 dBm signal across 600 Ohms will result in a voltage of 0.775 Volts. This is a value that is well known to every communications engineer.

The receiving section of the attenuation tester should be a voltmeter with a frequency-selectable filter, which can be adjusted to the received frequency. The reading scale should be in dBm and Volts, and the input impedance should be selectable to the same values as on the frequency generator.

A test is performed by either looping the transmission facility or placing another attended instrument at the other end of the transmission facility. If a one-way test is made, there should be voice communication between operators at both ends to coordinate their activities. A plot of measured attenuation at frequencies across the voice-band should give a good indication of the health of the circuit. Deviation from a flat line is called *attenuation distortion*. Comparison with the leased-line characteristics discussed in Chapter 13 will show any abnormalities.

Delay Distortion Measurements

Measurements of delay distortion are particularly important for data transmission. Excessive delay distortion smears the normally rectangular shape of data pulses such that they interfere with adjacent pulses. From the point of view of the operator, the delay distortion measuring set is somewhat similar to an attenuation tester. It has a variable frequency generator at the transmitting end, and allows you to select input and output impedance and select transmitting levels in dBm. The only visible difference to the operator is that, at the receiving end, there is also a scale that is calibrated in microseconds.

As shown in Figure 17-1, the transmitting section of a delay distortion measuring set sends a modulated wave, which is either looped back at the far end and returned and demodulated, or is measured at the far end by another delay-distortion measuring set. The zero crossings of the reference and received signals after demodulation are then compared. The time difference will vary as a function of the carrier frequency of the transmitted signal. The time difference of the zero crossings is then plotted as a function of the variable carrier frequency. This plot, when referring to a specific frequency (for example, 1000 Hz) is the relative delay distortion. The flatter the plot, the less delay distortion there is. A typical curve of delay distortion versus frequency of a switched voice telephone channel is shown in Figure 17-2. The minimum distortion is in the middle of the voice band between 1800 and 2000 Hz, and it increases at low and high frequencies to several milliseconds.

Figure 17-1 Delay distortion measuring set

AMPLITUDE

CARRIER
(A) FREQUENCY
fc

MODULATING
(B) FREQUENCY

RESULTANT
AMPLITUDE
(C) MODULATED
SIGNAL

ENVELOPE OF
(D) WAVEFORM (C)

ENVE-
LOPE
DELAY

THIS TIME
DIFFERENCE
MEASURED IN
MICROSEC

DELAYED
(E) ENVELOPE

TIME

ENVELOPE/GROUP DELAY:

* BELL: REQUIRES 4-WIRE LINE
            REQUIRES ONE OR 2 INSTRUMENTS
            USES EITHER RETURN OR FORWARD
            REFERENCE

*CCITT: REQUIRES 2-WIRE LINE
            REQUIRES 2 INSTRUMENTS
            IS AN END-TO-END MEASUREMENT

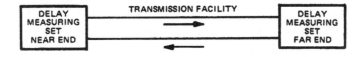

| DELAY MEASURING SET NEAR END | TRANSMISSION FACILITY → ← | DELAY MEASURING SET FAR END |

Figure 17-2 Delay distortion versus frequency

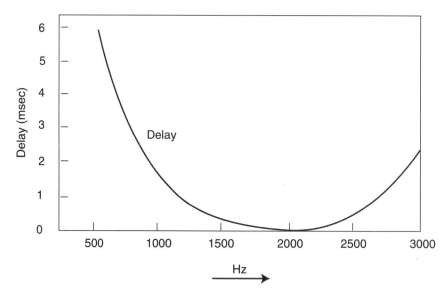

Noise Measurements

There are two kinds of noise that can affect data transmission: *steady* noise and *impulse* noise. Steady noise is determined by the characteristics of the transmitting and receiving equipment and by the modulation method. One of the components of steady noise is the thermal noise associated with each resistor, caused by the random movement of electrons. Steady noise can also be caused by interference from power lines. In general, the data communications user has no control over the steady noise, which ultimately determines the maximum data transmission rate.

The second type of noise is impulse noise, which consists of short bursts of noise, typically 10 to 15 dB above the floor of the steady noise. The impulse noise is caused by defective circuits, by crosstalk from other channels, by non-linearities in the modulation process, and by atmospheric interference. Because the design of transmitting and receiving modems takes the steady noise into consideration, it is not a problem for data transmission. However, impulse noise, being much higher than steady noise, can destroy the transmitted data during its brief occurrences. The solution to impulse noise is either an improvement to the circuit or use of error detection and correction through block re-transmission.

Steady noise is measured using a voltmeter that is bridged across the transmission line. The voltmeter should have a flat frequency response along the voice-band range. Some noise meters use the so-called C-weighting, shown in Figure 17-3, to measure noise on voice circuits. This weighting mimics the response of the average human ear and should not be used on data circuits.

Many telephone circuits include so-called companders and expanders. These circuits improve voice transmission by lowering circuit gain and therefore lowering the apparent noise during quiet intervals. The popular Dolby circuit, used for many high-fidelity recordings, is based on the same principle. Therefore, to obtain a true noise reading, the circuit should carry a signal of the same power as the expected data signal. The way to achieve this is to send a steady signal (for example, at 1000 Hz) over the transmission facility and filter it out with a sharp frequency-discrimination filter at the receiving end where noise is being measured.

Figure 17-3 C-weighting for noise measurements

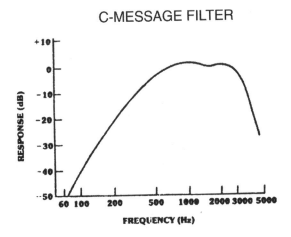

Impulse noise is measured with a voltmeter that is attached to an event counter. The measurement is performed over a certain time period, typically 15 to 30 minutes. The counter counts each noise burst that exceeds a specified level (in dBm). In general, a row of counters is preset to different noise levels, such as -15 dBm, -10 dBm and -5 dBm. All counters are started at the same time. Over a period of perhaps 30 minutes, the -15 dBm counter might indicate 35 counts, while the -10 dBm counter might indicate 10 counts, and the -5 dBm counter might indicate three counts. From this distribution, one can derive, by simple calculation, the expected bit error rate. If it is found that the expected error rate due to impulse noise is much lower than the actual error rate, then there must be some other cause for the errors. They may be caused by delay distortion or modem problems.

Phase Jitter Measurements

Phase jitter, as shown in Figure 17-4, affects the bit error rate, particularly in modems that use differential phase modulation and Quadrature amplitude modulation. As the bit value is determined by the relative phase of successive signal elements, phase jitter can affect signal recognition. This impairment is caused by the fading of microwave signals and by changes in carrier frequencies in the common carrier multiplex terminals. It is less common today, because most long-haul transmission is carried on digital circuits over optical-fiber facilities, which are less prone to fading. Phase jitter can also be caused by defective modems.

Phase jitter is measured by comparing the received signal with a steady signal generated by a local quartz-controlled oscillator. The phase difference between the received signal and the local signal is translated into a voltage, which is then measured and displayed.

Figure 17-4 Phase jitter effect on data signal components

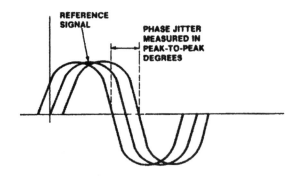

Eye Pattern Testing

The combined cumulative effect of various kinds of distortion and noise is best seen in the so-called eye pattern. The test setup to measure and display the eye pattern is shown in Figure 17-5. The setup consists of a pseudo-random bit generator and an oscilloscope. The oscilloscope derives its synchronization from the generator such that the oscilloscope trace starts at the beginning of each pulse.

Figure 17-5 Eye pattern measurement setup

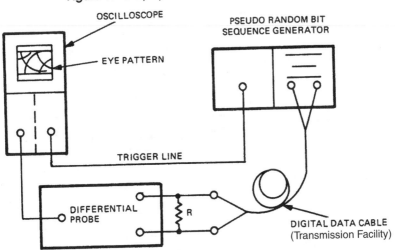

The eye pattern shows a superposition of random data pulses as they are seen by the receiving modem. Examples of observed eye patterns for two- and four-level modulation, and for different amounts of phase jitter, are shown in Figure 17-6. The reason for the smeared look of the patterns is that—due mainly to delay distortion—each data pulse leaves behind a decaying tail, which interferes with the successive data pulses. The height of this tail depends on the value of previous pulses. Therefore the sequence 01010 will add a different tail to a subsequent pulse than the sequence 11110. A random sequence of bits, which approximates the actual data transmission stream, will result in random interference and a smeared look of the eye pattern. As the center of the eye pattern closes, it becomes more and more difficult and costly for the receiving modem to distinguish between a binary 0 and a binary 1. When the eye closes completely, this distinction becomes impossible to detect and transmission fails.

Figure 17-6 Typical eye patterns with 5% and 50% phase jitter

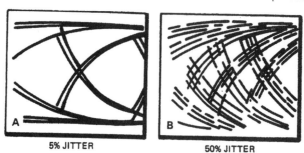

TWO LEVEL "EYE"

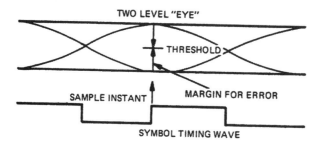

FOUR LEVEL "EYE"

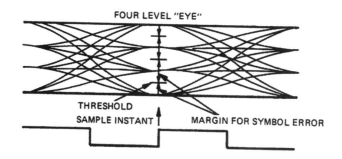

Specialized Transmission Measuring Sets

In addition to measurements of the basic transmission parameters—such as attenuation, delay distortion, phase jitter and noise—with separate measuring sets, there are instruments that combine several of these functions into one unit. It is also possible to make a single measurement with a single resulting value that is closely related to the actual data transmission quality. An instrument that was quite popular a number of years ago, and can still be found in some telephone companies today, is the so-called Peak-to-Average Ratio or P/AR meter. The function of the P/AR meter is now included in a specialized measuring set such as the Hewlett Packard 4934A Transmission Impairment Measuring Set (TIMS) described in the next section.

The P/AR signal is a complex waveform consisting of 16 harmonically-unrelated sine waves at different amplitudes. The power spectrum of the signal is shown in Figure 17-7 and the periodic P/AR signal as a function of time is shown in Figure 17-8. The receiver portion of the instrument computes the ratio of the received peak energy to average energy. This ratio is related to both attenuation and delay distortion and provides an indication of the circuit quality.

This simple measurement is particularly useful for comparisons between similar circuits. A measured P/AR value that is between 0 and 100 is a quick benchmark reading. It is most sensitive to the attenuation and delay distortion, and noise, but it will not indicate which particular parameter is at fault. A P/AR reading, which differs by at least four units from a previous reading, indicates some problem on the transmission line and should be followed by more detailed testing. This quick test can therefore detect a problem before it becomes serious.

Figure 17-7 P/AR power spectrum

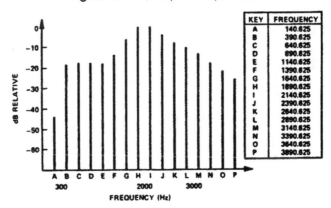

KEY	FREQUENCY
A	140.625
B	390.625
C	640.625
D	890.625
E	1140.625
F	1390.625
G	1640.625
H	1890.625
I	2140.625
J	2390.625
K	2640.625
L	2890.625
M	3140.625
N	3390.625
O	3640.625
P	3890.625

Figure 17-8 P/AR test signal

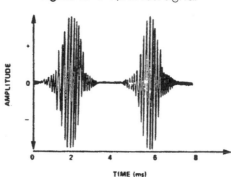

Transmission Impairment Measuring Sets

The majority of the previously-described analog tests can also be performed with a single instrument. The so-called Transmission Measuring Sets (TIMS) measure attenuation and delay distortion, phase jitter,
noise, and P/AR. Many of the tests are performed automatically, with
results plotted, recorded or printed.

An example of such a kitchen-sink instrument is the Hewlett
Packard HP4934A Transmission Impairment Measuring Set. It measures attenuation and delay distortion, flat and C-message circuit
noise, impulse noise, signal-to-noise (S/N) ratio, peak-to-average
ratio (P/AR), gain and phase hits, phase jitter, and inter-modulation
distortion. Plots and printouts that show the measured parameters
can be generated via a built-in RS232 serial interface. The S/N ratio
measurements are made by sending a single frequency signal and
measuring the received signal with a narrow-band filter, and at the
same time measuring the broad-band noise level with a band-rejec-
tion filter, which excludes the original single frequency signal. It
should be noted that the broad-band noise level must be measured at
the same time as the single frequency signal is being transmitted.
Otherwise an erroneous result might be obtained due to automatic
gain control and expanders present in many short-haul transmission
systems.

The HP4934A TIMS can also monitor the circuit while data is
being transmitted. The instrument can be bridged on a two-wire data
circuit and measure various transmission parameters while exchanging data with the remote modem. Thanks to the high-impedance
input of the instrument, testing can continue without interrupting
normal operation. The instrument can also simulate specific modems
and measure modem signals. In this mode, the instrument will measure the amplitude and frequency of the modem output in dBm and

Hz, respectively, the amplitude modulation and jitter in percent and phase jitter in degrees. The cost of this type of instrument is in the range of $5000 to $10,000.

Bit Error Rate Test Sets

Another popular instrument, the bit error rate test set (BERT), sends a pseudo-random sequence of bits over a given period of time to the remote terminal. The data is looped back, compared with the outgoing stream, and errors are counted. This type of instrument is particularly useful if the modem does not have a built-in error tester, as some do. Although there are some stand-alone BERT instruments on the market, the BERT function is usually included in most modem and line analyzers, such as in the modem testing instruments discussed in Chapter 16.

This short review of test methods and test instruments will assist the corporate user in analyzing most data-transmission problems.

Telephone Line and Modem Testing for Everyone

The previously-described methods of line testing are obviously not designed for the average personal-computer user. However, a simple method of line and modem testing is currently being offered free to the general public by U.S. Robotics. The main reason for this offer is to advise consumers buying the high-speed 56 kbps x2 modems if their telephone-line quality will support it. U.S. Robotics performs this service by establishing a special BBS and installing equipment to analyze a caller's telephone line and modem characteristics.

To access this special U.S. Robotics BBS, start your terminal-emulation program (refer to Chapter 10 if you need instructions). Select the standard parameters—8-1-N (8 data bits, 1 stop bit, no parity). The toll-free number to call, as of May 1997, is 1-888-877-9248. At the "first name?" prompt, type the words "Line Test." The remote computer will then analyze the connection and return a table of measured parameters and a graph of attenuation versus frequency. It will also tell you if your connection can support an x2 modem.

When I recently tested my telephone line and modem by means of the U.S. Robotics BBS, I received the following message:

```
x2 is possible on this connection, but you are likely to
experience reduced performance.

Since line conditions can change from call to call, we recommend
that you make several calls to this test system to ensure an
```

accurate diagnosis. If you have multiple phone lines available
for use, we recommend that you try calling this system from each
line.

Performance degradation is typically caused by noise in your
local phone lines.

It's important to remember that telephone network configurations
can change, so your chances of making high-speed connections may
improve over time.

The call to the BBS also generated a list of measured parameters,
as shown in Table 17-1. On a color display, the good values appear in
green, the so-so values appear in yellow, and the bad values appear in
red.

Table 17-1: Analysis of the Connection

Elapsed Time	00:00:19	Modulation	V.34+
Blocks Received	4	Speed	26400/28800
Blers	0	Symbol Rate	3200/3200
Blocks sent	51	Carrier Frequency	1829/1829
Link Naks	0	Trellis Code	64S-4D/64S-4D
Blocks resent	0	Nonlinear Encoding	ON/ON
Link Timeouts	0	Precoding	ON/OFF
Chars sent	1536	Shaping	OFF/OFF
Octets sent	1388	Preemphasis	4/4
Chars lost	0	Rx Lev/TX Lev/SNR	23.4/10.9/40.6
Chars Received	0	Echo Loss	Near Far
Octets Received	27	Roundtrip Delay	28
Protocol	LAPM	Retrains Request/Grant	0/0
Block Size	128	Fallback	Enabled
Window Size	15	HST Line Reversals	0
Compression	V42BIS	HST Equalization	Long
Dictionary Size	2048	SV: 03/21/97	DSP: 03/17/97
String Length	32	Reason:	Online

Notice in particular that no blocks of data were received in error (Blers = 0) and that the modem supports V.42 bis data compression protocol with a dictionary size of 2048 words. This method of data compression is described in Chapter 4.

The analysis also shows that I have an excellent telephone line with a data receive level of -23.4 dBm (Rx Lev = 23.4), a data transmit level of -10.9 dBm (TX Lev = 10.9), and a signal-to-noise (S/N) ratio of 40.6 dB (SNR = 40.6). A data receive level greater than -25 dBm and a S/N ratio greater than 35 dB is considered excellent.

The attenuation versus frequency plot obtained during the same test is shown in Figure 17-9. I would also consider it excellent. When analyzing your connection, watch in particular for sudden drop-offs below 3000 Hz, or for any dips in the attenuation versus frequency characteristics.

Figure 17-9 Line attenuation versus frequency

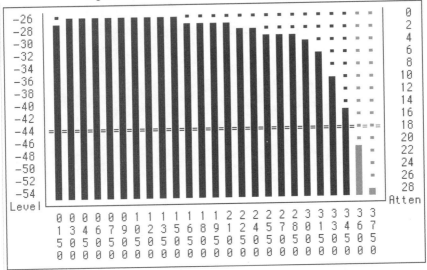

You can also experiment with the test facility by taking one of your extension phones off-hook. Just make sure that the extension is on the same line as your modem. You should do it after establishing the connection to the U.S. Robotics BBS. You can then play a radio next to the extension phone and see how the connection deteriorates. When I tried it, the connection speed dropped to 9,600 bps, the received level dropped to -30 dBm, and the S/N ratio dropped to 20 dB.

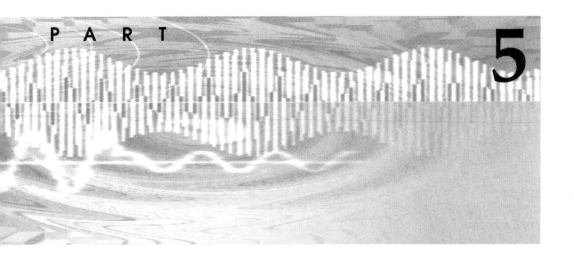

Look Into the Future

One hardly needs a crystal ball to anticipate many future developments in data transmission in general, and modem technology in particular. All one has to do is to extrapolate current trends towards lower prices, higher transmission rates and further miniaturization. It is quite certain that the future generation of PCs will have built-in modems as just another standard interface. Upgrades would then be performed as software patches. This trend was started by the development of the modem data pump (MDP) discussed in Chapter 12, which is basically a universal modem processor driven by software. To help in component miniaturization, the current two- and three-chip modem sets will certainly shrink into single-chip modems.

Other developments that loom on the horizon, which are discussed in Chapter 18, include improved subscriber loops, growth of the telephone network, the so-called killer applications, and the current debate over the Network Computer (NC) versus the Personal Computer (PC).

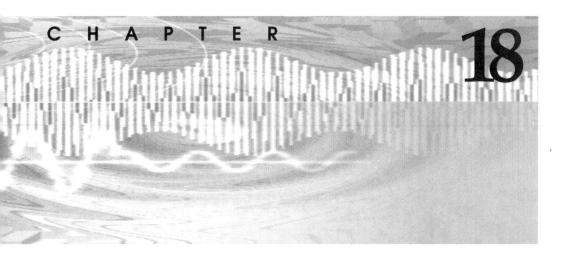

18

What Will the Future Bring?

The continuous increase in the computing power and memory of the personal computer will certainly improve the ease of use and user friendliness as the PC acquires the ability to talk, see and learn.

As far as transmission rate is concerned, a few years ago the 300 bps modem was prevalent for data transmission over the PSTN. By 1987 the 1200 bps full-duplex modem became the prevalent standard, with 2400 bps V.22 bis modems becoming more and more popular. The 9600 bps V.32 full-duplex modem made a short appearance in the early 1990s, and was soon replaced by the V.34 and V.Fast modems, which operated at 14,400 and 28,800 bps.

Improvements in long-haul telephone circuits, which picked up the pace with the deregulation of the telecommunications industry in the United States, resulted in the wholesale replacement of analog telephone long-haul facilities with digital facilities. As demand for end-to-end digital connections develops, the need for modems as we know them today may one day disappear. With replacement of the analog-loop by digital-loop, modems will be ultimately replaced by other data-communications devices such as Data Service Units, Channel Service Units (DSU/CSU), Terminal Adapters (TAs), and multiplexers. A multiplexer could combine data streams originated by computers, voice phones and assorted entertainment systems and send them over a fiber-optics link to their destination.

On a more prosaic tone, improvements in the local telephone offices—namely the replacement of compromise impedance-matching hybrid circuits, which terminate each local loop with active adaptive circuits—would provide a higher return loss on each line. This would in turn result in a much better echo performance and a lower circuit loss. Combined with use of low-noise long-haul digital facilities based on optical fiber, there would be a dramatic improvement in the signal-to-noise ratios on both voice and data circuits. Still, higher transmission speeds and higher throughput due to fewer errors would then be possible.

This is just a few comments on what the future will probably bring. If in doubt, be sure to check the next edition of this book. What follows is a description, in more detail, of more recent developments that will affect the communications industry in the coming years.

Improved Subscriber Loops

There are two current contenders for the replacement of the analog subscriber loop: the Asymmetric Digital Subscriber Line (ADSL), and the Very High Rate Digital Subscriber Line (VDSL). The following monographs that describe these two technologies were furnished by the ADSL Forum, an organization that promotes the development of these new methods of transmission. Though the tutorials use a lot of telephone jargon, they provide a great deal of information about the progress of these new technologies.

Asymmetric Digital Subscriber Line

Asymmetric Digital Subscriber Line (ADSL) converts existing twisted-pair telephone lines into access paths for multimedia and high-speed data communications. ADSL transmits more than 6 Mbps to a subscriber on one channel, and 640 kbps from and to the subscriber on a second channel. Such rates expand existing access capacities by a factor of 50 or more without new cabling. ADSL can literally transform the existing public information network from one limited to voice, text and low-resolution graphics to a powerful, ubiquitous system capable of bringing multimedia—including full-motion video—to telephone subscribers.

ADSL will possibly play a crucial role over the next 10 or more years as telephone companies enter new markets for delivering information in video and multimedia formats. New broad-band cabling will take decades to reach all prospective subscribers. But success of these new services will depend upon reaching as many subscribers as possible during the first few years. By bringing movies, television, video catalogs, remote CD-ROMs, corporate LANs, and the Internet into homes and small businesses, ADSL will make these markets viable, and profitable, for telephone companies and application suppliers alike.

Capabilities

An ADSL circuit connects an ADSL modem on each end of a twisted-pair telephone line, creating three information channels: a high-speed downstream channel, a medium-speed duplex channel, and a POTS (plain old telephone service) channel. The POTS channel is split off from the digital modem using filters, therefore guaranteeing uninterrupted POTS, even if ADSL fails. The high-speed downstream channel ranges from 1.5 to 6.1 Mbps, while duplex rates on the medium-speed channel range from 16 to 640 kbps. Each channel can be sub-multiplexed to create multiple, lower-rate channels.

ADSL modems provide data rates consistent with North American and European digital hierarchies and can be purchased with various speed ranges and capabilities. The minimum configuration provides 1.5 or 2.0 Mbps downstream and a 16 kbps duplex channel; others provide rates of 6.1 Mbps and 64 kbps duplex channels. Products with downstream rates up to 9 Mbps and duplex rates up to 640 kbps will be available in the near future. As the Asynchronous Transmission Mode (ATM) technology and market requirements mature, ADSL modems will accommodate ATM transport with variable rates and compensation for ATM overhead.

Downstream data rates depend on a number of factors, including the length of the copper line, its wire gauge, presence of bridged taps, and cross-coupled interference. Line attenuation increases with line length and frequency, and decreases as wire diameter increases. Ignoring bridged taps, the expected ADSL performance is shown in Table 18-1.

Table 18-1: Expected ADSL Performance

Data Rate	Wire Gauge	Distance	Wire Size	Distance
1.5 or 2 Mbps	24 AWG	18,000 ft	0.5 mm	5.5 km
1.5 or 2 Mbps	26 AWG	15,000 ft	0.4 mm	4.6 km
6.1 Mbps	24 AWG	12,000 ft	0.5 mm	3.7 km
6.1 Mbps	26 AWG	9,000 ft	0.4 mm	2.7 km

While these figures vary from telco to telco (telco is short for telephone company), these capabilities can cover up to 95% of a local loop's plant, depending on the desired data rate. Locations beyond these distances can be reached using fiber-optics-based digital-loop carrier systems.

One application that is envisioned for ADSL involves digital compressed video. As a real-time signal, digital video cannot use link- or network-level error control procedures commonly found in data communications systems. ADSL modems therefore incorporate forward error correction, which reduces errors caused by impulse noise. Error correction on a symbol-by-symbol basis also reduces errors caused by continuous noise coupled into a line.

At present, ADSL models support T1/E1 and V.35 digital interfaces for continuous bit-rate (CBR) signals. Future versions will offer LAN interfaces for direct connection to a personal computer and ATM interfaces for variable bit-rate (VBR) signals. Over time, ADSL units will be built directly into access node concentrators and so-called *premise service modules*, such as TV set-top boxes and personal computer interface cards.

Technology

To create multiple channels, ADSL modems divide the available bandwidth of a telephone line by means of frequency division multiplexing (FDM) and time division multiplexing (TDM). FDM assigns one band for upstream data and another band for downstream data. The downstream path is then divided by time-division multiplexing into one or more high-speed channels and one or more low-speed channels. The upstream path is also multiplexed into corresponding low-speed channels. Echo cancellation assigns the upstream band to overlap the downstream band, and separates the two by means of local echo cancellation, a technique well known in V.32 and V.34 modems. Echo cancellation uses bandwidth more efficiently, but at the expense of complexity and cost. ADSL splits off a 4 kHz region for POTS at the low-frequency end of the band.

An ADSL modem organizes the aggregate data stream—created by multiplexing downstream channels, duplex channels, and maintenance channels—together into blocks, and attaches an error-correction code to each block. The receiver then corrects errors that occur during transmission up to the limits implied by the code and the block length. The unit may, at the users option, also create super blocks by interleaving data within sub blocks; this allows the receiver to correct any combination of errors within a specific span of bits. The typical ADSL modem interleaves 20 ms of data, and can thereby correct error bursts as long as 0.5 ms. ADSL modems can therefore tolerate impulses of arbitrary magnitude whose effect on the data stream lasts no longer than 0.5 ms. Initial trials indicate that this level of correction will create effective error rates suitable for MPEG-2 and other digital video compression schemes.

Standards and Associations

The American National Standards Institute (ANSI), working group T1E1.4, recently approved an ADSL standard at rates up to 6.1 Mbps (ANSI Standard T1.413). The European Technical Standards Institute (ETSI) contributed an Annex to T1.413 to reflect the European requirements. T1.413 currently embodies a single terminal interface at the premise (the subscriber's) end. Issue II, now under study by T1E1.4, will expand the standard to include a multiplexed interface at the premise end, protocols for configuration and network management, and other improvements.

The ATM Forum has recognized ADSL as a physical-layer transmission protocol for unshielded twisted pair (UTP) media.

The ADSL Forum was formed in December of 1994 to promote the ADSL concept and facilitate the development of ADSL system architectures, protocols, and interfaces for major ADSL applications. The Forum has more than 60 members who represent service providers, equipment manufacturers, and semiconductor companies throughout the world.

Market Status

ADSL modems have been tested successfully by as many as 30 telephone companies, and hundreds of lines have been installed in various technology trials in North America and Europe. Several telephone companies plan market trials using ADSL, principally for video-on-demand, but including such applications as personal shopping, interactive games, and educational programming. Interest in personal computer applications grows, particularly for high-speed access to Internet resources.

Semiconductor companies have introduced transceiver chip sets that are already being used in market trials. These initial chip sets combine off-the-shelf components, programmable digital signal processors (DSPs), and custom ASICs (application specific integrated circuits). Continued investment by these semiconductor companies will increase functionality and reduce chip count, power consumption, and cost, enabling mass deployment of ADSL-based services in the near future.

Note: The ADSL Forum takes no position on particular implementations of ADSL, or specific vendor features, pricing, or performance. This monograph therefore omits any discussion of line code (the basic modulation system) or the various tradeoffs between performance and costs. The Forum does maintain a roster of vendors who are more than happy to address these areas. The Forum also has more detailed papers on technology and market aspects of ADSL. For further information, send an e-mail message to:

```
adslForum@adsl.com.
```

Very High Rate Digital Subscriber Line (VDSL)

It is becoming increasingly clear that telephone companies around the world are making decisions to include existing twisted-pair loops in their next-generation broad-band access networks. Hybrid fiber coax (HFC), a shared-access medium that is well suited to analog and digital broadcast, comes up somewhat short when asked to carry voice telephony, interactive video, and high-speed data communications at the same time. Fiber to the home (FTTH) is still prohibitively expensive in a marketplace soon to be driven by competition and cost. An attractive alternative, soon to be commercially practical, is a combination of fiber-optic cables that feed neighborhood optical network units (ONUs) and last-leg premises connections by existing or new copper. This topology, which can be called fiber to the neighborhood (FTTN), encompasses fiber to the curb (FTTC) with short drops and fiber to the basement (FTTB), serving tall buildings with vertical drops.

One of the enabling technologies for FTTN is very high-rate digital subscriber line, or VDSL. In simple terms, VDSL transmits high-speed data over short distances using twisted-pair copper telephone lines, with a range of speeds depending upon actual line length. The maximum downstream rate under consideration is between 51 and 55 Mbps over lines up to 1000 feet (300 meters) in length. Downstream speeds of 13 Mbps over lengths beyond 4000 feet (1500 meters) are also in the picture.

Upstream rates in early models will be asymmetric (meaning the upstream rate is different than the downstream rate), just like ADSL, at speeds from 1.6 to 2.3 Mbps. Both data channels will be separated in frequency from bands used for POTS and ISDN, enabling service providers to overlay VDSL on existing services. At present, the two high-speed channels will also be separated in frequency. As needs arise for higher-speed upstream channels or symmetric rates, VDSL systems may need to use echo cancellation.

This monograph presents VDSL in terms of projected capabilities, underlying technology, and outstanding issues. It is followed with a survey of standards activity and concludes with a suggestion that VDSL and ADSL together provide network providers an excellent combination for evolving a full-service network while offering virtually ubiquitous access to most PC applications and interactive TV applications as the network develops.

VDSL Projected Capabilities

While VDSL has not achieved the degree of definition that ADSL has, it has advanced far enough to discuss realizable goals, beginning with data rates and range. Downstream rates are 51.84 Mbps, 25.92 Mbps and 12.96 Mbps. Each rate has a corresponding target range, as shown in Table 18-2.

Table 18-2: Downstream and Upstream Rates for VDSL

Downstream Rates	Target Range (ft.)	Target Range (m)
12.96 - 13.8 Mbps	4500 ft	1500 meters
25.92 - 27.6 Mbps	3000 ft	1000 meters
51.84 - 55.2 Mbps	1000 ft	300 meters
Upstream Rates		
1.6 - 2.3 Mbps	4500 ft	1500 meters
19.2 Mbps	3000 ft	1000 meters

Early versions of VDSL will almost certainly incorporate the slower asymmetric rate. Higher upstream and symmetric configurations may only be possible for very short lines.

Like ADSL, VDSL must transmit compressed video, which is a real-time signal unsuited to error re-transmission schemes used in data communications. To achieve error rates compatible with compressed video, VDSL will have to incorporate forward error correction (FEC) with sufficient interleaving to correct all errors created by impulse noise of some specified duration. Interleaving introduces delay, in the order of 40 times the maximum-length correctable impulse.

Data in the downstream direction will be broadcast to every customer premise equipment (CPE) on a premise, or will be transmitted to a logically-separated hub that distributes data to the addressed CPE based upon cell, or time division multiplexing (TDM) within the data stream itself. Upstream multiplexing is more difficult. Systems must insert data onto a shared medium, possibly by a token-control method called *cell grants* that are passed in the downstream direction.

Migration and inventory considerations dictate VDSL units that can operate at various (preferably all) speeds with automatic recognition of a newly-connected device to a line or a change in speed. Passive network interfaces must support hot insertion, where a new VDSL premises unit can be connected to a line without having to turn off power and without interfering with the operation of other modems.

VDSL Technology

VDSL technology will resemble ADSL to a large degree, although ADSL must face much larger dynamic ranges and is considerably more complex as a result. VDSL must be lower in cost and lower in power, and premises VDSL units may have to implement a physical layer media access control (MAC) for multiplexing upstream data.

Line code candidates (modulation schemes)

Four line codes have been proposed for VDSL:

CAP

Carrier-less AM/PM, a version of suppressed-carrier QAM. For passive configurations, CAP would use QPSK upstream and a type of TDM for multiplexing (although CAP does not preclude an FDM approach to upstream multiplexing).

DMT

Discrete multi-tone, which is a multi-carrier system using discrete Fourier transforms to create and demodulate individual carriers. For passive configurations, DMT would use FDM for upstream multiplexing (although DMT does not preclude a TDM multiplexing strategy).

DWMT

Discrete wavelet multi-tone, which is a multi-carrier system using wavelet transforms to create and demodulate individual carriers. DWMT also uses FDM for upstream multiplexing, but also allows TDM.

SLC

Simple line code, a version of four-level base-band signaling that filters the base-band and restores it at the receiver. For passive configurations, SLC would most likely use TDM for upstream multiplexing, although FDM is possible.

Channel Separation

Early versions of VDSL will use FDM to separate downstream from upstream channels, and both of them from POTS and ISDN. Echo cancellation may be required for later-generation systems that feature symmetric data rates. A rather substantial distance, in frequency, will be maintained between the lowest data channel and POTS in order to enable very simple and cost-effective POTS splitters. The normal practice would be to locate the downstream channel above the upstream channel.

Upstream Multiplexing

If the premises VDSL unit comprises the network termination (an active NT), then the means of multiplexing upstream cells or data channels from more than one CPE into a single upstream becomes the responsibility of the premise network. The VDSL unit simply presents raw data streams in both directions. One type of premise network involves a star connecting each CPE to a switching or multiplexing hub; such a hub could be integral to the VDSL premise unit.

In a passive NT configuration, each CPE has an associated VDSL unit. A passive NT does not conceptually preclude multiple CPE per VDSL, but the question of active versus passive NT becomes a matter of ownership, not a matter of wiring topology and multiplexing strategies.

Now the upstream channels for each CPE must share a common wire. While a collision-detection system could be used, the desire for guaranteed bandwidth indicates one of two solutions. One invokes a cell-grant protocol in which downstream frames, which are generated at the ONU or further up the network, contain a few bits that grant access to specific CPE during a specified period subsequent to receiving a frame. A granted CPE can send one upstream cell during this period. The transmitter in the CPE must turn on, send a preamble to condition the ONU receiver, send the cell, then turn itself off.

A second method divides the upstream channel into frequency bands and assigns one band to each CPE. This method has the advantage of avoiding any media access control with its associated overhead (although a multiplexer must be built into the ONU), but either restricts the data rate available to any one CPE, or imposes a dynamic inverse-multiplexing scheme that lets one CPE send more than its share for a period. The latter would look a great deal like a media access control (MAC) protocol, but without the loss of bandwidth associated with carrier-detect-and-clear for each cell.

VDSL Issues

VDSL is still in the definition stage. Some preliminary products exist, but not enough is known yet about telephone-line characteristics, RFI emissions and susceptibility, upstream multiplexing protocols, and information requirements to frame a set of definitive, standardizable properties. One large unknown is the maximum distance that VDSL can reliably operate for a specified data rate. The reason this is unknown is because actual line characteristics at the frequencies required for VDSL are speculative, and items such as short bridged taps or un-terminated extension lines in homes, which have no affect on telephony, ISDN or ADSL, may have very detrimental affects on VDSL in certain configurations. Furthermore, VDSL invades the frequency ranges of amateur radio, and every above-ground telephone wire is an antenna that both radiates and attracts energy in the amateur radio bands. Balancing low signal levels—to prevent emissions that might interfere with amateur radio—with higher signals needed to combat interference by amateur radio, could be the dominant factor in determining line reach.

A second dimension of VDSL that is far from clear is the services environment. It can be assumed that VDSL will carry information in ATM cell format for video and asymmetric data communications, although optimum downstream and upstream data rates have not been ascertained. What is more difficult to assess is the need for VDSL to carry information in non-ATM formats (such as conventional PDH structures) and the need for symmetric channels at broad-band rates (above T1/E1). VDSL will not be completely independent of upper-layer protocols, particularly in the upstream direction where multiplexing data from more than one CPE may require knowledge of link-layer formats.

A third difficult subject is premises distribution and the interface between the telephone network and customer premises equipment (CPE). Cost considerations favor a passive network interface with premises VDSL installed in CPE, and upstream multiplexing handled much like local area network (LAN) busses. System management, reliability, regulatory constraints, and migration favor an active network termination, just like ADSL and ISDN, that can operate like a hub, with point-to-point or shared media distribution to multiple CPE on premises wiring that is independent and physically isolated from network wiring.

But costs cannot be ignored. Small telcos must spread common equipment costs, such as fiber-optic links, interfaces, and equipment cabinets, over a small number of subscribers. VDSL therefore has a much lower cost target than ADSL, which may connect directly from a wiring center, or cable modems, which also have much lower common equipment costs per user. Furthermore, VDSL for passive NTs may (only may) be more expensive than VDSL for active NTs, but the elimination of any other premises network electronics may make it the most cost-effective solution—and highly desired, despite the obvious benefits of an active NT.

Standards Status

At present five standards organizations/forums have begun work on VDSL: ANSI group T1E1.4, ETSI, DAVIC, The ATM Forum, and The ADSL Forum.

T1E1.4

The U.S. ANSI standards group T1E1.4 has just begun a project for VDSL, making a first attack on system requirements that will evolve into a system and protocol definition.

ETSI

The European Telecommunications Standards Institute (ETSI) has a VDSL standards project under the title High Speed (metallic) Access Systems (HSAS), and has compiled a list of objectives, problems, and requirements. Among its preliminary findings are the need for an active NT, and payloads in multiples of SDH Virtual Container VC-12, or 2.3 Mbps. ETSI works very closely with T1E1.4 and the ADSL Forum, with significant overlapping attendees.

DAVIC

The Digital Audio-Visual Council (DAVIC) has taken the earliest position on VDSL. Its first specification will define a line code for downstream data, another for upstream data, and a media access control for upstream multiplexing based on TDMA over shared wiring. As of its September 1995 meeting, DAVIC is only specifying VDSL for a single downstream rate of 51.84 Mbps and a single upstream rate of 1.6 Mbps over 300 meters or less of copper. The proposal assumes, and is driven to a large extent by a passive NT, and further assumes premises distribution from the NT over new coaxial cable or new copper wiring.

The ATM Forum

The ATM Forum has defined a 51.84 Mbps interface for private network UNIs and a corresponding transmission technology. It has also taken up the question of premises distribution and delivery of ATM all the way to the premises over various access technologies described previously.

The ADSL Forum

The ADSL Forum has begun consideration of VDSL. In keeping with its charter, the Forum will address network, protocol, and architectural aspects of VDSL for all prospective applications, leaving line code and transceiver protocols to T1E1.4 and ETSI, and higher-layer protocols to organizations such as the ATM Forum and DAVIC.

Relationship with ADSL

VDSL has an odd technical resemblance to ADSL. VDSL achieves data rates nearly 10 times greater than ADSL, but ADSL is the more complex transmission technology, in large part because ADSL must contend with much larger dynamic ranges than VDSL. However, the two are essentially cut from the same cloth. ADSL employs advanced transmission techniques and forward error correction to realize data rates from 1.5 to 9 Mbps over twisted-pair ranging to 18,000 feet. VDSL employs the same advanced transmission techniques and forward error correction to realize data rates from 13 to 55 Mbps over twisted-pair wiring ranging to 4500 feet. Indeed, the two can be considered a continuum—a set of transmission tools that delivers about as much data as theoretically possible over varying distances of existing telephone wiring.

VDSL is clearly a technology suitable for a full-service network (assuming full-service does not imply more than two HDTV channels over the highest rate VDSL). It is equally clear that telephone companies cannot deploy ONUs overnight, even if all the technology were available. ADSL may not be a full-service network technology, but it has the singular advantage of offering service over lines that exist today, and ADSL products are closer in time than VDSL.

Many new services being contemplated today can be delivered at speeds at or below T1/E1 rates: video conferencing, Internet access, video on demand, and remote LAN access. For such services, ADSL/VDSL provides an ideal combination for network evolution. On the longest lines, ADSL delivers a single channel. As line length shrinks, either from natural proximity to a central office or deployment of fiber-based access nodes, ADSL and VDSL simply offer more channels and capacity for services that require rates above T1/E1 (such as digital live television or virtual CD-ROM access).

Network Growth

Moore's Law about doubling capacities of silicon chips every 18 months while keeping their prices the same, is still holding. This famous "law" was first formulated by Gordon Moore, one of the founders of Intel Corporation. A similar law seems to apply to the growth of communications networks. Although Internet usage grows at about 10% per month, it has not yet reached its full capacity, and new technological developments keep adding new capacity to the network. Copper wires are being replaced with coaxial cable and with optical fiber, while the improved modulation schemes squeeze more and more bandwidth from the existing transmission media. All one can say at this point is that the communications network will expand via satellites, such as Motorola's Iridium project, and optical fiber, to eventually cover the entire globe.

Killer Applications and More Power on the Desktop

Once in a while a new application comes along and starts a technological revolution. Such applications cannot be anticipated—they come from the genius of one or more individuals. The first spreadsheet program, Visicalc, moved the personal computer from the hobby status to a professional tool. Similarly, the graphical user interface (GUI) of the World Wide Web (the Web) opened the Internet to those who were completely unfamiliar with computer technology. These are the real "killer applications."

Technological advances make such new applications possible. The Web would not have come along if we were still using the text-oriented TRS-80 computers or VT-100 terminals. As the speed of microprocessors increases and the price of memory drops, I am sure that more killer applications will arrive in the next few years. It may be a new virtual reality game or a new financial application that will capture people's imagination and make a few more billionaires.

Network Computers—PC versus NC

There is currently a debate going on between two computer gurus—Bill Gates of Microsoft Corporation, the proponent of the full-fledged Personal Computer (PC), and Larry Ellison of Oracle Corporation, the proponent of the bare-bones Internet-connected Network Computer (NC).

Larry Ellison, whose company is a major developer of large centralized database systems, argues that computer users should store all their data in large centralized databases accessible via the Internet—not on their local hard disks and diskettes. There is a certain logic in this argument. People are willing to keep their cash in centralized banks—it is less likely to be lost or stolen from there than at home. Storing data on a PC requires frequent backups, and leads to occasional data loss due to crashes, mistakes and device failure. It also requires a higher level of sophistication and computer knowledge to manage data on the local PC than to put it in the care of professionals at some central location.

Another argument in favor of the NC is that even in the wealthiest countries such as the United States, the market penetration of the personal computer in 1996 was only 30%. People are scared by both the complexity and cost of the product. By contrast, TV and phones are found in over 90% of American homes. Both a TV set and a phone are simple, easy to use and inexpensive appliances connected to large, expensive and extremely complicated networks handled by professionals. Following this analogy, an inexpensive and simple to use network computer connected to the complicated Internet network would make the information available to everyone. One would no longer need word processors, spreadsheets and data on the local computer—instead, at the click of the mouse, one could tap the latest and greatest version of each program from some Internet site.

Well, this is one side of the debate. The argument of the PC proponents is that all is fine and well and, in due time with a few good programmers, it will be possible to make computers easy to use and fun for the masses. The crippled PC, called NC, would be hardly cheaper but would rely solely on a "big brother" for proper operation. Based on experience of many people, a large corporation with famous professionals can also mess up things in a big way. People who get million-dollar bills from the gas company for the last month of home usage can attest to this. Also, the administration of the network of software and storage providers would be quite difficult and expensive, with the cost borne by subscribers.

I still remember the days of remote-computer timesharing in the early 1970s. I would dial in to a remote computer, which would execute programs and store my data. The local storage medium was just a smelly waxed paper tape with lots of holes that was being spilled out from my teletypewriter. The monthly bill for the service was typically $1000.

And finally, there are those who believe that both sides are right and that PCs and NCs will find their own place in the computing universe, and will live in harmony for ever after. A recent development—the Web TV, a set-top box that is connected to the phone line and TV, provides a limited but easy-access connection to the Internet. Web TV did not take the world by storm, but sold a few hundred thousand units nevertheless in its first year of operation.

Epilogue

With Moore's Law still holding, you can be sure that by the time you finish reading this book most of your computer hardware will be obsolete, while newer and faster modems will take over the market. Still, the knowledge you ave gained about data Communications will not become obsolete and will give you a good understanding of the latest developments.

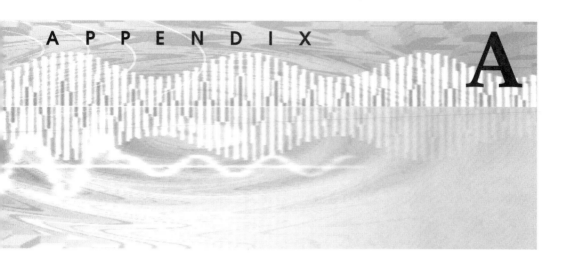

Modem-Related URL Internet Addresses

The following URLs contained modem-related information as of May 1997. They are listed in alphabetical order.

3C562 ETHERLINK III LAN+MODEM PC CARD
> 3C562 ETHERLINK III LAN+MODEM PC CARD by 3Com Corporation. Date: 10 JUN 1996 Bulletin #: L-3910 Faxback Document ID: #27176. The Novell Labs Faxback can...
> http://labs.novell.com/yes/l3910.htm - size 2K - 26 Nov 96

Application: Modem Design Guide
> Click on any item for design advice. Navigation Options: Top of Page. Design Guide Text. Application Page. Design Phase | Application | Procurement |...
> http://www.designphase.com/application/design_guides/modem_gui.html - size 2K - 5 Nov 96

Boca Modem-Init Strings
> Boca Modem-Init Strings. model and initialization string. port speed and notes. Complete Communicator Gold Internal...
> http://www.eaglesnest.net/modems/boca.html - size 2K - 21 Oct 96

Buying a Business Modem
> Buying a Business Modem. by John Kaufeld. Question: Our insurance brokerage relies on a couple of dial-up information services for customer policy quotes..
> http://www.datatech.com/hot/f96_4.htm - size 7K - 21 Jan 97

Card Mounted Sync-Async Short Haul Modem
> Model 77 - Card Mounted Sync-Async Short Haul Modem. Full Duplex Sync/Async 19.2 KBPS. Clocking from External, Internal or Received Data. Compatible with..
> http://telebyteusa.com/catalog/77.htm - size 2K - 5 Oct 96

Compaq Online - Quick Reference Guide -- Appendix D, Modem Configuration Infor
> Appendix D, Modem Configuration Information. Document Number: 107315-027, Volume 2 Publication Date: July 1996. 9600 BPS Fax/Data Modem Spare Part...
> http://www.compaq.com/support/techpubs/qrg/volume2/APPD.html - size 8K - 10 Dec 96

DTR - Wireless - Radio Modem
> Conexco Sem Fio. Wireless - acabou de encontrar a solugco! Fazemos sua conexco atravis de Radio Modem com velocidades ati 64k smncrono ou assmncrono....
> http://www.merconet.com.br/dtr/radio.htm - size 1K - 6 Dec 96

Data Communications - Modems : Lab 8
> Data Communications : Modems. Laboratory Exercise 8. BLPR.L8. Modem installation testing and diagnostics.   Back to PRACTICAL ACTIVITIES...
> http://www.tafe.sa.edu.au/institutes/torrens-valley/programs/eit/datacoms/lab8.htm - size 4K - 28 Feb 97

Dual Short Haul Modem - Rack Mounted
> Model 75 - Dual Short Haul Modem - Rack Mounted. Two Modems Per Module. Plugs into Rack Mountable Card Cage Model 76-2. Contains Self-Check Capability....
> http://telebyteusa.com/catalog/75.htm - size 2K - 5 Oct 96

External Modem Installation
> MODEM INSTALLATION GUIDE. EXTERNAL TYPE. 1. The first in any installation is to choose the type of Modem you want. The two preferred   types of..
> http://home.eznet.net/~beamer/helpdocs/modemext.htm - size 4K - 17 Aug 96

Fast Wire. Short Haul Modem
> Model 214 - Fast Wire. Short Haul Modem. Full Duplex on Single Pair. 0 to 38.4 KBPS. Adaptive Echo Cancellation. ISDN Technology. Up to 3 Miles....
> http://telebyteusa.com/catalog/214.htm - size 4K - 5 Oct 96

Fax Modem Testing Products Overview
> Fax Modem Testing Products Overview. Conformance Testing of Class 1/2.0 Implementations. Addresses the Class 1 and recently-approved Class 2.0 protocol...
> http://www.gentech.com/faxmodem/faxmodem.html - size 7K - 28 Aug 96

GLB RADIO MODEM DATA SYSTEMS
> designer & manufacturer of RF Radio and Radio Modem, Radio Modem. Supplier of modules and complete RF systems for a wide variety of data
> http://www.glb.com/products.html - size 3K - 4 Mar 97
> http://www.glb.com/services.html - size 2K - 4 Mar 97
> http://www.glb.com/company.html - size 4K - 28 Feb 97
> http://www.glb.com/mail.html - size 4K - 28 Feb 97

http://www.glb.com/whatsnew.html - size 5K - 27 Feb 97
http://www.glb.com/4wire.html - size 4K - 27 Feb 97
http://www.glb.com/apps.html - size 6K - 26 Feb 97
http://www.glb.com/ - size 6K - 26 Feb 97

GTE Cable Modem HomePage

Welcome to GTE Cable Modem Service Tired of waiting an eternity to download files? Stuck in another Internet rush-hour traffic jam? Fed up with ISP busy...
http://www.centripedus.com/gte/gte.index.html - size 2K - 26 Feb 97

Hayes Modem-Init Strings

Hayes Modem-Init Strings. OPTIMA Series .. ULTRA Series ... ACCURA Series. model and initialization string. port speed and notes. Hayes OPTIMA 144 +...
http://www.eaglesnest.net/modems/hayes.html - size 5K - 21 Oct 96

High-Speed Modem Technology

High-Speed Modem Technology. The Advantages Of A Standards-Based, V.34 Solution. This White Paper examines the forthcoming V.34 communications standard...
http://www.megahertz.com/technology/wpv34.html - size 10K - 6 May 96

Intertex Data AB's Modem HomePage

Welcome to Intertex Data AB. On our homepage you can: get detailed information on our New IX35 Modem lineIncluding SVD, DSVD and MORE ! sign up on the...
http://www.intertex.se/ - size 2K - 4 Feb 97

Mall-By-Modem Index Page

On-Line Shopping and Much, Much More!! Mall-By-Modem. hopes to create an atmosphere where shopping can be fun and easy. You don't have to fight to...
http://www.mallbymodem.com/ - size 3K - 11 Feb 97

Modem & Multimedia Solutions

Modem and Multimedia Solutions. In this era of widespread access to Internet and on-line services, modems have become ubiquitous. With the advancement of...
http://www.lucent.com/micro/wam/modem.html - size 4K - 16 Oct 96

Modem Compatability Matrix

Home] [Services Index] [Online Help Index] [Index] [Feedback] Modem Compatability Matrix. Page 1 of 3. Page 2 of 3. Page 3 of 3. [Home] [Services Index]...
http://www.verilink.com/byrum/service/miscserv/modem.htm - size 1K - 29 May 96

Modem Connection - Pilot Organizer

The Pocket Ethernet Adaptor III. The Xircom Pocket Ethernet Adaptor III is the easiest way to connect any PC user to an Ethernet LANa. The Xircom Pocket...
http://www.modemconnection.com.au/combo.htm - size 5K - 12 Sep 96

Modem Design Considerations

Modem Design Considerations. Maximizing data throughput under all line conditions is a complex process that is overly simplified by most product...
http://www.microcom.com/modems/wpmodem.htm - size 6K - 5 Feb 97

Modem Initialization Strings

Modem Initialization Strings. This is a list of modem initialization strings designed as a customer support aid for NeboNet Information Services. If you...
http://www.nebonet.com/ts/modem/comppc.html - size 718 bytes - 26 Nov 95

Modem Installation

> Modem Installation. SigQuest will install, configure, and test your new new modem, and save you the trouble of having to worry about COM port and IRQ...
> http://206.242.196.3/sigquest/modem.htm - size 700 bytes - 16 Jan 97

Modem Standards

> WELCOME !!!!!!! to the Modem Standards Homepage, a site dedicated to explaining the mystery of the MODEM... Made possible by: Siao Ly. Ji Ma. Trey...
> http://www2.gsu.edu/~gs02sel/modem.html - size 7K - 26 Feb 97

Modem String/CCL Database

> Modem String/CCL Database. 4. by Modem Files Contained in. Database only. MODEMS.INI. MODEMS.SNM. Shiva ARA 2.0.1 CCLs. SNM Modems. [Home] [Help] [...
> http://www.shiva.com/prod/ccl/4.html - size 1K - 14 Jan 97

Modem US Robotics Courier

> Repleto de indicadores. Modem US Robotics Courier. Este msdem externo destaca sin duda por su gran tamaqo y peso, que hace recordar a los modems de hace...
> http://www.pcactual.esegi.es/han021.html - size 2K - 20 Mar 96

Modem University, Havenport University and Modem College Home Page

> Modem University, Havenport University and Modem College. Connecting the world's learners with the world's experts. Contents. General Information. Program.
> http://www.modemu.com/ - size 2K - 21 Nov 95

ModemSurfer Modem Surfer Motorola V.34 Internal Desktop Data Fax Modem FAQs Fr

> ModemSURFR. TM. V.34 Internal Desktop/Data/Fax Modem. E-mail Sales Assistance or call Toll Free 1-800-646-6415 for your Discount Price Quote. NEW $50...
> http://www.netlinkweb.com/modemsurfer.htm - size 21K - 26 Feb 97

Motorola Modem Settings

> Motorola Modem Settings. Last Modified: May 8, 1996. Here are a few useful settings for using modems with Xylogics terminal servers. Note that the AT...
> http://www.xylogics.com/support/appnotes/motorola_modems.html - size 1K - 8 May 96

Motorola Modem-Init Strings

> Motorola Modem-Init Strings. model and initialization string. port speed and notes. Motorola BitSURFR Pro AT%A2=95. 115200. Motorola/Codex 3220 Plus...
> http://www.eaglesnest.net/modems/motorola.html - size 2K - 21 Oct 96

Navas 28800 Modem FAQ

> Navas 28800 Modem FAQ. TM. (Answers to Frequently Asked Questions) Copyright 1995-1997 The Navas Group. SM. , All Rights Reserved. Permission is granted...
> http://web.aimnet.com/~jnavas/modem/faq.htm - size 15K - 5 Mar 97
> http://www.aimnet.com/~jnavas/modem/faq.html - size 15K - 3 Mar 97

Online with a high speed cable modem

> Intro | Hardware | Setup | Surfing | Links | FAQ. Appearance. The modem is 2.6" tall by 6.6" wide by 10" deep and weighs 6 pounds. It has a.
> http://www.infowest.com/cable/hardware.htm - size 5K - 13 Aug 96

Rockwell modem Custom Electronic Design

> Digital Acoustics provides advanced electronic design for OEM's worldwide. Please contact us with you requirements. We provide .... Computer Modem designs.
> http://www.digac.com/webrwp.html - size 2K - 10 May 97

Spider Modem Quick Installation Check List
APPLICATION NOTE. TITLE: Spider Modem Quick Installation Check List. Number: 2. Revision: a. 1.0 SCOPE. This application note lists the steps necessary to.
http://www.inetinc.com/spider/AP002.html - size 6K - 1 Dec 96

Telecom Analysis Systems, Inc. -- FAQ's for Modem/Fax Testing
Telecom Analysis Systems, Inc. Commonly Asked Questions- Modem/Fax Test Equipment. 1. Can I plug my 100 Series or Series II Telephone Network Emulator...
http://www.taskit.com/faq_vb.html - size 6K - 17 Nov 96

TestingModem
Testing your Modem. FreePrint will attempt to automatically detect and configure your modem. For print shops: If FreePrint can detect and configure your...
http://www.freemail.com/00000016.htm - size 2K - 26 Aug 96

The Hayes AT Modem Commands
The Hayes AT Modem Commands. This is a description of common modem AT commands. General | Ampersand (&) | S Registers. Dial Commands. 0-9 Digits to Dial *.
http://help.unc.edu/help-desk/modem/hayes.html - size 5K - 9 Nov 95

The Modem Superstore Home Page to. Modem Superstore.
Welcome to our site. We hope you enjoy your visit. To view these pages we recommend using Netscape Navigator version 3.0. The...
http://www.modemsuperstore.com.au/ - size 2K - 2 Dec 96

The ROLM "modem"
Using the built in serial port or 'modem' on your ROLM phone. More About the Dataline. The data line allows you terminal access to any campus hosts that...
http://roundtable.cif.rochester.edu/users/tparker/rolm/modem.html - size 7K - 18 Dec 95

The ROLM "modem"
Using the built in serial port or 'modem' on your ROLM phone. More About the Dataline. The data line allows you terminal access to any campus hosts that...
http://roundtable.cif.rochester.edu/users/tparker/rolm/modem.html - size 7K - 18 Dec 95

U.S. Robotics Modem Glossary
A] [B] [C] [D] [E] [F] [G] [H] [I] [J] [K] [L] [M] [N] [O] [P] [Q] [R] [S] [T] [U] [V] [W] [X] [Y] [Z] Analog Signals Continuous, varying waveforms such...
http://www.usr.com/home/gloss1.html - size 16K - 8 Mar 97

U.S. Robotics Modem Info
U.S. Robotics 28.8 kbps Sportster V.34. U.S. Robotics has corrected a bug in it's 28.8 kbps Sportster V.34 chip set that caused the modem to pause when...
http://ascenture.net/maininfo.HTM - size 736 bytes - 1 Dec 96

US Robotics Modem Drivers
US Robotics Drivers and Utilities. Tagram System Corporation updates all of its files weekly. There may be instances when there are newer drivers...
http://www.tagram.com/usrmdm.htm - size 2K - 29 Oct 96

US Robotics Modem Settings
US Robotics Modem Settings. Last Modified: May 8, 1996. Here are a few useful settings for using modems with Xylogics terminal servers. Note that the AT...
http://www.xylogics.com/support/appnotes/usr_modems.html - size 2K - 8 May 96

UniPro Products US Robotics 56800 Modem
>US Robotics 56800bps Modem. The ideal dial up modem for internet access. US Robotics Sportster Features. runs at 28800/33600 now. upgradable to 56800 bps..
>http://196.7.97.1/products/usr56.htm - size 1K - 4 Nov 96

Windows 95 Modem Configuration
>Windows 95 Modem Configuration. If you have not already installed and configured a modem, click the Modem icon in Control Panel. Click the Add... button...
>http://www.dcrt.nih.gov/csb/training/pw95/sld009.htm - size 1K - 24 Oct 96

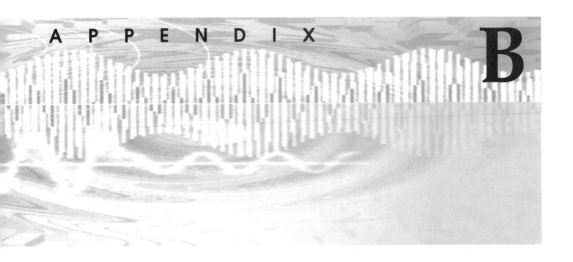

List of Selected Modem and Software Manufacturers

Manufacturer	Information	Tech Support	Bulletin Board
ATI Technologies	(416) 756-0718	(416) 756-0711	(416) 756-4591
Cardinal	(800) 233-0187	(717) 293-3124	(717) 293-3074
Compucom	(800) 228-6648	(408) 732-4500	(408) 738-4990
Hayes	(404) 441-1617	(404) 441-1617	(800) 874-2937
Image Communications	(201) 935-8880	(201) 935-8880	
Intel	(800) 538-3373	(503) 629-7000	(503) 645-6275
VocalTEC (Internet Phone)	(201) 768-9400		
Microcom	(800) 822-8224	(617) 551-1313	(617) 551-1655
Multi-Tech	(800) 328-9717	(800) 328-9717	(612) 785-9875
Practical Peripherals	(800) 442-4774	(818) 991-8200	(818) 706-2467
Prometheus	(800) 477-3473	(503) 624-0571	(503) 691-5199
Supra	(800) 727-8772	(503) 967-2440	(503) 967-2444
Telebit	(800) 835-3248	(800) 835-3248	n/a
U.S. Robotics	(800) 342-5877	(800) 982-5151	(708) 982-5092
Zoom	(800) 666-6191	(617) 423-1076	(617) 451-5284

Support BBS for Communications Programs	Phone Number
Procomm Plus (Datastorm Technologies, Inc.)	(314) 875-0523
Telix (Exis Inc.)	(416) 439-9399
Qmodem (The Forbin Project, Inc.)	(319) 233-6157
HyperAccess 5 (Hilgraeve Inc.)	(313) 243-5915
Crosstalk for Windows (DCA)	(404) 740-8428
MicroPhone II (Software Ventures)	(415) 849-1912

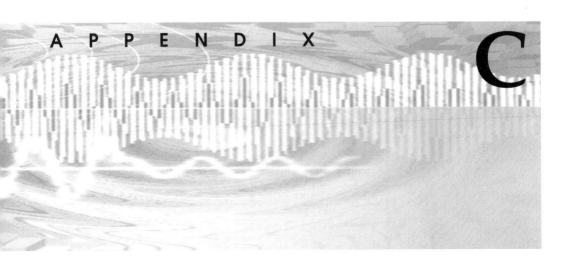

ASCII Codes

Binary	Hex	Character	Binary	Hex	Character	Binary	Hex	Character
0000000	00H	<NUL>	0001111	0FH	<SI>	0011110	1EH	<RS>
0000001	01H	<SOH>	0010000	10H	<DLE>	0011111	1FH	<US>
0000010	02H	<STX>	0010001	11H	<DC1>	0100000	20H	space
0000011	03H	<ETX>	0010010	12H	<DC2>	0100001	21H	!
0000100	04H	<EOT>	0010011	13H	<DC3>	0100010	22H	"
0000101	05H	<ENQ>	0010100	14H	<DC4>	0100011	23H	#
0000110	06H	<ACK>	0010101	15H	<NAK>	0100100	24H	$
0000111	07H	<BELL>	0010110	16H	<SYN>	0100101	25H	%
0001000	08H	<BKSP>	0010111	17H	<ETB>	0100110	26H	&
0001001	09H	<TAB>	0011000	18H	<CAN>	0100111	27H	`
0001010	0AH	<LF>	0011001	19H	<EM>	0101000	28H	(
0001011	0BH	<VT>	0011010	1AH	<SUB>	0101001	29H	)
0001100	0CH	<FF>	0011011	1BH	<ESC>	0101010	2AH	*
0001101	0DH	<CR>	0011100	1CH	<FS>	0101011	2BH	+
0001110	0EH	<SO>	0011101	1DH	<GS>	0101100	2CH	,

Binary	Hex	Character	Binary	Hex	Character	Binary	Hex	Character
0101101	2DH	-	1001001	49H	I	1100101	65H	e
0101110	2EH	.	1001010	4AH	J	1100110	66H	f
0101111	2FH	/	1001011	4BH	K	1100111	67H	g
0110000	30H	0	1001100	4CH	L	1101000	68H	h
0110001	31H	1	1001101	4DH	M	1101001	69H	I
0110010	32H	2	1001110	4EH	N	1101010	6AH	j
0110011	33H	3	1001111	4FH	O	1101011	6BH	k
0110100	34H	4	1010000	50H	P	1101100	6CH	l
0110101	35H	5	1010001	51H	Q	1101101	6DH	m
0110110	36H	6	1010010	52H	R	1101110	6EH	n
0110111	37H	7	1010011	53H	S	1101111	6FH	o
0111000	38H	8	1010100	54H	T	1110000	70H	p
0111001	39H	9	1010101	55H	U	1110001	71H	q
0111010	3AH	:	1010110	56H	V	1110010	72H	r
0111011	3BH	;	1010111	57H	W	1110011	73H	s
0111100	3CH	<	1011000	58H	X	1110100	74H	t
0111101	3DH	=	1011001	59H	Y	1110101	75H	u
0111110	3EH	>	1011010	5AH	Z	1110110	76H	v
0111111	3FH	?	1011011	5BH	(	1110111	77H	w
1000000	40H	@	1011100	5CH	\	1111000	78H	x
1000001	41H	A	1011101	5DH	)	1111001	79H	y
1000010	42H	B	1011110	5EH	^	1111010	7AH	z
1000011	43H	C	1011111	5FH	_	1111011	7BH	{
1000100	44H	D	1100000	60H	`	1111100	7CH		
1000101	45H	E	1100001	61H	a	1111101	7DH	}
1000110	46H	F	1100010	62H	b	1111110	7EH	~
1000111	47H	G	1100011	63H	c	1111111	7FH	Δ
1001000	48H	H	1100100	64H	d			

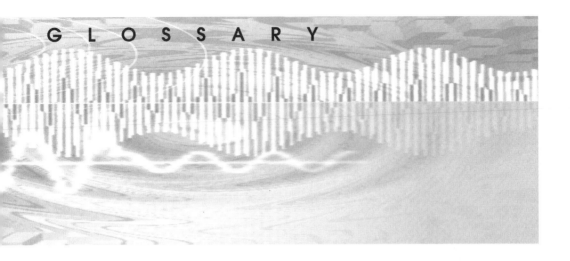

G L O S S A R Y

This short glossary contains terms that relate to modems, data and voice communications. Engineers and military personnel have always had a propensity for acronyms. Here are some of them:

ARQ
Automatic repeat request. A general term for error control protocols featuring hardware detection and re-transmission of defective data.

ASCII
American Standard Code for Information Exchange, a binary code representation of letters, numbers and special characters. It is universally supported for transfer of computer data.

Asynchronous
Data transmission in which the actual data is preceded by a start bit and followed by a stop bit to separate transmitted characters.

Auto Answer
Modem feature that enables the detection of a ring and answers without manual assistance.

Baud Rate
The number of discrete signal events per second occurring on a communications channel. Baud is often incorrectly used instead of bits per second (bps).

BBS

Bulletin board system. A host system into which callers may dial with their modems to read and send electronic mail, upload and download files, and chat online with other callers.

Bit

Binary digit. A single basic unit of computer signaling, consisting of a value of 0 or 1, meaning off or on, or mark or space, respectively.

Buffer

A memory area used for temporary storage during input/output (I/O) operations.

Byte

A group of eight bits acted upon as a group, which may have a readable ASCII value such as a letter or number, or some other coded meaning to the computer. 1 kilobyte = 1024 bytes; 64kbytes = 65,536 bytes (or characters).

Carrier

A continuous frequency capable of being modulated (impressed with another information-carrying signal). Carriers are generated by modems through the transmission lines of telephone companies.

CCITT

An acronym for the French name of the international telecommunications organization called the International Telephone and Telegraph Consultative Committee. This international organization used to define standards for telephone and data equipment. It is now called the ITU-T.

Conference

An area of public messages on a Bulletin Board System, usually with a particular topic and frequently a conference host or moderator to guide the discussion. Also called a Special Interest Group (SIG) or Echo.

cps

Characters per second. A transfer rate estimated from the bit rate and length of each character. If each character is 8 bits long and includes a start and stop bit for asynchronous transmission, each character needs 10 bits to be sent. At 28,800 bps, data is transmitted at approximately 2800 cps.

CRC

Cyclical redundancy checking. An error-detection algorithm performed on each block of data at the sending end and then included in the block that is sent. A CRC is again performed at the receiving end of the transmission. As each block is received, the CRC value is checked against the CRC value sent. If they match, the block is assumed to be error-free. Many file transfer protocols will request that a block be resent until the block is received without error.

DTE

Data terminal equipment. The device that is the originator or destination of the data sent by a modem.

Data Compression Protocols

Protocols that compress data during its transmission (on the fly). Compression of data by a modem allows more information to be transferred in a shorter time. Protocols used for data compression include V.42bis and MNP 5.

Data Transmission Protocols

Standards for modulation and transmission of data at various speeds. The standards are Bell 103 and V.21 for 300 bps, Bell 212A and V.22 for 1200 bps, V.22 bis for 2400 bps, V.32 for 9600 bps, and V.34 for 28,800 bps. Proprietary protocols are also used extensively for higher rates.

DTR

Data Terminal Ready. A signal generated by most modems, indicating a connection between the DTE (computer) and the modem. When DTR is high, the computer is connected.

Error Control Protocols

Modem-based techniques that check the reliability of characters or blocks of data at the hardware level. Examples include MNP 2-4 and V.42.

Flow Control

A mechanism that controls the flow of data to and output from a modem and computer. Either hardware or software can be used for this control to prevent data loss. Hardware flow-control, which uses the data control lines of a serial interface, is more accurate than using software to control the flow of information through a serial port. Flow control is necessary if the communications port is locked at a higher rate than the connection rate.

Freeware

Computer software that may be distributed on Bulletin Board Systems or the Internet, and for which the author requests no license fee or registration fee. Not to be confused with Shareware.

Full Duplex

Signals flow in both directions at the same time. It is sometimes used to refer to the suppression of online LOCAL ECHO and allows the remote system to provide a REMOTE ECHO.

Half Duplex

Signals flow in both directions, but only one way at a time. It is sometimes used to refer to activation of LOCAL ECHO, which causes a copy of data that is sent to the sending system's display to provide a visual indication for the operator.

Host System

Another name for a bulletin board system (BBS) or ISP server.

ISP

Internet service provider. A computer system that is connected to the Internet network for the purpose of providing access to subscribers.

K56 Flex

A high-speed modem transmission method introduced by a consortium of modem manufacturers.

LAN

Local area network. A group of computers that are interconnected with cables and software, allowing hard disks and other devices to be shared among many users.

MNP

Microcom Networking Protocols. A set of hardware and software error detection and correction protocols and data compression techniques developed by Microcom Corporation.

NVRAM

Non-volatile random access memory. A user-programmable memory chip whose data is retained when power to the chip is turned off. NVRAM is used in many modems to store default settings.

On/Off-Hook

A term referring to older telephone sets where you lifted a telephone receiver off the hook to place or answer a call, and replaced it back on the hook to hang up.

Packer

A program that compresses multiple files into a single file, such as PKZIP, ARC, or LHARC.

Packet

A mail packet (with a .QWK extension) from a host system. The term is also used for a packet of data that travels over a packet-switching network such as the Internet.

Parity

An error-detection method used in both communications and computer memory checking to determine character validity. Communications systems now make use of the more efficient block checking, although parity must still be matched in a communications session for a transfer to take place correctly. Host communications in BBS environments omit parity checking (no parity).

Protocol

A system of rules and procedures governing communication between two devices. For example, file-transfer protocols in your communications program refer to a set of rules governing how error checking will be performed on blocks of data.

Public Domain

Computer software on which no copyright exists (usually by a specific statement to that effect by the author), and which may be freely used and distributed.

Remote Echo

A copy of the data being received is returned to the sending system that sent it, for display on the original sender's screen. This provides verification to the sending operator that the data was actually received by the recipient. See Full/Half duplex.

Shareware

Computer software that is distributed on the honor system, which may be freely copied and distributed to other potential users (not for commercial use), but for which payment (sometimes called a registration fee) is required for continued use beyond an initial evaluation period. The same as "try before you buy."

SysOp

System operator of a BBS. The person responsible for setting up and maintaining a BBS.

Thread

A group of BBS messages and replies that are organized and sorted by topic.

Unpacker

A program that uncompresses a file from a packer, such as PKUNZIP.

x2©

A high-speed modem transmission method introduced by U.S. Robotics.

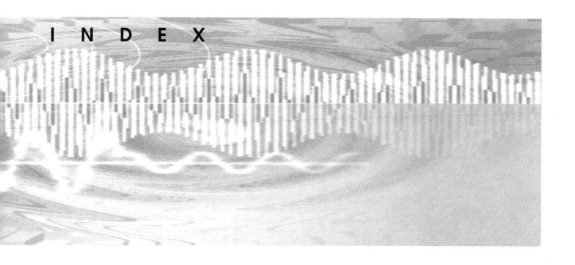

INDEX

LICENSE AGREEMENT AND LIMITED WARRANTY

READ THE FOLLOWING TERMS AND CONDITIONS CAREFULLY BEFORE OPENING THIS CD PACKAGE, *HANDS-ON NETSCAPE CD—WINDOWS 3.1*. THIS LEGAL DOCUMENT IS AN AGREEMENT BETWEEN YOU AND PRENTICE-HALL, INC. (THE "COMPANY"). BY OPENING THIS SEALED CD PACKAGE, YOU ARE AGREEING TO BE BOUND BY THESE TERMS AND CONDITIONS. IF YOU DO NOT AGREE WITH THESE TERMS AND CONDITIONS, DO NOT OPEN THE CD PACKAGE. PROMPTLY RETURN THE UNOPENED CD PACKAGE AND ALL ACCOMPANYING ITEMS TO THE PLACE YOU OBTAINED THEM FOR A FULL REFUND OF ANY SUMS YOU HAVE PAID.

1.   **GRANT OF LICENSE:** In consideration of your purchase of this book, and your agreement to abide by the terms and conditions of this Agreement, the Company grants to you a nonexclusive right to use and display the copy of the enclosed software program (hereinafter the "SOFTWARE") on a single computer (i.e., with a single CPU) at a single location so long as you comply with the terms of this Agreement. The Company reserves all rights not expressly granted to you under this Agreement.

2.   **OWNERSHIP OF SOFTWARE:** You own only the magnetic or physical media (the enclosed CD) on which the SOFTWARE is recorded or fixed, but the Company and the software developers retain all the rights, title, and ownership to the SOFTWARE recorded on the original CD copy(ies) and all subsequent copies of the SOFTWARE, regardless of the form or media on which the original or other copies may exist. This license is not a sale of the original SOFTWARE or any copy to you.

3.   **COPY RESTRICTIONS:** This SOFTWARE and the accompanying printed materials and user manual (the "Documentation") are the subject of copyright. The individual programs on the CD are copyrighted by the authors of each program. Some of the programs on the CD include separate licensing agreements. If you intend to use one of these programs, you must read and follow its accompanying license agreement. You may not copy the Documentation or the SOFTWARE, except that you may make a single copy of the SOFTWARE for backup or archival purposes only. You may be held legally responsible for any copying or copyright infringement which is caused or encouraged by your failure to abide by the terms of this restriction.

4.   **USE RESTRICTIONS:** You may not network the SOFTWARE or otherwise use it on more than one computer or computer terminal at the same time. You may physically transfer the SOFTWARE from one computer to another provided that the SOFTWARE is used on only one computer at a time. You may not distribute copies of the SOFTWARE or Documentation to others. You may not reverse engineer, disassemble, decompile, modify, adapt, translate, or create derivative works based on the SOFTWARE or the Documentation without the prior written consent of the Company.

5.   **TRANSFER RESTRICTIONS:** The enclosed SOFTWARE is licensed only to you and may not be transferred to any one else without the prior written consent of the Company. Any unauthorized transfer of the SOFTWARE shall result in the immediate termination of this Agreement.

6.   **TERMINATION:** This license is effective until terminated. This license will terminate automatically without notice from the Company and become null and void if you fail to comply with any provisions or limitations of this license. Upon termination, you shall destroy the Documentation and all copies of the SOFTWARE. All provisions of this Agreement as to warranties, limitation of liability, remedies or damages, and our ownership rights shall survive termination.

7.   **MISCELLANEOUS:** This Agreement shall be construed in accordance with the laws of the United States of America and the State of New York and shall benefit the Company, its affiliates, and assignees.

8.   **LIMITED WARRANTY AND DISCLAIMER OF WARRANTY:** The Company warrants that the SOFTWARE, when properly used in accordance with the Documentation, will operate in substantial conformity with the description of the SOFTWARE set forth in the Documentation. The Company does not warrant that the SOFTWARE will meet your requirements or that the operation